STABILITY ANALYSIS OF EARTH SLOPES

STABILITY ANALYSIS OF EARTH SLOPES

Yang H. Huang
University of Kentucky

VAN NOSTRAND REINHOLD COMPANY
NEW YORK CINCINNATI TORONTO LONDON MELBOURNE

Library of Congress Catalog Card Number: 82-8432
ISBN: 0-442-23689-1

Manufactured in the United States of America

Published by Van Nostrand Reinhold Company Inc.
135 West 50th Street, New York, N.Y. 10020

Van Nostrand Reinhold Publishing
1410 Birchmount Road
Scarborough, Ontario M1P 2E7, Canada

Van Nostrand Reinhold
480 Latrobe Street
Melbourne, Victoria 3000, Australia

Van Nostrand Reinhold Company Limited
Molly Millars Lane
Wokingham, Berkshire, England

15 14 13 12 11 10 9 8 7 6 5 4 3 2 1

Library of Congress Cataloging in Publication Data
Huang, Yang H. (Yang Hsien), 1927-
    Stability analysis of earth slopes.

    Includes bibliographical references and index.
    1. Slopes (Soil mechanics)   I. Title.
TA710.H785   1982   624.1'51   82-8432
ISBN 0-442-23689-1   AACR2

Preface

During the past several years I have been engaged in applied research related to the stability analysis of slopes. This research was supported by the Institute for Mining and Minerals Research, University of Kentucky, in response to the Surface Mining Control and Reclamation Act of 1977, which requires stability analysis for refuse dams, hollow fills, and spoil banks created by surface mining. The results of the research have been published in several journals and reports and also presented in a number of short courses. Both the simplified and the computerized methods of stability analysis, as developed from this research, have been widely used by practicing engineers throughout Kentucky for the application of mining permits. The large number of out-of-state participants in the short courses indicates that the methods developed have widespread applications.

This book is a practical treatise on the stability analysis of earth slopes. Special emphasis is placed on the utility and application of stablity formulas, charts, and computer programs developed recently by the author for the analysis of human-created slopes. These analyses can be used for the design of new slopes and the assessment of remedial measures on existing slopes. To make the book more complete as a treatise on slope stability analysis, other methods of stability analysis, in addition to those developed by the author, are briefly discussed. It is hoped that this book will be a useful reference, classroom text, and users' manual for people interested in learning about stability analysis.

This book is divided into four parts and 12 chapters. Part I presents the fundamentals of slope stability and consists of five chapters. Chapter 1 describes slope movements and discusses some of the more well-known methods for stability analysis. Chapter 2 explains the mechanics of slope failures and defines the factor of safety for both cylindrical and plane failures. Chapter 3 discusses both the laboratory and the field methods for determining the shear strength of soils used for stability analysis. Chapter 4 illustrates some methods for estimating the location of the phreatic surface and determining the pore pressure ratio. Chapter 5 outlines remedial measures for correcting slides. Part

II presents simplified methods of stability analysis and consists of two chapters. Chapter 6 derives some simple formulas for determining the factor of safety of plane failure, while Chap. 7 compiles a number of stability charts for determining the factor of safety of cylindrical failure. These methods can be applied by using a simple pocket calculator without the service of a large computer. Part III presents computerized methods of stability analysis and consists of three chapters. The SWASE (Sliding Wedge Analysis of Sidehill Embankments) computer program for plane failure is presented in Chap. 8, and the REAME (Rotational Equilibrium Analysis of Multilayered Embankments) program for cylindrical failure is presented in Chap. 9. Applications of these programs to a number of examples are illustrated in Chap. 10. Part IV presents several methods of stability analysis, other than those used in SWASE and REAME, and consists of two chapters. Chapter 11 describes two methods used exclusively for homogeneous slopes: the friction circle method and the logarithmic spiral method. Chapter 12 discusses six more methods used for either homogeneous or nonhomogeneous slopes involving cylindrical or noncylindrical failure surfaces: the earth pressure method, Janbu's method, Morgenstern and Price's method, Spencer's method, the finite element method, and the probabilistic method. Complete listings of SWASE and REAME in both FORTRAN and BASIC are presented in the Appendices. These programs are designed for small computers and should be found useful to those who have only limited computing capability.

The author should like to thank the Institute for Mining and Minerals Research for the support of this research, which makes possible the publication of this book. The helpful comments by Dr. Donald H. Gray, Professor of Civil Engineering, University of Michigan, are gratefully acknowledged.

Yang H. Huang
Professor of Civil Engineering
University of Kentucky

Contents

PART IV.   OTHER METHODS OF STABILITY ANALYSIS

STABILITY ANALYSIS OF EARTH SLOPES

Part I
Fundamentals of Slope Stability

1
Introduction

1.1 SLOPE MOVEMENTS

The stability analysis of slopes plays a very important role in civil engineering. Stability analysis is used in the construction of transportation facilities such as highways, railroads, airports, and canals; the development of natural resources such as surface mining, refuse disposal, and earth dams; as well as many other human activities involving building construction and excavations. Failures of slopes in these applications are caused by movements within the human-created fill, in the natural slope, or a combination of both. These movement phenomena are usually studied from two different points of view. The geologists consider the moving phenomena as a natural process and study the cause of their origin, their courses, and the resulting surface forms. The engineers investigate the safety of construction based on the principles of soil mechanics and develop methods for a reliable assessment of the stability of slopes, as well as the controlling and corrective measures needed. The best result of stability studies can be achieved only by the combination of both these approaches. The quantitative determination of the stability of slopes by the methods of soil mechanics must be based on a knowledge of the geological structure of the area, the detailed composition and orientation of strata, and the geomorphological history of the land surface. On the other hand, geologists may obtain a clearer picture of the origin and character of movement process by checking their considerations against the results of engineering analyses based on soil mechanics. For example, it is well known that one of the most favorable settings for landslides is the presence of permeable or soluble beds overlying or interbedded with relatively impervious beds. This geological phenomenon was explained by Henkel (1967) using the principles of soil mechanics.

Slope failures involve such a variety of processes and disturbing factors that they afford unlimited possibilities of classification. For instance, they can be divided according to the form of failures, the kind of materials moved, the age, or the stage of development.

One of the most comprehensive references on landslides or slope failures is a special report published by the Transportation Research Board (Schuster and Krizek, 1978). According to this report, the form of slope movements is divided into five main groups: falls, topples, slides, spreads, and flows. A sixth group, complex slope movements, includes combination of two or more of the above five types. The kind of materials is divided into two classes: rock and soil. Soil is further divided into debris and earth. Table 1.1 shows the classification of slope movements.

Recognizing the form of slope movements is important because they dictate the method of stability analysis and the remedial measures to be employed. They are described as follow:

In falls, a mass of any size is detached from a steep slope or cliff, along a surface on which little or no shear displacement takes place, and descends mostly through the air by free fall, leaping, bounding, or rolling. Movements are very rapid and may or may not be preceded by minor movements leading to progressive separation of the mass from its source.

In topples, one or more units of mass rotate forward about some pivot point, below or low in the unit, under the action of gravity and forces exerted by adjacent unit or by fluids in cracks. In fact, it is tilting without collapse.

In slides, the movement consists of shear strain and displacement along one or several surfaces that are visible or may reasonably be inferred, or within a relatively narrow zone. The movement may be progressive; that is, shear failure may not initially occur simultaneously over what eventually be-

Table 1.1   Classification of Slope Movements.

Type of Movement			TYPE OF MATERIAL		
				Engineering Soils	
			Bedrock	Predominantly Coarse	Predominantly Fine
Falls			Rock fall	Debris fall	Earth Fall
Topples			Rock topple	Debris topple	Earth topple
Slides	Rotational		Rock slump	Debris slump	Earth slump
	Translational	Few units	Rock block slide	Debris block slide	Earth block slide
		Many units	Rock slide	Debris slide	Earth slide
Lateral spreads			Rock spread	Debris spread	Earth spread
Flows			Rock flow (deep creep)	Debris flow (soil creep)	Earth flow (soil creep)
Complex			combination of two or more principal types of movement		

(After Varnes, 1978)

comes a defined surface of rupture, but rather it may propagate from an area of local failure. This displaced mass may slide beyond the original surface of rupture onto what had been the original ground surface, which then becomes a surface of separation. Slides are divided into rotational slides and translational slides. This distinction is important because it affects the methods of analysis and control.

In spreads, the dominant mode of movement is lateral extension accommodated by shear or tensile fractures. Movements may involve fracturing and extension of coherent material, either bedrock or soil, owing to liquefaction or plastic flow of subjacent material. The coherent upper units may subside, translate, rotate, or disintegrate, or they may liquify and flow. The mechanism of failure can involve elements not only of rotation and translation but also of flows; hence some lateral spreading failures may be regarded as complex. The sudden spreading of clay slopes was discussed by Terzaghi and Peck (1967).

Many examples of slope movement cannot be classed as falls, topples, slides, or spreads. In unconsolidated materials, these generally take the form of fairly obvious flows, either fast or slow, wet or dry. In bedrock, the movements are extremely slow and distributed among many closely spaced, non-interconnected fractures that result in folding, bending, or bulging.

According to age, slope movements are divided into contemporary, dormant, and fossil movements. Contemporary movements are generally active and relatively easily recognizable by their configuration, because the surface forms produced by the mass movements are expressive and not affected by rainwash and erosion. Dormant movements are usually covered by vegetation or disturbed by erosion so that the traces of their last movements are not easily discernible. However, the causes of their origin still remain and the movement may be renewed. Fossil movements generally developed in the Pleistocene or earlier periods, under different morphological and climatic conditions, and cannot repeat themselves at present.

According to stage, slope movements can be divided into initial, advanced, and exhausted movements. At the initial stage, the first signs of the disturbance of equilibrium appear and cracks in the upper part of the slope develop. In the advanced stage, the loosened mass is propelled into motion and slides downslope. In the exhausted stage, the accumulation of slide mass creates temporary equilibrium conditions.

Chowdhury (1980) classified slides according to their causes: (1) landslides arising from exceptional causes such as earthquake, exceptional precipitation, severe flooding, accelerated erosion from wave action, and liquefaction; (2) ordinary landslides, or landslides resulting from known or usual causes which can be explained by traditional theories; and (3) landslides which occur without any apparent cause.

It can be seen from the above discussion that the stability of slopes is a complex problem which may defy any theoretical analysis. In this book only

the slide type of failures will be discussed, not only because it is more amenable to theoretical analysis but also because it is the predominant type of
failures, particularly in the human-created slopes.

1.2 LIMIT PLASTIC EQUILIBRIUM

Practically all stability analyses of slopes are based on the concept of limit
plastic equilibrium. First, a failure surface is assumed. A state of limiting
equilibrium is said to exist when the shear stress along the failure surface is
expressed as

$$\tau = \frac{s}{F} \tag{1.1}$$

in which τ is the shear stress, s is the shear strength, and F is the factor of
safety. According to the Mohr-Coulomb theory, the shear strength can be
expressed as

$$s = c + \sigma_n \tan \phi \tag{1.2}$$

in which c is the cohesion, σ_n is the normal stress, and ϕ is the angle of
internal friction. Both c and ϕ are known properties of the soil. Once the
factor of safety is known, the shear stress along the failure surface can be
determined from Eq. 1.1.

In most methods of limit plastic equilibrium, only the concept of statics
is applied. Unfortunately, except in the most simple cases, most problems in
slope stability are statically indeterminate. As a result, some simplifying assumptions must be made in order to determine a unique factor of safety. Due
to the differences in assumptions, a variety of methods, which result in different factors of safety, have been developed, from the very simple wedge
method (Seed and Sultan, 1967) to the very sophisticated finite-element
method (Wang, Sun, and Ropchan, 1972). Between these two extremes are
the methods by Fellenius (1936), by Bishop (1955), by Janbu (1954, 1973), by
Morgenstern and Price (1965), and by Spencer (1967). As the purpose of this
book is to present the most practical methods that can be readily used by
practicing engineers, only the wedge, Fellenius, and simplified Bishop methods will be discussed in detail, while all other methods will only be briefly
described. To emphasize basic concepts in these other methods, only the most
simple cases with no pore water pressure will be considered, and no mathematical derivations of complex formulas will be attempted. Readers interested
in the detailed procedure should refer to the original publications.

1.3 STATICALLY DETERMINATE PROBLEMS

Figure 1.1 shows three example problems in which the factor of safety can be determined from statics.

In Fig. 1.1a, a fill is placed on a sloping ground. The failure surface is assumed to be a plane at the bottom of the fill along the sloping ground. The weight of the fill is W, the force normal to the failure plane is N, and the shear force, T, along the failure plane can be expressed as

$$T = \frac{C + N \tan \phi}{F} \tag{1.3}$$

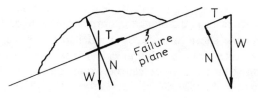

(a) PLANE FAILURE

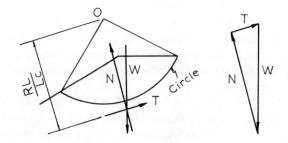

(b) CYLINDRICAL FAILURE ($\phi = 0$)

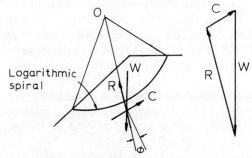

(c) LOGARITHMIC SPIRAL FAILURE

FIGURE 1.1. Statically determinate problems.

in which C is the total cohesion resistance which is equal to the unit cohesion, c, multiplied by the area of failure surface. There are a total of three unknowns; viz. the factor of safety, F, the magnitude of N, and the point of application of N. According to statics, there are also three equilibrium equations; specifically, the sum of forces in the normal direction is zero, the sum of forces in the tangential direction is zero, and the sum of moments about any given point is zero. The moment requirement implies that W, T, and N must meet at the same point. Knowing the magnitude and direction of W and the direction of N and T, the magnitude of N and T can be determined from the force diagram shown in the figure, and the factor of safety from Eq. 1.3.

In Fig. 1.1b, a failure circle, or a cylinder in three dimensions, is assumed to cut through a soil with $\phi = 0$. When $\phi = 0$, the shear resistance is dependent on the cohesion only, independent of the normal force. By assuming that the cohesion, c, is distributed uniformly along the failure arc, the line of application of the shear force, T, is parallel to the chord at a distance of RL/L_C from the center, where R is the radius, L is the arc length, and L_C is the chord length. The shear force can be expressed as

$$T = \frac{cL_c}{F} \tag{1.4}$$

The problem is statically determinate because there are only three unknowns: the factor of safety, F; the magnitude of N; and the point of application of N. The location of these three forces and the force diagram are shown in the figure. The factor of safety can be determined simply by taking moment at the center of circle to determine T. The factor of safety can then be determined from Eq. 1.4.

Figure 1.1c shows a logarithmic spiral failure surface. The three forces shown are the weight, W, the cohesion force, C, and the resultant of normal and frictional forces, R. Even though ϕ is not zero, the problem is still statically determinate because the resultant of normal and frictional forces along a logarithmic spiral always passes through the origin. The three unknowns are the factor of safety, F, the magnitude of R, and the point of application of R. The location of these three forces and the force diagram are shown in the figure. If only the factor of safety is required, it can be easily determined by taking the moment at the origin, as will be fully described in Sect. 11.2

1.4 STATICALLY INDETERMINATE PROBLEMS

Except for the simple cases in Fig. 1.1, most problems encountered in engineering practice are statically indeterminate. Figure 1.2 shows a fill with two failure planes. This problem is statically indeterminate because there are five unknowns but only three equations. The five unknowns are the factor of safety, F, the magnitude and point of application of N_1, and the magnitude

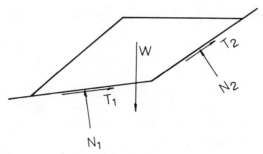

FIGURE 1.2. Statically indeterminate problem.

and point of application of N_2. To make the problem statically determinate, it is necessary to divide the fill into two blocks and arbitrarily assume the forces acting between the two blocks. Of course, different assumptions will result in different factors of safety.

The case of circular failure surface shown in Fig. 1.1b is also statically indeterminate if the angle of internal friction of the soil, ϕ, is not zero. Because the frictional force along the failure arc is indeterminate, there are six unknowns but only three equations. The six unknowns are the factor of safety, F, the magnitude and point of application of N, and the magnitude, direction, and point of application of T. To make the problem statically determinate, it is necessary to assume the distribution of normal stress along the failure surface, thus relating T to N and eliminating the three unknowns on T. This concept is used in the friction circle method, as will be discussed in Sect. 11.1

A very powerful method, which can be applied to either circular or non-circular failure surface, is the method of slices. Figure 1.3 shows an arbitrary failure surface which is divided into a number of slices. The forces applied on a slice are shown in the free body diagram. If the failure mass is divided into

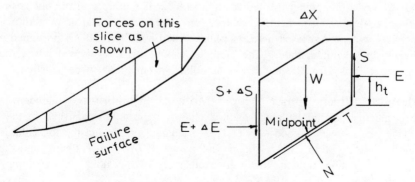

FIGURE 1.3. Method of slices.

a sufficient number of slices, the width, Δx, will be small and it is reasonable to assume that the normal force, N, is applied at the midpoint of the failure surface. In the free body diagram, the known forces are the weight, W, and the shear force, T, which can be written as

$$T = \frac{C + N \tan \phi}{F} \tag{1.5}$$

in which C is the cohesion force, which is equal to the unit cohesion multiplied by the area of failure surface at the bottom of the slice. The unknowns are the factor of safety, F, the shear force on the vertical side, S, the normal force on the vertical side, E, the vertical distance, h_t, and the normal force, N. If there are a total of n slices, the number of unknowns is $4n - 2$, as tabulated below:

UNKNOWN	NUMBER
F (related to T)	1
N	n
E	$n - 1$
S	$n - 1$
h_t	$n - 1$
Total	$4n - 2$

Since each slice can only have three equations by statics, two with respect to forces and one with respect to moments, the number of equations is $3n$. Therefore, there is an indeterminacy of $n - 2$. The problem can be solved statically only by making assumptions on the forces on the interface between slices.

1.5 METHODS OF STABILITY ANALYSIS

Due to the large number of methods available, it is neither possible nor desirable to review each of them. Therefore, only the most popular or well-known methods will be discussed here. Hopkins, Allen and Deen (1975) presented a review on several methods. The methods can be broadly divided into three categories, based on the number of equilibrium equations to be satisfied.

Methods Which Satisfy Overall Moment Equilibrium. Included in this category are the Fellenius method (1936), the simplified Bishop method (1955), the $\phi = 0$ method (Taylor, 1937; Huang, 1975), and the logarithmic spiral method (Taylor, 1937; Huang and Avery, 1976). The Fellenius and the simplified Bishop methods are used exclusively in this book for developing stability charts and the REAME computer program.

The Fellenius method has been used extensively for many years because it is applicable to nonhomogeneous slopes and is very amenable to hand calculation. This method is applicable only to circular failure surfaces and considers only the overall moment equilibrium; the forces on each side of a slice are ignored. Where high pore pressures are present, a modified version of the Fellenius method, based on the concept of submerged weight and hereafter called the normal method (Bailey and Christian, 1969), is available. The Fellenius and the normal methods were used in this book to generate stability charts for practical use. The factor of safety obtained by the normal method is usually slightly smaller than that given by the simplified Bishop method.

In the simplified Bishop method, overall moment equilibrium and vertical force equilibrium are satisfied. However, for individual slices, neither moment nor horizontal force equilibrium is satisfied. Although equilibrium conditions are not completely satisfied, the method is, nevertheless, a satisfactory procedure and is recommended for most routine work where the failure surface can be approximated by a circle. Bishop (1955) compared the factor of safety obtained from the simplified method with that from a more rigorous method in which all equilibrium conditions are satisfied. He found that the vertical interslice force, S, could be assumed zero without introducing significant error, typically less than one percent. Hence, the simplified procedure, which set the vertical interslice force to zero, gives approximately the same result as the rigorous procedure, which satisfies all the equilibrium conditions.

The $\phi = 0$ method involving circular failure surfaces was used to develop the stability charts for the total stress analysis, as will be described in Sect. 7.7. The logarithmic spiral method will be described in Sect. 11.2

Force Equilibrium Methods.   Several methods have been proposed which satisfy only the overall vertical and horizontal force equilibrium as well as the force equilibrium in individual slices or blocks. Although moment equilibrium is not explicitly considered, these methods may yield accurate results if the inclination of the side forces is assumed in such a manner that the moment equilibrium is implicitly satisfied. Arbitrary assumptions on the inclination of side forces have a large influence on the factor of safety. Depending on the inclination of side forces, a range of safety factors may be obtained in many problems. Force equilibrium methods should be used cautiously, and the user should be well aware of the particular side force assumptions employed.

The procedure most amenable to hand calculation is the sliding wedge method in which the active or passive earth pressure theory is employed to determine the side forces. This procedure has been used by Méndez (1971) and the Department of Navy (1971) and will be described in Sect. 12.1. Another procedure, as used by Seed and Sultan (1967), is to assume the shear stress along the failure surface as a quotient of shear strength and factor of safety, as indicated by Eq. 1.1, and determine the factor of safety from force equilibrium. This method, which requires iterations, is more cumbersome but is used in this book to develop equations for plane failure and the SWASE computer program.

Moment and Force Equilibrium Methods. Included in this category are the methods by Janbu (1954, 1973), Morgenstern and Price (1965), and Spencer (1967). The basic concept in these methods is the same; the difference lies in the assumption of the interslice forces. If both moment and force equilibrium are satisfied, the assumption on interslice forces should have only small effect on the factor of safety obtained. All these methods can be applied to both circular and noncircular failure surfaces.

In Janbu's method, the location of the interslice normal force, or the line of thrust, must be arbitrarily assumed. For cohesionless soils, the line of thrust should be selected at or very near the lower third point. For cohesive soils, the line of thrust should be located above the lower third point in a compressive zone (passive condition) and somewhat below it in an expansive zone (active condition). This method is very easy to use and does not require as much individual judgment as Morgenstern and Price's method. Janbu's method will be described in Sect. 12.2.

In Morgenstern and Price's method, a simplifying assumption is made regarding the relationship between the interslice shear force, S, and the normal force, E. Having made the assumption and obtained the output from the computer, all the computed quantities must be examined to determine whether they seem reasonable. If not, a new assumption must be made. Bishop (1955) indicated that the range of equally correct values of safety factor might be quite narrow and that any assumption leading to reasonable stress distributions and magnitudes would give practically the same factor of safety. Whitman and Bailey (1967) solved several problems using Morgenstern and Price's procedure and the simplified Bishop method and found the resulting difference was seven percent or less, usually less than two percent. Morgenstern and Price's method will be described in Sect. 12.3

In Spencer's method, it is assumed that the interslice forces be parallel; i.e., the angle of inclination, δ, or $\tan^{-1} (S/E)$, is the same at every vertical side. By considering the force and moment equilibrium for each slice, two recursive formulas can be derived for determining the two unknowns F and δ. Spencer's method will be described in Sect. 12.4.

In contrast to all the above methods, the finite element method determines the normal and shear stresses along a failure surface by considering the elastic properties of soils in terms of Young's modulus and Poisson's ratio. This method will be described in Sect. 12.5.

All the above methods are deterministic in that the shear strength of soils, the loadings applied to the slope, and the required factor of safety are assumed to be known. In reality, a large variation in shear strength, and possibly also in loading, exists. The probabilistic method, which provides information on the probability of failure, will be described in Sect. 12.6.

2
Mechanics of Slides

2.1 TYPES OF FAILURE SURFACES

The purpose of stability analysis is to determine the factor of safety of a potential failure surface. The factor of safety is defined as a ratio between the resisting force and the driving force, both applied along the failure surface. When the driving force due to weight is equal to the resisting force due to shear strength, the factor of safety is equal to 1 and failure is imminent. Figure 2.1 shows several types of failure surfaces.

The shape of the failure surface may be quite irregular, depending on the homogeneity of the material in the slope. This is particularly true in natural slopes where the relic joints and fractures dictate the locus of failure surfaces. If the material is homogeneous and a large circle can be formed, the most critical failure surface will be cylindrical, because a circle has the least surface area per unit mass, the former being related to the resisting force and the latter to the driving force. If a large circle cannot be developed, such as in the case of an infinite slope with depth much smaller than length, the most crit-

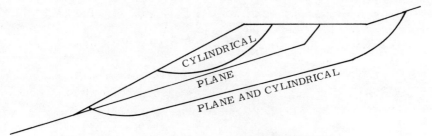

$$\text{FACTOR OF SAFETY} = \frac{\text{RESISTING FORCE}}{\text{DRIVING FORCE}}$$

FIGURE 2.1. Types of failure surfaces.

ical failure surface will be a plane parallel to the slope. If some planes of weakness exist, the most critical failure surface will be a series of planes passing through the weak strata. In some cases, a combination of plane, cylindrical, and other irregular failure surfaces may also exist.

In this book only the case of plane or cylindrical failure will be discussed in detail. If the actual failure surface is quite irregular, it must be approximated by either a series of planes or a cylinder before a stability analysis can be made. Methods involving the use of irregular failure surfaces will be briefly reviewed in Chap. 12.

2.2 PLANE FAILURE SURFACES

Figure 2.2 shows a plane failure surface along the bottom of a sidehill embankment. The reason that a plane failure exists is because the original hillside is not properly scalped and a layer of weak material remains at the bottom.

The resisting force along the failure surface can be determined from the Mohr-Coulomb theory as indicated by Eq. 2.1

$$s = c + \sigma_n \tan \phi \qquad (2.1)$$

in which s is the shear strength or resisting stress, c is the cohesion, σ_n is the stress normal to failure surface, and ϕ is the angle of internal friction. Note that c and ϕ are the strength parameters of the soil along the failure surface. For a failure plane of length L and unit width, the resisting force is equal to $cL + W \cos \alpha \tan \phi$, where W is the weight of soil above failure plane and α

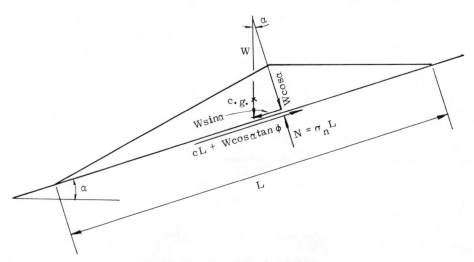

FIGURE 2.2. Plane failure — simple case.

is the degree of natural slope. Note that $W \cos \alpha$ is the component of weight normal to the failure plane. The driving force is the component of weight along the failure plane and is equal to $W \sin \alpha$. The factor of safety is a ratio between the resisting force and the driving force, or

$$F = \frac{C + N \tan \phi}{T} = \frac{cL + W \cos \alpha \tan \phi}{W \sin \alpha} \tag{2.2}$$

The case becomes more complicated if the failure surface is comprised of two or more planes.

Figure 2.3 shows a failure surface along two different planes. The sliding mass is divided into two blocks. Both the normal and the shear forces on each plane depend on the interacting force between the two blocks and can only be determined by considering the two blocks jointly. The lower block has a weight, W_1, and a length of failure plane, L_1; the upper block has a weight, W_2, and a length of failure plane, L_2.

Figure 2.4 shows the free-body diagram for each block. The shear force along the failure plane is equal to the shear resistance divided by the factor of

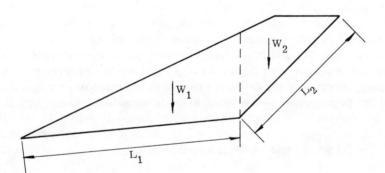

FIGURE 2.3.   Plane failure — complex case.

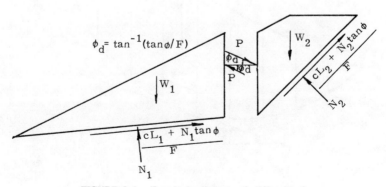

FIGURE 2.4.   Free body diagram of sliding blocks.

safety. For the two blocks shown, there are four unknowns: N_1, N_2, P, and F, where N_1 and N_2 are the forces normal to the failure plane at the lower and upper blocks, respectively; P is the force between the two blocks; and F is the factor of safety. Based on statics, there are also four equations; i.e., in each block the summation of forces in the horizontal and the vertical directions is equal to 0. Since there are four equations and four unknowns, one can solve for the factor of safety.

If the weaker planes are not continuous or the location of the critical planes is not known apriori, it is necessary to try different failure planes until a minimum factor of safety is obtained.

2.3 CYLINDRICAL FAILURE SURFACES

To find the minimum factor of safety for a cylindrical failure surface, a large number of circles must be tried to determine which is most critical. Figure 2.5 shows one of the many circles for which the factor of safety is to be determined. The sliding mass is divided into n slices. The ith slice has a weight, W_i, a length of failure surface, L_i, an angle of inclination, θ_i, and a normal force, N_i. The factor of safety is a ratio between the resisting force and the driving force. According to the Mohr-Coulomb theory, the resisting force in slice i is $cL_i + N_i \tan \phi$. Note that N_i depends on the forces on the two sides of the slice and is statically indeterminate unless some simplifying assumptions are made. The driving force is equal to $W_i \sin \theta_i$, which is the component of weight along the failure surface. The driving force is independent of the forces on the two sides of the slice because whenever there is a force on one side of the slice, there is a corresponding force, the same in magnitude but opposite in direction on the adjacent side, thus neutralizing their effect. The factor of safety can be determined by

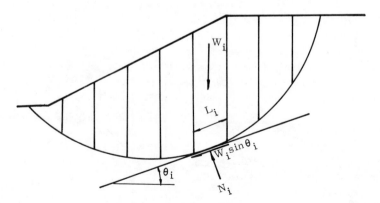

FIGURE 2.5. Cylindrical failure surface.

$$F = \frac{\sum_{i=1}^{n} (cL_i + N_i \tan \phi)}{\sum_{i=1}^{n} (W_i \sin \theta_i)} \tag{2.3}$$

When the failure surface is circular, the factor of safety can be defined as the ratio between two moments. Because both the numerator and the denominator in Eq. 2.3 can be multipied by the same moment arm, which is the radius of the circle, it makes no difference whether the force or the moment is used. In the Fellenius method, it is assumed that the forces on the two sides of a slice are parallel to the failure surface at the bottom of the slice, so they have no effect on the force normal to the failure surface, or

$$N_i = W_i \cos \theta_i \tag{2.4}$$

Thus Eq. 2.3 becomes

$$F = \frac{\sum_{i=1}^{n} (cL_i + W_i \cos \theta_i \tan \phi)}{\sum_{i=1}^{n} (W_i \sin \theta_i)} \tag{2.5}$$

2.4 TOTAL STRESS VERSUS EFFECTIVE STRESS

There are two methods for analyzing the stability of slopes: total stress analysis and effective stress analysis. Total stress analysis is based on the undrained shear strength, and is also called s_u-analysis. Effective stress analysis is based on the drained shear strength and is also called $\bar{c}$, $\bar{\phi}$-analysis. The undrained shear strength is usually used for determining short-term stability during or at the end of construction, and the drained shear strength for long-term stability. Since undrained strength is determined by the initial conditions prior to loading, it is not necessary to determine the effective stress at the time of failure.

If a soil undergoing undrained loading is saturated, ϕ can be assumed as zero so $\phi = 0$ analysis, which is a special case of s_u-analysis, may be used; otherwise, a c_u, ϕ_u-analysis should be performed. In the s_u-analysis, pore pressure should be taken as zero along any failure surface where undrained strength is used. This step does not imply that pore pressures actually are zero, but rather is done to be consistent with the assumption that undrained strength can be expressed independently of the effective stress at failure.

The major difference between a total stress analysis and an effective stress analysis is that the former does not require a knowledge of the pore pressure while the latter does. In principle, short-term stability can also be analyzed in terms of effective stress, and long-term stability in terms of total

Table 2.1   Choice of Total Versus Effective Stress Method.

SITUATION	PREFERRED METHOD	COMMENT
1. End-of-construction with saturated soils, construction period short compared to consolidation time	S_u − analysis with $\phi = 0$ and $c = S_u$	$\bar{c}, \bar{\phi}$ − analysis permits check during construction using actual pore pressures
2. Long term stability	$\bar{c}, \bar{\phi}$ − analysis with pore pressure given by equilibrium ground-water conditions	
3. End-of-construction with partially saturated soil; construction period short compared to consolidation time	Either method: c_u, ϕ_u from undrained tests or $\bar{c}, \bar{\phi}$ plus estimated pore pressures	$\bar{c}, \bar{\phi}$ − analyses permits check during construction using actual pore pressures
4. Stability at intermediate times	$\bar{c}, \bar{\phi}$ − analysis with estimated pore pressures	Actual pore pressure must be checked in field

(After Lambe and Whitman, 1969)

stress. However, this would require extra testing efforts and is therefore not recommended. Table 2.1 shows the choice of total versus effective stress analysis.

For fill slopes constructed of saturated fine-grained soils or placed on saturated foundations, the total stress analysis for short-term stability is more critical because of the increase in pore pressure due to loading and the decrease in pore pressure, or the increase in effective stress, with time. For cut slopes in saturated soils, the effective stress analysis for long-term stability is more critical because of the decrease in pore pressure due to unloading and the increase in pore pressure, or the decrease in effective stress, with time. In certain cases, a failure surface may pass partly through a free draining soil, where strength is appropriately expressed in terms of effective stress, and partly through a clay, where undrained strength should be used. In such cases, the parameters $\bar{c}$ and $\bar{\phi}$, together with appropriate pore pressures, apply along one portion of the surface and the $\phi = 0$ or c_u, ϕ_u-analysis with zero pore pressure applies along the other part.

Equation 2.3 is based on total stress analysis, which does not include the effect of seepage. If there is any seepage, it can be represented by a phreatic surface as shown in Fig. 2.6. A pore pressure equal to the depth below the phreatic surface, h_{iw}, multiplied by the unit weight of water, γ_w, will exist along the failure surface. Therefore, the effective normal force, $\bar{N}_i$, is equal to the total normal force, N_i, minus the neutral force $\gamma_w h_{iw} b_i \sec \theta_i$.

In terms of effective stress, the Mohr-Coulomb theory can be represented by

$$s = \bar{c} + \bar{\sigma}_n \tan \bar{\phi} \qquad (2.6)$$

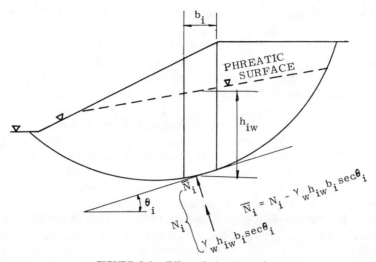

FIGURE 2.6. Effect of phreatic surface.

in which $\bar{c}$ is the effective cohesion, $\bar{\sigma}_n$ is the effective normal stress, and $\bar{\phi}$ is the effective angle of internal friction. Therefore, Eq. 2.3 for cylindrical failure can be written in terms of the effective stress as

$$F = \frac{\sum\limits_{i=1}^{n}(\bar{c}L_i + \bar{N}_i \tan \bar{\phi})}{\sum\limits_{i=1}^{n}(W_i \sin \theta_i)} \qquad (2.7)$$

One simple method for effective stress analysis is by assuming

$$\bar{N}_i = (1 - r_u) W_i \cos \theta_i \qquad (2.8)$$

in which r_u is the pore pressure ratio, which will be discussed in Chap. 4. Comparing Eq. 2.8 with Eq. 2.4, it can be seen that the normal stress for effective stress analysis is reduced by a factor of $1 - r_u$. By substituting Eq. 2.8 into Eq. 2.7

$$F = \frac{\sum\limits_{i=1}^{n}[\bar{c}L_i + (1 - r_u)W_i \cos \theta_i \tan \bar{\phi}]}{\sum\limits_{i=1}^{n}(W_i \sin \theta_i)} \qquad (2.9)$$

The use of Eq. 2.9 for determining the factor of safety is called the normal method. Another method, which has received more popularity, is the simplified Bishop method, in which the forces on the two sides of a slice are assumed to be horizontal. In the total stress analysis where $\phi = 0$, both methods result in the same factor of safety, whereas in the effective stress analysis, the normal method generally results in a smaller factor of safety and is therefore on the safe side.

In effective stress analysis, three cases need to be considered: steady-state seepage, rapid drawdown, and earthquake. The case of steady-state seepage is shown in Fig. 2.6. The phreatic surface outside the slope is along the face of the slope and the ground surface.

Rapid drawdown is usually the most critical situation in the design of earth dams. The downstream slope is controlled by the case of steady-state seepage, but the upstream slope is controlled by the case of rapid drawdown. As shown in Fig. 2.7, the phreatic surface under rapid drawdown is along the dashed line and the surface of both slopes. As more of the sliding mass is under water, the upstream slope of the dam may be made flatter than the downstream slope.

In the case of earthquake, a horizontal seismic force is applied at the centroid of each slice, as shown in Fig. 2.8. The seismic force is equal to

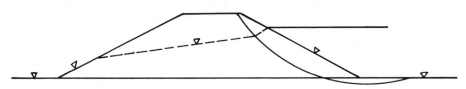

FIGURE 2.7. Phreatic surface for rapid drawdown.

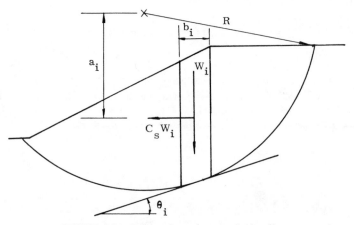

FIGURE 2.8. Driving force due to seismic effect.

C_sW_i, where C_s is the seismic coefficient and ranges from 0.03 to 0.27 or more depending on the geographic location. It is assumed that this force has no effect on the normal force, $\bar{N}_i$, or the resisting moment, so only the driving moment is affected. The factor of safety can be determined by

$$ F = \frac{\sum\limits_{i=1}^{n} (\bar{c}b_i \sec \theta_i + \bar{N}_i \tan \bar{\phi})}{\sum\limits_{i=1}^{n} (W_i \sin \theta_i + C_sW_ia_i/R)} \qquad (2.10) $$

in which b_i is the width of slice, a_i is the moment arm, and R is the radius of circle. By using the normal method

$$ F = \frac{\sum\limits_{i=1}^{n} [\bar{c}b_i \sec \theta_i + (1-r_u)W_i \cos \theta_i \tan \bar{\phi}]}{\sum\limits_{i=1}^{n} (W_i \sin \theta_i + C_sW_ia_i/R)} \qquad (2.11) $$

Figure 2.9 shows the seismic zone map of continental United States (Algermissen, 1969). The seismic coefficient, C_s, to be used for each zone is shown in Table 2.2. In Table 2.2, Eq. 2.12, based on an average epicentral distance of 15 miles (24 km), is used to determine the seismic coefficient (Neumann, 1954).

$$ C_s = \frac{\log^{-1} [0.267 + (MM - 1) \times 0.308]}{980} \qquad (2.12) $$

in which MM indicates the Modified Mercalli intensity scale.

The earthquake analysis presented above is called the psuedo-static method. For high risk dams in seismically active regions, more sophisticated dynamic analyses as suggested by Seed et al. (1975a, 1975b) should be used.

In effective stress analysis for rapid drawdown or earthquake, it is assumed that the pore pressure after drawdown or earthquake is the same as before. If these excitations cause a significant change in pore pressure, total stress analysis may have to be used.

2.5 GUIDELINES FOR STABILITY ANALYSIS

Figure 2.10 shows the guidelines for analyzing fills and dams. Because the short-term stability is usually more critical, total stress analysis at the end of construction should be performed first. Effective stress analysis can also be used to check stability at the end of construction, replacing the total stress

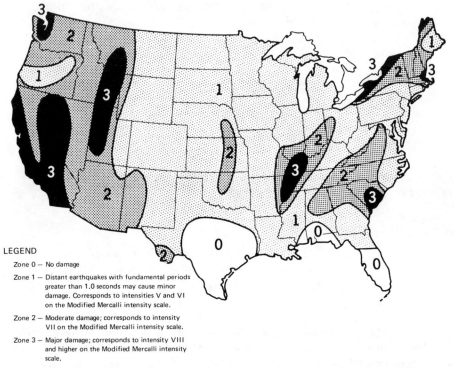

LEGEND

Zone 0 — No damage

Zone 1 — Distant earthquakes with fundamental periods greater than 1.0 seconds may cause minor damage. Corresponds to intensities V and VI on the Modified Mercalli intensity scale.

Zone 2 — Moderate damage; corresponds to intensity VII on the Modified Mercalli intensity scale.

Zone 3 — Major damage; corresponds to intensity VIII and higher on the Modified Mercalli intensity scale.

FIGURE 2.9.   Seismic zone map of continental United States (After Algermissen, 1969)

analysis, if the pore pressure is known or measured. Effective stress analyses can then be made on steady-state seepage, rapid drawdown, and earthquake, respectively.

Figure 2.11 shows guidelines for analyzing cut slopes. Because the long-term stability is usually more critical, an effective stress analysis for long-term stability should be made first. If the required factor of safety for steady-state seepage is the same as that for rapid drawdown, only the case of rapid drawdown need be considered because it is more critical than the steady-state seepage. After the long-term stability is satisfied, the short-term stability at the end of construction should be checked.

It should be noted that the case of rapid drawdown cannot occur in

Table 2.2   Seismic Coefficients Corresponding to Each Zone.

ZONE	INTENSITY OF MODIFIED MERCALLI SCALE	AVERAGE SEISMIC COEFFICIENT	REMARK
0	—	0	No damage
1	V and VI	0.03 to 0.07	Minor damage
2	VII	0.13	Moderate damage
3	VII and higher	0.27	Major damage

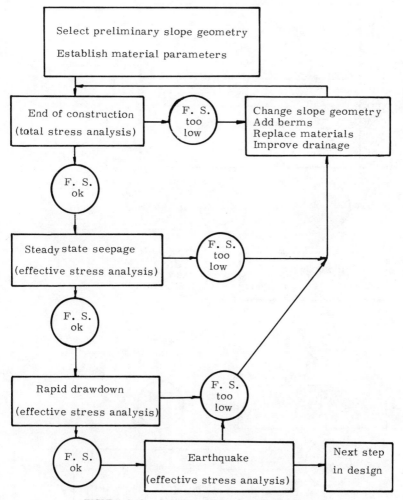

FIGURE 2.10. Stability analysis for fills and dams.

coarse-grained soils because the pore pressure in the slope will dissipate as the pool level lowers. Rapid drawdown also cannot occur in fine-grained soils unless the cut or fill slope is used to store water for a long period of time, such as in a reservoir. Occasional flooding for a few days or weeks cannot saturate the slope to effect the detrimental drawdown conditions.

2.6 FACTORS OF SAFETY

In the stability analysis of slopes, many design factors cannot be determined with certainty. Therefore, a degree of risk should be assessed in an adopted design. The factor of safety fulfills this requirement. The factor should take into account not only the uncertainties in design parameters but also the con-

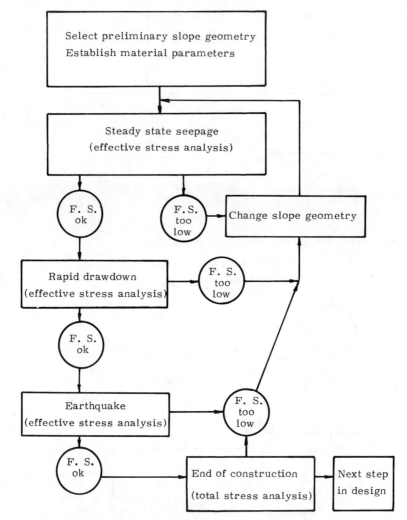

FIGURE 2.11. Stability analysis for permanent cut slopes.

sequences of failure. Where the consequences of failure are slight, a greater risk of failure or a lower factor of safety may be acceptable.

The potential seriousness of failure is related to many factors other than the size of project. A low dam located above or close to inhabited buildings can pose a greater danger than a high dam in a remote location. Often, the most potentially dangerous types of failure involve soils that undergo a sudden release of energy without much warning. This is true for soils subjected to liquefaction and that have a low ratio between the residual and peak strength.

Table 2.3 shows the factors of safety suggested by various sources for mining operations (D'Appolonia Consulting Engineers, 1975; Federal Register, 1977; Mine Branch, Canada, 1972; National Coal Board, 1970). All of these

Table 2.3 Factors of Safety Suggested for Mining Operations.

UNITED STATES (FEDERAL REGISTER, 1977)		MINIMUM SAFETY FACTOR
I	End of construction	1.3
II	Partial pool with steady seepage saturation	1.5
III	Steady seepage from spillway or decant crest	1.5
IV	Earthquake (cases II and III with seismic loading)	1.0

UNITED STATES (D'APPOLONIA CONSULTING ENGINEERS, INC., 1975)	SUGGESTED MINIMUM FACTORS OF SAFETY WITH HAZARD POTENTIAL		
	HIGH	MODERATE	LOW
Designs based on shear strength parameters measured in the laboratory	1.5	1.4	1.3
Designs that consider maximum seismic acceleration expected at the site	1.2	1.1	1.0

BRITAIN (NATIONAL COAL BOARD, 1970)	FACTOR OF SAFETY	
	I*	II**
(1) For slip surfaces along which the peak shear stress is used.	1.5	1.25
(2) For slip surfaces passing through a foundation stratum which is at its residual shear strength (slip circles wholly within the bank should satisfy (1)).	1.35	1.15
(3) For slip surfaces passing along a deep vertical subsidence crack where no shear strength is mobilized and which is filled with water (slip surfaces wholly within intact zones of bank and foundations should satisfy (1)).	1.35	1.15
(4) For slip surfaces where both (2) and (3) apply.	1.2	1.1

CANADA (MINES BRANCH, CANADA, 1972)	FACTOR OF SAFETY	
	I*	II**
Design is based on peak shear strength parameters	1.5	1.3
Design is based on residual shear strength parameters	1.3	1.2
Analyses that include the predicted 100-year return period accelerations applied to the potential failure mass	1.2	1.1
For horizontal sliding on base of dike in seismic areas assuming shear strength of fine refuse in impoundment reduced to zero	1.3	1.3

*where there is a risk of danger to persons or property
**where no risk of danger to persons or property is anticipated

stipulations are based on the assumptions that the most critical failure surface is used in the analysis, that strength parameters are reasonably representative of the actual case, and that sufficient construction control is ensured.

For earth slope composed of intact homogeneous soils, when the strength parameters have been chosen on the basis of good laboratory tests and a careful estimate of pore pressure has been made, a safety factor of at least 1.5 is commonly employed (Lambe and Whitman, 1969). With fissured clays and for nonhomogeneous soils, larger uncertainties will generally exist and more caution is necessary.

3
Shear Strength

3.1 SUBSURFACE INVESTIGATIONS

The shear strength of soils can be determined by field or laboratory tests. No matter what tests are used, it is necessary to conduct an overall geologic appraisal of the site, followed by a planned subsurface investigation. The purpose of the subsurface investigation is to determine the nature and extent of each type of material that may have an effect on the stability of the slope. A detailed knowledge of the slope from toe to crest is essential. Fills situated over a deep layer of clays and silts may merit expensive drilling. Auger holes, pits, or trenches will suffice for smaller fills or those with bedrock only a short distance below the surface.

The log of boring forms the permanent record used for design. Both disturbed and undisturbed samples can be taken while boring. To obtain reliable results, the strength parameters should be determined from undisturbed or remolded samples. However, the effective strength parameters of saturated granular soils and silty clays are not affected significantly by the moisture content and density, so disturbed samples may be used for the direct shear test to determine the effective strength. It is difficult to generalize the appropriate number, depth, and spacing of borings required for a project since these will depend upon a variety of factors, such as site conditions, size of the project, etc. Often the final location of borings should be made in the field and additions must be made to the boring program based on the information from the borings already completed. During boring the depth to the ground water table can also be determined.

3.2 FIELD TESTS

There are a variety of field tests for determining the shear strength of soils. However, only the standard penetration test and the Dutch cone test, which are used in conjunction with subsurface explorations, and the vane shear test will be discussed here. These tests are applicable to soils free from substantial gravel or cobble-sized particles.

For sands, some attempts have been made to correlate the effective friction angle, $\bar{\phi}$, with results from both the standard penetration test and the Dutch cone test. Curves for determining $\bar{\phi}$ are given in Fig. 3.1 for the standard penetration test (Schmertmann, 1975) and Fig. 3.2 for the Dutch cone test (Trofimenkov, 1974). Both figures require the determination of the overburden pressure in terms of effective stress.

To determine the effective friction angle, it is necessary to calculate the effective overburden pressure during the field test. If there is no water table, the effective overburden pressure is equal to the depth below ground surface multiplied by the total unit weight of soil. If part of the overburden is below the water table, the submerged unit weight should be used for the overburden below the water table.

Figures 3.1 and 3.2 are designed for cohesionless sands but are also applicable to fine-grained soil with a small cohesion. If the material tested has some cohesion, the effective friction angle determined from the figures should be reduced a few degrees to compensate for the extra cohesion. By keeping the total strength the same, i.e., $\bar{c} + \bar{p} \tan \bar{\phi}_c = \bar{p} \tan \bar{\phi}$, the corrected angle of internal friction can be determined by

$$\tan \bar{\phi}_c = \tan \bar{\phi} - \frac{\bar{c}}{\bar{p}} \tag{3.1}$$

in which $\bar{\phi}_c$ is the corrected effective friction angle, $\bar{\phi}$ is the effective friction angle obtained from Figs. 3.1 and 3.2, $\bar{c}$ is the effective cohesion, which can be arbitrarily assumed a small value, say 0.05 to 0.1 tsf (4.8 to 9.6 kPa); and $\bar{p}$ is the effective overburden pressure.

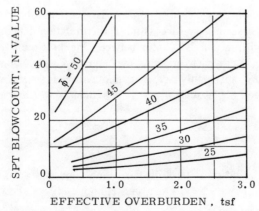

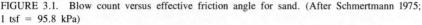

FIGURE 3.1. Blow count versus effective friction angle for sand. (After Schmertmann 1975; 1 tsf = 95.8 kPa)

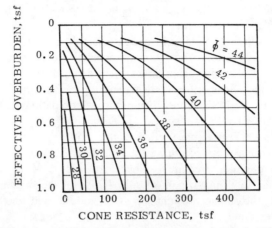

FIGURE 3.2.   Cone resistance versus effective friction angle for sand. (After Trofimenkov, 1974; 1 tsf = 95.8 kPa)

Figure 3.3 shows the correlation between the standard penetration test and the unconfined compressive strength for clays (Department of Navy, 1971). The undrained shear strength, or the cohesion for total stress analysis with $\phi = 0$, is equal to one half of the unconfined compressive strength. If the clay is normally consolidated with $N <$ depth in feet/5 or $N <$ depth in meters/1.5, Schmertmann (1975) suggests

$$s_u \ (\text{tsf}) \geq N/15 \tag{3.2}$$

or

$$s_u \ (\text{kPa}) \geq 6.4N \tag{3.3}$$

in which s_u is the undrained shear strength and N is the blow count per foot of penetration.

The correlation between the cone resistance and the undrained shear strength for clays is not very successful. The correlation depends on the over-consolidation ratio, which is a ratio between the maximum precompression and the existing overburden pressure. An equation used by many engineers for clays that are not highly sensitive and with an overconsolidation ratio smaller than 2 and a plasticity index greater than 10 is (Schmertmann, 1975)

$$s_u = \frac{q_c - \gamma z}{16} \tag{3.4}$$

in which q_c is the bearing stress on the Dutch cone and γz is the total over-burden pressure. Drnevich, Gorman and Hopkins (1974) show that

$$s_u = 0.8 \times \text{friction sleeve resistance} \tag{3.5}$$

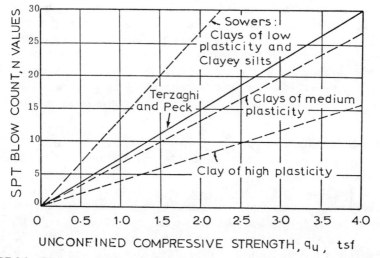

FIGURE 3.3.   Blow count versus unconfined compressive strength. (After Department of Navy, 1971; 1 tsf = 95.8 kPa)

The vane shear test is commonly used for determining the undrained strength of clays in situ. The test basically consisting of placing a four-blade vane in the undisturbed soil and rotating it from the surface to determine the torsional force required to cause a cylindrical surface to be sheared by the vane. This force is then converted to the undrained shear strength. Both the peak and residual undrained strength can be determined by measuring the maximum and the postmaximum torsional forces. The standard method for the field vane shear test is described in ASTM (1981). Because of the difference in failure mode, the results of vane tests do not always agree with other shear tests. Although empirical corrections are available, such as the correction curves proposed by Bjerrum (1972), the vane strength should be used with caution in highly plastic or very sensitive soils, where experience shows that the vane strength usually exceeds other results.

3.3 LABORATORY TESTS

Laboratory tests complement field tests to give a more complete picture of the materials within the slope and their engineering properties. Furthermore, it is possible in the laboratory to establish the changes in soil behavior due to the changes in environment. For example, the construction of an embankment will certainly affect the shear strength in the foundation soils. Field tests before construction cannot establish these changes, while laboratory tests can simulate these changes as they occur in the field.

The major laboratory tests for determining the shear strength of soils include the direct shear test, the triaxial compression test, the unconfined compression test, and the laboratory vane shear test. The direct shear test is very easy to operate because of its simplicity. Due to the thin specimen used, drained conditions exist for most materials except for the highly plastic clays. Therefore, the direct shear strength is usually in terms of the effective stress. The triaxial compression test can be used for determining either the total strength or the effective strength. The unconfined compression and the vane shear test can only be used for determining the undrained shear strength.

Direct Shear Test. Figure 3.4 shows a schematic diagram of the direct shear box. The sample is placed between two porous stones to facilitate drainage. The normal load is applied to the sample by placing weights in a hanger system. The shear force is supplied by the piston driven by an electric motor. The horizontal displacement is measured by a horizontal dial and the shear force by a proving ring and load dial, which are not shown in the figure.

For granular materials and silty clays, such as coal refuse and mine spoils, Huang (1978b) found that their effective strength can be easily determined by the direct shear tests, which check closely with the results of triaxial compression tests with pore pressure measurements. His suggested procedure is as follows:

The soil is air-dried and sieved through a no. 4 sieve (4.75 mm). The material retained on no. 4 is discarded because the specimen is only 2.5 in. (63.5 mm) in diameter and is not adequate for large particles. The material passing the no. 4 sieve is mixed with a large amount of water to make it very plastic and then is placed in the direct shear box. To prevent the sample from squeezing out, a Teflon ring is used to separate the two halves of the shear box.

After a given normal stress is applied for about 10 min, the shear stress is

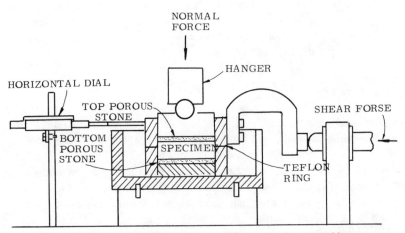

FIGURE 3.4. Schematic diagram of direct shear box assembly.

applied at a rate of 0.02 in. (0.5 mm) per min until the specimen fails, as indicated by a decrease in the reading of the proving ring dial. If the specimen does not fail, the test is stopped at 25 min, or a horizontal deformation of 0.5 in. At least three tests involving three different normal stresses must be performed. Figure 3.5 shows the stress–displacement curves of a fine refuse under three different normal stresses: 0.52, 1.55, and 2.58 tsf (49.8, 148.5, and 247.2 kPa). In all three curves the shear stress increases with the horizontal displacement up to a peak and then decreases until a nearly constant value is obtained. The shear stress at the peak is called the peak shear strength, while that at the constant value is called the residual shear strength. Because of progressive failure, the average shear strength actually developed along a failure surface is somewhere between the peak and the residual strength. If the two strengths are not significantly different, as is the case shown in Fig. 3.5, the peak strength can be used. Figure 3.6 shows a plot of peak shear stress versus normal stress for fine coal refuse. A straight line passing through the three points is drawn. The vertical intercept at zero normal stress is the effective cohesion, $\bar{c}$, and the angle of the straight line with the horizontal is the effective friction angle, $\bar{\phi}$. The figure shows that the fine coal refuse has an effective cohesion of 0.1 tsf (9.6 kPa) and an effective friction angle of 35.4°.

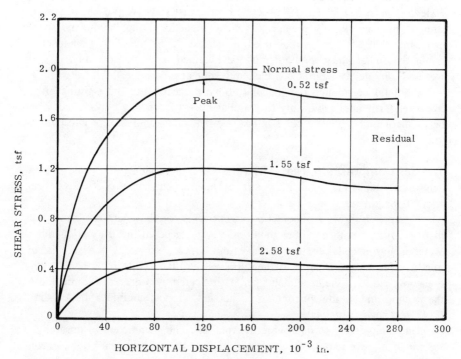

FIGURE 3.5.   Stress-displacement curves of fine refuse. (1 in = 2.54 cm, 1 tsf = 95.8 kPa)

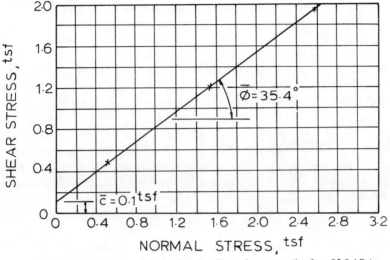

FIGURE 3.6.   Strength of fine refuse by direct shear test. (1 tsf = 95.8 kPa)

It is believed that the strength parameters determined from the above procedure are quite conservative because: (1) Only the fraction passing the no. 4 sieve is used in the test. If sufficient coarse materials are present, the shear strength may be slightly greater. (2) No compaction is applied to the specimen other than the normal load used in the test. If the material is compacted, a slightly higher shear strength may be obtained. (3) The specimen is very wet and completely saturated which may not occur in the field.

It should be pointed out that the above procedure for determining effective strength is applicable only to granular materials or silty clays. It may not be used for highly plastic clays unless the rate of loading is kept exceedingly low.

Triaxial Compression Test.   The triaxial compression test can be used to determine either the total peak strength parameters or the effective peak strength parameters.

Figure 3.7 shows a schematic diagram of a triaxial chamber. The specimen is covered with a rubber membrane and placed in the triaxial chamber. Water is introduced into the chamber and a given confining pressure is applied. A vertical axial stress is then applied, and the deformation and loading dials are read until the specimen fails, as indicated by a decrease in reading of the loading dial. If the specimen does not fail, the test continues until a strain of 15 percent is obtained.

One simple way to determine the total strength parameters of unsaturated soils is to prepare two identical specimens, and then subject one to the unconfined compression test and the other to the triaxial compression test. The confining pressure used for the triaxial test should nearly equal the maximum

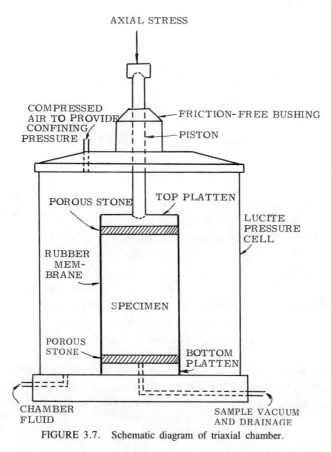

AXIAL STRESS

COMPRESSED
AIR TO PROVIDE ──── FRICTION-FREE BUSHING
CONFINING
PRESSURE ──── PISTON

POROUS STONE   TOP PLATTEN

LUCITE
PRESSURE
CELL

RUBBER
MEM-
BRANE

SPECIMEN

POROUS
STONE

BOTTOM
PLATTEN

CHAMBER
FLUID

SAMPLE VACUUM
AND DRAINAGE

FIGURE 3.7.   Schematic diagram of triaxial chamber.

overburden pressure expected in the field. The procedure for the unconfined compression test is similar to the triaxial test, except that the specimen is not enclosed in the rubber membrane and that no confining pressure is applied. To prevent any drainage in the triaxial test, the drainage valves must be closed. After both tests are completed, two Mohr's circles are drawn and a straight line tangent to these two circles is the Mohr's envelope. The vertical intercept of the envelope at zero normal stress is the cohesion, and the angle of the envelope with the horizontal is the friction angle as shown in Fig. 3.8. The total strength parameters generally exhibit a large cohesion and a small friction angle. If the specimen is completely saturated, the envelope will be horizontal with an angle of internal friction equal to zero.

The effective strength parameters can be determined by a consolidated drained test or consolidated undrained test with pore pressure measurements. Instead of using the total normal stress as shown in Fig. 3.8, the shear stress is plotted versus the effective normal stress, and a Mohr's envelope in terms of effective stress is obtained. The vertical intercept of the envelope at zero

BROWNISH SANDY MINE SPOIL

MOISTURE CONTENT   18.7 %

DRY UNIT WEIGHT   106 pcf

OPTIMUM MOISTURE CONTENT   13.5 %

MAXIMUM DRY DENSITY   115 pcf

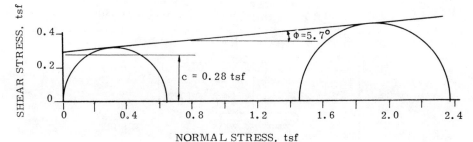

FIGURE 3.8. Total strength parameters of compacted specimen. (1 tsf = 95.8 kPa, 1 pcf = 157.1 N/m³)

effective normal stress is the effective cohesion, and the angle of the envelope with the horizontal is the effective friction angle. The effective strength parameters always exhibit a small effective cohesion and a large effective friction angle.

Another procedure to obtain the effective strength parameters is by the use of stress path method (Lambe and Whitman, 1969). Figure 3.9 shows the $\bar{p}$ versus q diagram of the fine coal refuse used in the direct shear test shown in Fig. 3.6. Note that

$$\bar{p} = \frac{\bar{\sigma}_1 + \bar{\sigma}_3}{2} \qquad (3.6)$$

$$q = \frac{\sigma_1 - \sigma_3}{2} = \frac{\bar{\sigma}_1 - \bar{\sigma}_3}{2} \qquad (3.7)$$

in which σ_1 is the major principal stress and σ_3 is the minor principal stress. The corresponding effective stresses, $\bar{\sigma}_1$ and $\bar{\sigma}_3$, can be determined by substracting the measured pore pressure from the total stresses, σ_1 and σ_3.

The triaxial compression tests with pore pressure measurements were made on two specimens, one subject to an effective confining pressure of 0.6 tsf (58 kPa) and the other to 1.4 tsf (134 kPa). To saturate the specimens, filter strips were placed around the sample and a back pressure was applied. A straight line tangent to the failure part of the stress path is called the K_f-line. The angle of the K_f-line with the horizontal is called $\bar{\alpha}$, which is related to $\bar{\phi}$ by

$$\sin \bar{\phi} = \tan \bar{\alpha} \qquad (3.8)$$

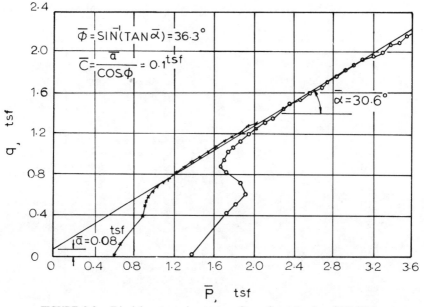

FIGURE 3.9. Triaxial compression test on fine refuse. (1 tsf = 95.8 kPa)

The intercept of K_f-line with the q-axis is called $\bar{a}$, which is related to $\bar{c}$ by

$$\bar{c} = \frac{\bar{a}}{\cos \bar{\phi}} \qquad (3.9)$$

Figure 3.9 shows that the fine coal refuse has an effective cohesion of 0.10 tsf (7.7 kPa) and an effective friction angle of 36.3°, which checks closely with the result of direct shear test.

An advantage of using the stress path method is that the effective cohesion and the effective friction angle may be estimated from a single test by approximating a straight line through the stress points at the latter part of the test, whereas two tests are required if the Mohr's circle is used.

3.4 TYPICAL RANGES AND CORRELATIONS

Table 3.1 shows the average effective shear strength of soils compacted to proctor maximum dry density at optimum moisture content (Bureau of Reclamation, 1973). The shear strength tabulated is the effective shear strength, or

$$s = \bar{c} + (\sigma_n - u) \tan \bar{\phi} \qquad (3.10)$$

in which u is the pore water pressure. If the soil is subject to saturation, $\bar{c} = c_{sat}$. If the soil is at the optimum moisture content and the maximum density, $\bar{c} = c_o$.

Table 3.1 Average Effective Shear Strength of Compacted Soils.

UNIFIED CLASSIFICATION	SOIL TYPE	PROCTOR COMPACTION		AS COMPACTED COHESION c_o tsf	SATURATED COHESION c_{sat} tsf	FRICTION ANGLE $\bar{\phi}$ deg
		MAXIMUM DRY DENSITY pcf	OPTIMUM MOISTURE CONTENT %			
GW	well graded clean gravels, gravel-sand mixture	>119	<13.3	*	*	>38
GP	poorly graded clean gravels, gravel sand mixture	>110	<12.4	*	*	>37
GM	silty gravels, poorly graded gravel-sand-silt	>114	<14.5	*	*	>34
GC	clayey gravels, poorly graded gravel-sand-clay	>115	<14.7	*	*	>31
SW	well graded clean sands, gravelly sands	119±5	13.3±2.5	0.41±0.04	*	38±1
SP	poorly graded clean sands, sand-gravel mixture	110±2	12.4±1.0	0.24±0.06	*	37±1
SM	silty sands, poorly graded sand-silt mixture	114±1	14.5±0.4	0.53±0.06	0.21±0.07	34±1
SM-SC	sand-silt-clay with slightly plastic fines	119±1	12.8±0.5	0.21±0.07	0.15±0.06	33±3
SC	clayey sands, poorly graded sand-clay mixture	115±1	14.7±0.4	0.78±0.16	0.12±0.06	31±3
ML	inorganic silts and clayed silts	103±1	19.2±0.7	0.70±0.10	0.09±*	32±2
ML-CL	mixtures of inorganic silts and clays	109±2	16.8±0.7	0.66±0.18	0.23±*	32±2
CL	inorganic clays of low to medium plasticity	108±1	17.3±3	0.91±0.11	0.14±0.02	28±2
OL	organic silts and silty clays of low plasticity	*	*	*	*	*
MH	inorganic clayey silts, elastic silts	82±4	36.3±3.2	0.76±0.31	0.21±0.09	25±3
CH	inorganic clays of high plasticity	94±2	25.5±1.2	1.07±0.35	0.12±0.06	19±5
OH	organic clays and silty clays	*	*	*	*	*

*denotes insufficient data, > is greater than, < is less than
(After Bureau of Reclamation, 1973; 1 pcf=157.1 N/m^3, 1 tsf=95.8 kPa)

The shear strength listed in Table 3.1 is for compacted soils. For natural soils, the effective cohesion may be larger or smaller than the listed values depending on whether the soil is overly or normally consolidated, but the effective angle of internal friction should not be much different. Kenney (1959) presented the relationship between sin $\bar{\phi}$ and the plasticity index for normally consolidated soils, as shown in Figure 3.10. Although there is considerable scatter, a definite trend toward decreasing $\bar{\phi}$ with increasing plasticity is apparent. Bjerrum and Simons (1960) presented a similar relationship for both undisturbed and remolded soil as shown in Fig. 3.11. The relationship by Kenney (1959) is plotted in dashed curve for comparison. Skempton (1964) presented a correlation between the residual effective angle of internal

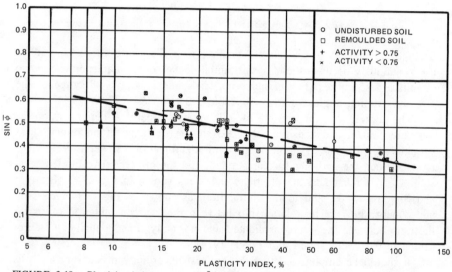

FIGURE 3.10. Plasticity index versus sinϕ for normally consolidated soils. (After Kenney, 1959)

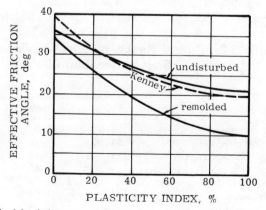

FIGURE 3.11. Plasticity index versus effective friction angle. (After Bjerrum and Simmons, 1960)

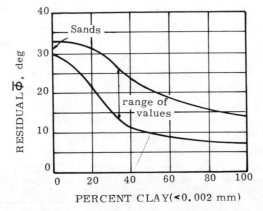

FIGURE 3.12.   Percent clay versus residual effective friction angle. (After Skempton, 1964)

friction and percent of clay as shown in Fig. 3.12. These curves should be used cautiously because there is substantial scatter in the data points used to establish these curves. For granular materials and silts, typical range of the effective friction angle is shown in Table 3.2 (Bowles, 1979).

The undrained shear strength of soils varies a great deal depending on the moisture content and density. Table 3.3 shows the range of undrained shear strength of soils and a simple method of identification (Sowers, 1979). The ratio between the undrained shear strength and the effective overburden pressure, $s_u/\bar{p}$, is related to the plasticity index, as shown in Fig. 3.13 (Bjerrum and Simons, 1960), and to the liquid limit as shown in Fig. 3.14 (Karlsson and Viberg, 1967). The relationship between the overconsolidation ratio and the ratio between the overconsolidated and normally consolidated $s_u/\bar{p}$ is presented in Fig. 3.15 (Ladd and Foott, 1974).

It should be emphasized that the greatest uncertainty in stability analysis arises in the determination of shear strength. The error associated with stability computations is usually small compared with that arising from the selection of strength parameters. Therefore, careful judgment in the selection of strength parameters is needed in the stability analysis of slopes.

Table 3.2 Typical Range of Effective Friction Angle for Soils Other than Clays.

	EFFECTIVE FRICTION ANGLE, deg	
SOIL	LOOSE	DENSE
Gravel, crushed	36–40	40–50
Gravel, bank run	34–38	38–42
Sand, crushed (angular)	32–36	35–45
Sand, bank run (subangular)	30–34	34–40
Sand, beach (well rounded)	28–32	32–38
Silty sand	25–35	30–36
Silt, inorganic	25–35	30–35

(After Bowles, 1979)

Table 3.3 Undrained Shear Strength of Soils.

CONSISTENCY	UNDRAINED SHEAR STRENGTH, tsf	FIELD TEST
Very soft	0–1	Squeezed between fingers when fist is closed
Soft	1–2	Easily molded by fingers
Firm	2–4	Molded by strong pressure of fingers
Stiff	4–6	Dented by strong pressures of fingers
Very stiff	6–8	Dented only slightly by finger pressure
Hard	8+	Dented only slightly by pencil point

(After Sowers, 1979; 1 tsf = 95.8 kPa)

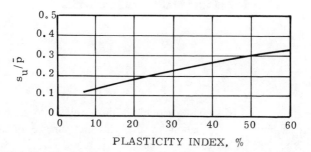

FIGURE 3.13. Plasticity index versus $s_u/\bar{p}$. (After Bjerrum and Simons, 1960)

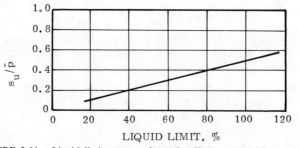

FIGURE 3.14. Liquid limit versus $s_u/\bar{p}$. (After Karlsson and Viberg, 1967)

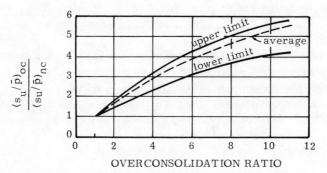

FIGURE 3.15. Relationship between overconsolidated and normally consolidated $s_u/\bar{p}$. (After Ladd and Foott, 1974)

4
Phreatic Surfaces

4.1 FLOW NETS

In the stability analysis of slopes, particularly those related to earth dams, it is necessary to estimate the location of the phreatic surface or the line of seepage. In the case of an existing slope, the phreatic surface can be determined from the subsurface investigation with adjustments for seasonal changes. If the slope has not been constructed and is quite complex in configuration, the easiest way to determine the phreatic surface is by drawing a flow net, as shown in Fig. 4.1. For a homogeneous and isotropic cross section, if the flow net satisfies the basic requirements that the flow lines and equipotential lines are perpendicular and form squares or rectangles of the same shape and that the vertical distance between equipotential lines along the phreatic surface are the same, the assumed phreatic surface is correct; otherwise, the phreatic surface must be changed until a satisfactory flow net is obtained. For an anisotropic cross section, a transformation based on the ratio between vertical and horizontal permeabilities must be made so that square flow nets can still be constructed. For a nonhomogeneous cross section, the flow nets must satisfy the continuity and interface conditions; as a result, the flow nets in some regions must be rectangular instead of square. Methods for constructing flow nets can be found in most textbooks in soil mechanics and also in Cedergren (1977).

It should be noted that the phreatic surface based on the construction of flow net or the measurement in the field during investigation will be different from the one at the time of failure. Because an adequate factor of safety is required in any design, the slope is not supposed to fail, so the phreatic surface prior to failure can be used for stability analyses. However, if a slope has failed and a stability analysis is made to back calculate the shear strength, the pore pressure at the time of failure should be measured or properly estimated.

FILTER

ROCK TOE

TOE DRAIN

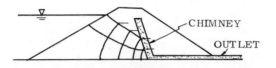

CHIMNEY

OUTLET

CHIMNEY DRAIN

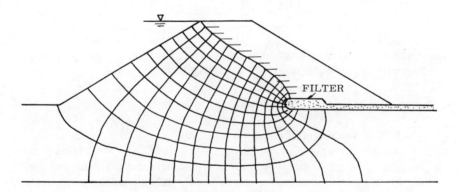

FILTER

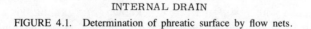

INTERNAL DRAIN

FIGURE 4.1. Determination of phreatic surface by flow nets.

4.2 EARTH DAMS WITHOUT FILTER DRAINS

Figure 4.2 shows an earth dam on an impervious base. The downstream face of the dam has a slope of S: 1 (horizontal to vertical). If no drainage system is provided, the downstream slope should be relatively flat, generally with a slope not steeper than 1.5:1 (horizontal to vertical). In such a case, Dupuit's assumption that in every point on a vertical line the hydraulic gradient is constant and equal to the slope, dy/dx, is valid. The seepage through the dam can be expressed by Darcy's law as

$$q = k(y - x \tan \alpha)\frac{dy}{dx} \qquad (4.1)$$

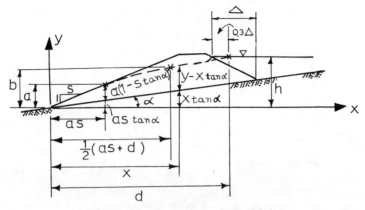

FIGURE 4.2.   Earth dam on inclined ledge.

in which q is the discharge per unit time, k is the permeability, α is the angle of inclination of base, and x and y are the coordinates. At the point of exit

$$q = \frac{ka(1 - S \tan \alpha)}{S} \qquad (4.2)$$

in which a is the y-coordinate of the exit point. Combining Eqs. 4.1 and 4.2 and integrating, an equation of the following form is obtained.

$$\text{Function } (x, y, c_1) = 0 \qquad (4.3)$$

in which c_1 is a constant of integration.

Following Casagrande's procedure, it is assumed that the theoretical line of seepage starts from the pool level at a distance of 0.3Δ from the dam, where Δ is the horizontal distance shown in Fig. 4.2. Therefore, when the toe of the downstream slope is used as the origin of coordinates, one point on the line of seepage with $x = d$ and $y = h$ is known. Substituting this x,y pair into Eq. 4.3 allows evaluation of the constant of integration, c_1. Assume that the x and y coordinates of the exit point are aS and a, respectively. Substituting this x,y pair into Eq. 4.3, an equation of the following form is obtained

$$\text{Function } (a) = 0 \qquad (4.4)$$

Equation 4.4 was solved by Huang (1981a) using a numerical method and the results are presented in Fig. 4.3. Next, assume that the x and y coordinates of the midpoint are $(aS + d)/2$ and b, respectively. Substituting this x,y pair into Eq. 4.3 and bearing in mind that the value of a has been determined form Eq. 4.4, an equation of the following form is obtained

$$\text{Function } (b) = 0 \qquad (4.5)$$

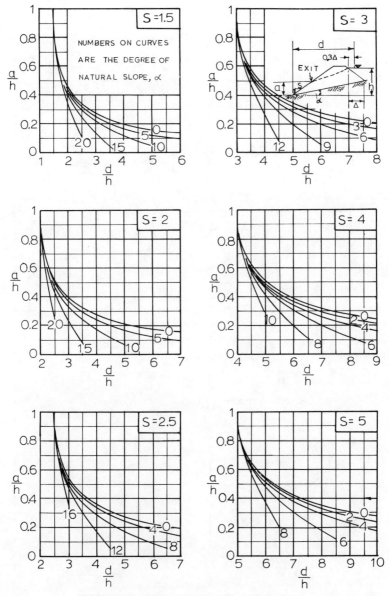

FIGURE 4.3. Chart for determining point of exit.

The solution of Eq. 4.5 is presented in Fig. 4.4.

Knowing three points on the phreatic surface, i.e., the starting point, the midpoint, and the exit point, a curve can be drawn, which is the theoretical line of seepage. Because the actual line of seepage must be perpendicular to the upstream slope and tangent to the downstream slope, the theoretical line can be slightly adjusted to fulfill these boundary requirements.

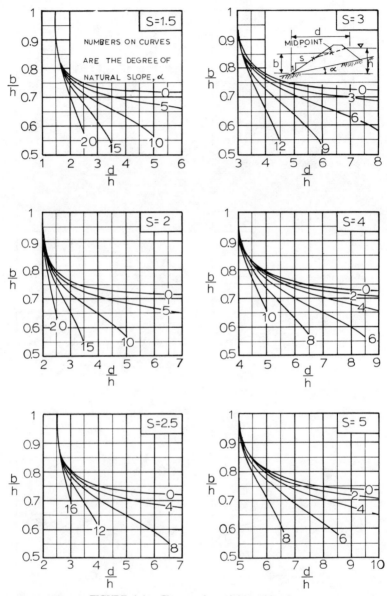

FIGURE 4.4. Chart or determining mid point.

When the dam is constructed on a horizontal base, i.e., $\alpha = 0$, the solution is the same as that developed by Schaffernak and Iterson in 1916 as described by Casagrande (1937).

The application of Figs. 4.3 and 4.4 can be illustrated by the following example:

Figure 4.5 shows the cross section of a dam. As shown in the figure, d

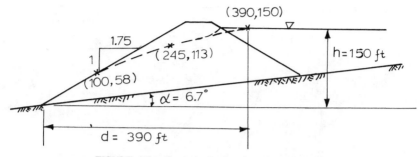

FIGURE 4.5. Example for locating phreatic surface.

$= 390$ ft (118.9 m) and $h = 150$ ft (45.7 m), thus $d/h = 2.6$. From Fig. 4.3, if $S = 1.5$, $d/h = 2.6$ and $\alpha = 6.7°$, then $a/h = 0.29$; if $S = 2$, then $a/h = 0.46$. Because the outslope of the dam is 1.75, $a/h = (0.29 + 0.46)/2 = 0.38$, or $a = 57$ ft (17.4 m). From Fig. 4.4, if $S = 1.5$, then $b/h = 0.73$; if $S = 2$, then $b/h = 0.77$; so $b/h = (0.73 + 0.77)/2 = 0.75$, or $b = 113$ ft (34.4 m). Knowing the coordinates of the three points, as indicated by the figure, the line of seepage can be determined.

4.3 EARTH DAMS WITH FILTER DRAINS

If a drainage system is provided within the dam, such as the use of porous shell, the line of seepage at the exit point may become pretty steep and the Dupuit's assumption is no longer valid. Based on Casagrande's method of assuming that the hydraulic gradient is dy/dl. where l is the distance along the line of seepage, Gilboy (1933) developed a simple chart for a dam on a horizontal base, as shown in Fig. 4.6. The insert in Fig. 4.6 shows only the more impervious part of the dam; the porous shell, if any, is not shown. When the slope, β, is less that 30°, the solution checks closely with Fig. 4.3 for $\alpha = 0$. The solution shown in Fig. 4.6 is very satisfactory for slopes up to 60°. If deviations of 25 percent are permitted, it may even be used up to 90°, i.e., for a vertical discharge face.

For $60° < \beta \le 180°$, Casagrande (1937) developed a method for sketching the line of seepage. Figure 4.7 shows an earth dam with an underfilter, or $\beta = 180°$. The equation for the line of seepage was derived by Kozeny and can be expressed as a basic parabola (Harr, 1962)

$$x = \frac{y^2 - y_0^2}{2y_0} \tag{4.6}$$

If the line of seepage is determined by the coordinates d and h of one known point, then

$$y_0 = \sqrt{d^2 + h^2} - d \tag{4.7}$$

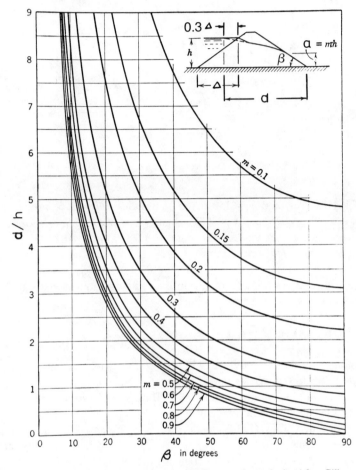

FIGURE 4.6. Location of phreatic surface by L. Casagrande method. (After Gilboy, 1933)

By comparing the line of seepage obtained from flow nets for various angles, β, with the basic parabola, Casagrande found the distance, Δl, between point A on the basic parabola and point B on the line of seepage, as shown in Fig. 4.8 for a toe drain with $\beta = 135°$. To adjust the basic parabola, a correction factor, c_f, must be obtained from Fig. 4.9. The correction factor is defined as

$$c_f = \frac{\Delta l}{l + \Delta l} \qquad (4.8)$$

in which $l + \Delta l$ is the distance from the origin to the basic parabola along the slope surface. Equation 4.8 can be used to determine Δl. Having plotted the basic parabola and determined the discharge point by scaling a distance of Δl

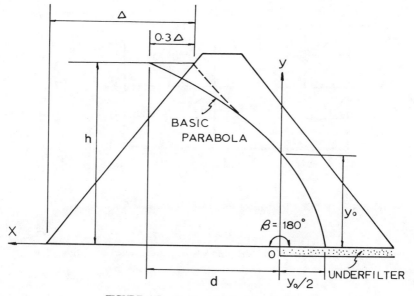

FIGURE 4.7.   Basic parabola for underfilter.

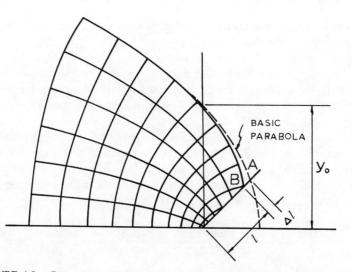

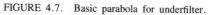

FIGURE 4.8.   Comparison of basic parabola and flow net. (After Casagrande, 1937)

from the basic parabola, the entire line of seepage can be easily sketched in. The following rules are convenient for sketching the line of seepage

1. When $\beta < 90°$, the line of seepage is tangent to the slope; when $\beta \geq 90°$, it is tangent to a vertical line.

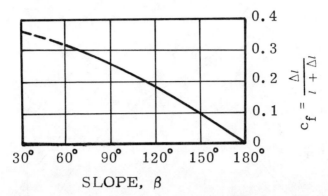

FIGURE 4.9. Correction factor for phreatic surface. (After Casagrande, 1937)

2. When $y = 0$, $x = = -y_0/2$.
3. When $x = 0$, $y = y_0$, the line of seepage makes an angle of $45°$ with horizontal.

4.4 PORE PRESSURE RATIO

In the computerized methods of stability analysis for steady-state seepage or rapid drawdown, it is desirable to determine the pore water pressure from a prescribed phreatic surface. However, this is usually not possible in the simplified methods. In order to include the effect of pore pressure, a soil parameter called the pore pressure ratio, r_u, is used instead. The pore pressure ratio is defined as a ratio between the total pore pressure and the total overburden pressure, or between the total upward force due to water pressure and the total downward force due to the weight or overburden pressure. According to the Archimedes' principle, the upward force is equal to the weight of water displaced, or the volume of sliding mass under water multiplied by the unit weight of water. The downward force is equal to the weight of sliding mass. Therefore, the pore pressure ratio can be determined by

$$r_u = \frac{\text{Volume of sliding mass under water} \times \text{unit weight of water}}{\text{Volume of sliding mass} \times \text{unit weight of soil}} \quad (4.9)$$

Since the unit weight of water is approximately equal to one half the unit weight of soil, the pore pressure ratio can be determined approximately by

$$r_u = \frac{\text{Cross section area of sliding mass under water}}{2 \times \text{total cross section area of sliding mass}} \quad (4.10)$$

Figure 4.10 shows the conversion of a phreatic surface to a pore pressure ratio for both the plane and the cylindrical surfaces. If the failure surface is

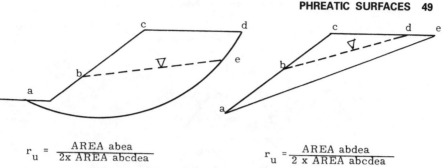

$$r_u = \frac{\text{AREA abea}}{2 \times \text{AREA abcdea}} \qquad\qquad r_u = \frac{\text{AREA abdea}}{2 \times \text{AREA abcdea}}$$

FIGURE 4.10. Determination of pore pressure ratio.

known, the pore pressure ratio can be determined by Eq. 4.9 or 4.10, as shown in the figure. If the location of the failure surface is not known, proper judgement or prior experience in estimating its location is needed in order to determine the pore pressure ratio.

In all the simplified methods of stability analysis, the pore pressure ratio defined by Eq. 4.9 is used to reduce the effective stress along the failure surface by a factor of $(1 - r_u)$. Figure 4.11 shows a sliding mass placed on a slope with an inclination of α. If W is the weight of the mass, the effective force, $\bar{N}$, normal to the failure plane at the base of the sliding mass is

$$\bar{N} = (1 - r_u)\, W \cos \alpha \qquad\qquad (4.11)$$

If the purpose of the pore pressure ratio is to reduce the weight of soil, Eq. 4.9 based on the Archimedes' principle should be the best to use. However, this method has the disadvantage that the location of the failure surface must be known or estimated. This definition of pore pressure ratio is slightly different from that proposed by Bishop and Morgenstern (1960), who defined the pore pressure ratio as

$$r_u = \frac{u}{\gamma h} \qquad\qquad (4.12)$$

in which u is the pore water pressure, γ is the unit weight of soil, and h is the depth of point below soil surface. Based on Eq. 4.12, the pore pressure along the failure plane at the bottom of the sliding mass is $r_u \gamma h$, or the neutral force within a horizontal distance, dx, is $r_u \gamma h \sec \alpha\, dx$. Since

$$W = \gamma \int h\, dx \qquad\qquad (4.13)$$

the neutral force normal to the failure plane is

$$U = r_u (\gamma \int h\, dx) \sec \alpha$$

or
$$U = r_u W \sec \alpha \qquad\qquad (4.14)$$

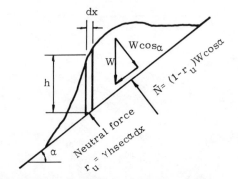

FIGURE 4.11. Use of pore pressure ratio to reduce normal force.

The total force normal to the failure plane is $W \cos \alpha$, so the effective force is

$$\bar{N} = W \cos \alpha - r_u W \sec \alpha \qquad (4.15)$$

The use of $r_u W \sec \alpha$ as the neutral force is not reasonable because when $r_u = 0.5$ and $\alpha \geq 45°$, the effective normal force $\bar{N}$ is negative. To overcome this difficulty, engineers have long used the concept of submerged weight for determining the effective stress, as indicated by Eq. 4.11.

As the pore pressure ratio based on Eq. 4.12 is generally not uniform throughout a slope, Bishop and Morgenstern (1960) suggested the use of an average pore pressure ratio in conjunction with their stability charts presented in Sect. 7.1. The average can be determined by weighing the pore pressure ratio over the area throughout the entire slope, irrespective of the location of the failure surface, or

$$r_u = \frac{\Sigma(\text{Pore pressure ratio} \times \text{area})}{\text{Total area}} \qquad (4.16)$$

For problems where the magnitude of pore pressure depends on the degree of consolidation, such as during or at the end of construction, Bishop (1955) suggested

$$r_u = \bar{B} \qquad (4.17)$$

in which $\bar{B}$ is a soil parameter, which can be determined by

$$\bar{B} = \frac{\Delta u}{\Delta \sigma_1} = B\left[\frac{\Delta \sigma_3}{\Delta \sigma_1} + A\left(1 - \frac{\Delta \sigma_3}{\Delta \sigma_1}\right)\right] \qquad (4.18)$$

where A and B are pore pressure coefficients, as described by Skempton (1954). The application of the undrained triaxial test with pore pressure measurements to determine these soil parameters was presented by Bishop and Henkel (1957).

Equations 4.12 and 4.16 are useful when the pore pressure is determined from field measurements or from the construction of flow nets. The advantage of this method is that it is not necessary to know the location of the circle apriori. If the failure surface is a shallow circle, the pore pressure ratio obtained from Eq. 4.9, based on the area of the sliding mass only, may be much smaller than that from Eq. 4.16, based on the area of the slope. If the failure surface lies midway between the base and the surface of the slope, the pore pressure ratio obtained from Eqs. 4.9 and 4.16 will not be too much different.

In view of the fact that the average pore pressure ratio is an approximate value which cannot be determined accurately, it is suggested that Eq. 4.9 be used in cases of steady-state seepage or rapid drawdown where a phreatic surface can be established.

5
Remedial Measures for Correcting Slides

5.1 FIELD INVESTIGATION

The scope of field investigations should include topography, geology, water, weather, and history of slope changes. If a slide has occurred, the shape of sliding surface should also be determined.

Topography. The topography or geometry of the ground surface is an overt clue to past landslide activity and potential instability. More detail than that shown on existing topographic or project design maps is usually required for landslide studies. Because the topography of a landslide is continually changing, the area must be mapped at different times, if possible, from several years before construction to several years after remedial measures are undertaken. Ultimately, the effectiveness of corrective measures is expressed by whether the topography changes.

If a detailed survey of the area and the preparation of a contour map are not possible due to the lack of time, at least several cross sections must be surveyed from the accumulated masses at the toe of the slide to above the head scarp. The cross sections must be long enough to cover part of the undisturbed area above and below the slide. The surface of the area should not be shown in a simplified form, but with as many topographical features as possible, such as all marked edges, swell and depressions, scarps, and cracks, etc. The surveyed sections are supplemented by the logs of borings.

Air photos are most useful for the investigation of landslides, because they offer a perfect three-dimensional view of the area. From an air photo, one can determine precisely the boundaries of a landslide, as the slope surface below the scarp is irregularly undulated with ponded depressions. Also the character of vegetation on the slope affected by the slide differs from that of the undisturbed adjacent slope. The amount of movement is easily determined from the offset of linear features, such as highways, railroads, alleys, etc., as soon as they continue to the undisturbed area.

Geology.   Geologic structure is frequently a major factor in landslides. Although this topic includes major large scale structural features such as folds and faults, the minor structural details, including joints, small faults, and local shear zones may be even more important. The landslide and the surrounding area should be mapped geologically in detail. On the map, the shape of the head scarp and the area of accumulation, outcrops of beds, offsets in strata, and changes in joint orientation, dips, and strikes should be identified.

An important characteristic of the sliding slope is the shape of cross section. If the slope was sculptured by erosion and covered with waste redeposited by rainwash, the profile forms a gentle curve at its transition into the flood plain. Even a very ancient landslide is recognizable from the curved bulged shape of the toe.

Water.   Water is a major factor in most landslides. The plan for corrective measures requires a good knowledge of the hydrogeological conditions of the slide itself and of its surroundings. The first task is to determine the depth of groundwater table and its fluctuation and to map all streams, springs, seeps, wet grounds, undrained depressions, aquiferous pressures, and permeable strata.

The changes of slope relief produced by sliding alter the drainage condition of surface water as well as the regime of groundwater. The seepage of groundwater has a significant effect on slope stability. Less pressure is built up when water is seeping out of the ground than when the exits for groundwater are blocked. For example, in one major slide area, landslide activity was always preceded by a stoppage of spring discharge near the toe; the cessation of movement was marked by an increase in spring discharge. Slip surfaces are generally impervious, retaining both surface and ground water. When slip surfaces approach the ground surface, new springs and wet grounds appear. In the boring logs, the depth and fluctuation of groundwater must be recorded. However, the pore pressure in clayey soils affected by sliding cannot be determined simply by observing the water level in a boring, because by filling the bore hole, the water loses the pressure in its vicinity. Therefore, the installation of piezometric instruments for pore pressure measurement is needed.

Weather.   The climate of the area, incuding rainfall, temperature, evaporation, wind, snowfall, relative humidity, and barometric pressure, is the ultimate dynamic factor influencing most landslides. The effects of these factors can seldom be evaluated analytically because the relations are too complex. Empirical correlations of one or more of these factors, particularly rainfall, snow, and melting temperatures, with episodes of movement or movement rates can point out their influences that must be controlled to minimize movements.

History of Slope Changes. Many clues can alert the investigator to past landslides and future risks. Some of these are hummocky ground, bulges, depressions, cracks, bowed and deformed trees, slumps, and changes in vegetation. The large features can be determined from large-scale maps and air photos; however, the evidence often is either hidden by vegetation or is so small that it can only be determined by direct observation. Even then, only one intimately familiar with the soil, geologic materials, and conditions in that particular area can recognize the potential hazards.

Among the most difficult kinds of slides to recognize and guard against are old landslides that have been covered by glacial till or other more recent sediments. Recent and active landslides can be easily recognized by their fresh appearance with steep and bared head scarp, open cracks and strung tree roots. The state of tree growth is indicative of the age of movements. Trees on unstable ground are tilted downslope but tend to return to a vertical position during the period of rest, so that the trucks become conspicuously bent. From the younger, vertically growing trunk segments, the date of the last sliding movement can be inferred.

Shape of Failure Surfaces. As noted in Sect. 1.1, slides are divided into two types, that is, rotational and translational. Rotational slides are characterized by rotation of the block or blocks of which they are composed, whereas translational slides are marked by lateral separation with very little vertical displacement and by vertical, rather than concave, cracks. Figure 5.1 is a schematic diagram of these two types of slides, which took place during the 1964 Alaska earthquake (Hansen, 1965).

The most common examples of rotational slides are little-deformed slumps, which are slides along a surface of rupture that is concave upward. The exposed cracks are concentric in plan and concave toward the direction of movement. In many rotational slides the underlying surface of rupture, together with the exposed scarps, is spoon-shaped. If the slide extends for a considerable distance along the slope perpendicular to the direction of movement, much of the rupture surface may approach the shape of a cylinder whose axis is parallel to the slope. In the head area, the movement may be almost wholly downward and have little apparent rotation. However, the top surface of each unit commonly tilts backward toward the slope, although some blocks may tilt forward. The classic purely rotational slide on a circular or cylindrical surface is relatively uncommon in natural slopes due to their internal inhomogeneities and discontinuities. Since rotation slides occur most frequently in fairly homogeneous materials, their incidence among constructed embankments and fills, and hence their interest to engineers, has been high relative to other types of failure. The stability charts in Chap. 7 and the

REAME computer program in Chap. 9 are based on the cylindrical failure surface.

In translational slides, the mass progresses down and out along a more or less planar or gently undulatory surface and has little of the rotational movement or backward tilting characteristics. The moving mass commonly slides out on the original ground surface. The simple equations in Chap. 6 and the SWASE computer program in Chap. 8 are based on the plane failure surface.

The distinction between rotational and translational slides is useful in planning control measures. The rotational slide, if the surface of rupture dips into the hill at the foot of the slide, tends to restore equilibrium in the unstable mass, the driving moment during movement decreases and the slide may stop moving. A translational slide, however, may progress indefinitely if the surface on which it rests is sufficiently inclined as long as the shear resistance along this surface remains lower than the more or less constant driving force. The movement of translational slides is commonly controlled structurally by surfaces of weakness, such as faults, joints, bedding planes, and variations in shear strength between layers of bedded deposits, or by the contact between firm bedrock and overlying detritus.

The location of the slip surface can be determined by an inclinometer. The inclinometer measures the change in inclination or tilt of a casing in a bore hole and thus allows the distribution of lateral movements to be determined as a function of depth below the ground surface and as a function of time. Inclinometers have undergone rapid development to improve reliability, provide accuracy, reduce weight and bulk of instruments, lessen data acquisition and reduction time, and improve versatility of operation under adverse conditions. Automatic data-recording devices, power cable reels, and other features are now available.

5.2 PRELIMINARY PLANNING

When a slide takes place, it is necessary to determine the causes of the slide, so that proper remedial measures can be taken to correct it. The processes involved in slides comprise a continuous series of events from cause to effect. Seldom, if ever, can a slide be attributed to a single definite cause. The detection of the causes may require continuous observations, and a final decision cannot be made within a short time. Since water is the major cause which may initiate a slide, or intensify a slide after it has occurred, the following initial remedial measures should be taken as soon as possible.

1. Capture and drain all the surface water which flows into the slide area.
2. Pump the groundwater out from all wells in the slide area and dewater all the drainless depressions.
3. Fill and tamp all open cracks to prevent the infiltration of surface water.

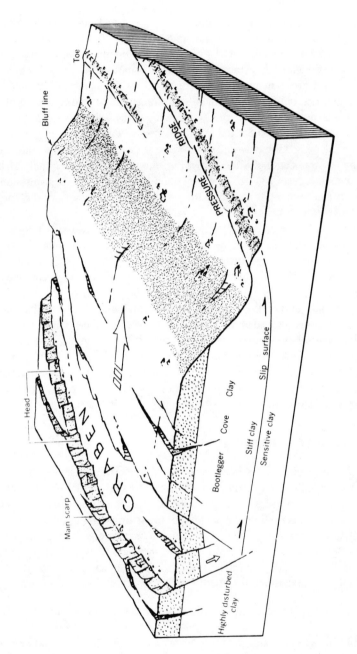

TRANSLATIONAL SLIDE

Toe

Bluff line

PRESSURE RIDGE

Head

Main scarp

GRABEN

Bootlegger Cove Clay

Stiff clay

Sensitive clay

Slip surface

Highly disturbed clay

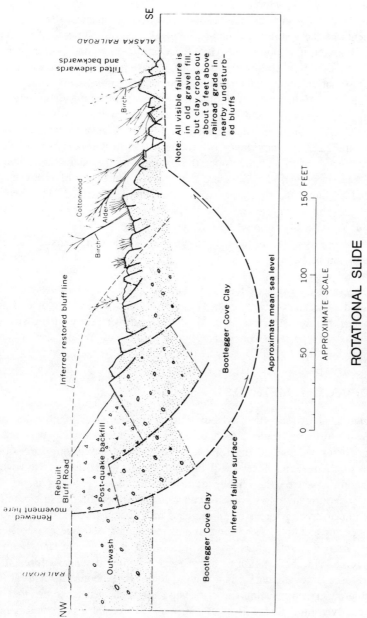

ROTATIONAL SLIDE

FIGURE 5.1. Translational and rotational slides. (After Hansen, 1965)

Only after the completion of the initial measures should other permanent and more expensive measures based on a detailed investigation be undertaken.

Peck (1967) described the catastrophic slide in 1966 on the Baker River north of Seattle, WA, which may be entitled "The Death of a Power Plant." He claimed that the current state of the art was still unable to make reliable assessments of the stability of many, if not all, natural slopes under circumstances of practical importance. After the destruction of the power plant by the slide, he asked the following questions: "Was spending the time necessary to get information about subsurface conditions and movements a tactical error? Could the unfavorable developments be prevented by providing extensive resloping, deep drainage, and other means rather than the use of the observational methods?" Although these questions cannot be answered satisfactorily, it clearly indicates the importance of prompt action in the correction of slides.

After the geometry of the slide, the location of the water table, and the soil parameters of various layers are determined, the factor of safety can be calculated by either the simplified or the computerized method. The factor of safety at the time of failure should be close to 1. If not, some of the parameters used in the analysis must be adjusted.

If the slope is homogeneous and there is only one soil, the shear strength of the soil can be back calculated by assuming a safety factor of 1. This shear strength can then be used for the redesign of the slope.

Based on the results of the investigation, a new slope is designed, and the method of stability analysis can be used to determine the factor of safety. If a strong retaining structure is used, the stability or safety of the structure should also be separately considered by the principles of soil and structural mechanics.

5.3 CORRECTIVE METHODS

Corrective methods can be used either to decrease the driving forces or to increase the resisting forces. As the factor of safety is a ratio between the resisting forces and the driving forces, a reduction of the driving forces or an increase of the resisting forces will increase the factor of safety. However, a classification based on the change of driving or resisting forces is not used here because in many cases it is difficult to tell whether the measure results in a reduction of driving forces or an increase of resisting forces. Take the effect of seepage for example. According to soil mechanics, seepage can be treated in two different ways: either a combination of total weight and boundary neutral force or a combination of submerged weight and seepage force. If the seepage force is considered, it is a driving force. If the boundary neutral force is considered, it is a resisting force, because the shear resistance changes with the boundary neutral force. Most of the practical examples presented here for illustrating the different corrective methods were reported by the Transportation Research Board (Schuster and Krizek, 1978).

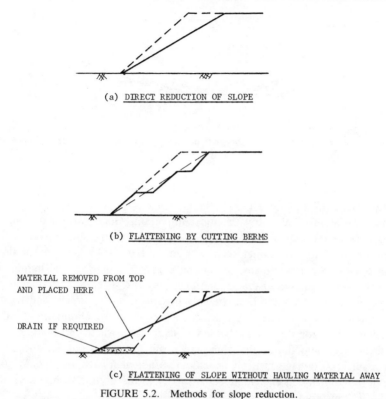

(a) <u>DIRECT REDUCTION OF SLOPE</u>

(b) <u>FLATTENING BY CUTTING BERMS</u>

MATERIAL REMOVED FROM TOP
AND PLACED HERE

DRAIN IF REQUIRED

(c) <u>FLATTENING OF SLOPE WITHOUT HAULING MATERIAL AWAY</u>

FIGURE 5.2. Methods for slope reduction.

Slope Reduction or Removal of Weight.

Figure 5.2 shows three methods for slope reduction; that is, direct reduction, flattening by cutting berms, and flattening without hauling material away. Although the third method is most economical, care must be taken to ensure that the material to be placed on the toe is of good quality. If necessary, a drainage blanket should be placed to minimize the effect of water.

Figure 5.3 shows a slope flattening which was used effectively on a 320 ft (98 m) cut for a southern California freeway (Smith and Cedergren, 1962). A failure took place during construction on a 1:1 benched cut slope composed predominantly of sandstone and interbedded shales. After considerable study and analysis, the slope was modified to $3h:1v$, and the final roadway grade was raised some 60 ft (18 m) above the original ground elevation. Moreover, to provide additional stability, earth buttresses were placed from roadway levels to a height of 70 ft (21 m).

Figure 5.4 shows the stabilization of the Cameo slide above a railroad in the Colorado River Valley by partial removal of the weight (Peck and Ireland, 1953). Stability analyses determined that the removal of volume B was more effective than the removal of volume A, as expected.

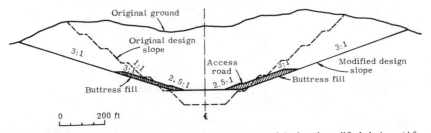

FIGURE 5.3. Typical section of Mulholland cut showing original and modified design. (After Smith and Cedergren, 1962; 1 ft = 0.305 m)

Surface Drainage. Of all possible methods for correcting slides, proper drainage of water is probably the most important. Good surface drainage is strongly recommended as part of the treatment for any slide. Every effort should be made to ensure that surface waters are carried away from a slope. The surface of the area affected by sliding is generally uneven and hummocky and traversed by unnoticed cracks and deep fissures. Therefore, reshaping the surface of a slide mass can be extremely beneficial in that cracks and fissures are sealed and water-collecting surface depressions are eliminated. This is particularly true for the cracks behind a scarp face where large volumes of water can seep into the failure zone and result in serious consequences. Although surface drainage in itself is seldom sufficient for the stabilization of a slope in motion, it can contribute substantially to the drying of the material in the slope, thus controlling the slide.

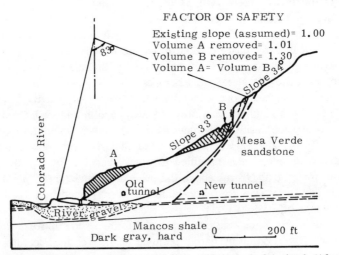

FIGURE 5.4. Stabilization of the Cameo slide by partial removal of the head. (After Peck and Ireland, 1953; 1 ft = 0.305 m)

Subsurface Drainage. As groundwater is one of the major causes of slope instability, the subsurface drainage is a very effective remedial measure. Methods frequently used are the installation of horizontal drains, vertical drainage wells, and drainage tunnels. A horizontal drain is a small diameter well drilled into a slope on approximately a 5 to 10 percent grade and fitted with a perforated pipe. Pipes should be provided to carry the collected water to a safe point of disposal to prevent surface erosion. A vertical drainage well can be either a gravity drain or a pumped well, depending on whether there is an outlet for the water to drain by gravity. In many cases, a horizontal drain can be drilled to intercept the vertical drain at the bottom. A drainage tunnel is a deep and large structure, usually about 3 ft (1 m) wide by 6 ft (2 m) high in cross section, constructed for the purpose of discharging a large amount of water. The effectiveness of a drainage tunnel may be increased by short or long drainage borings in the walls, floor, or roof of the tunnel.

Figure 5.5 shows the use of both surface and subsurface drainage for correcting the slide on the Castaic-Alamos Creek in California (Dennis and Allan, 1941). The surface water is collected by the intercepting trench connected to the perforated pipe and gravel subdrain, which are also used for subdrainage. Figure 5.6 shows the use of both vertical and horizontal drains for correcting an active landslide that occurred at San Marcos Pass near Santa Barbara, CA (Root, 1958). The vertical wells were about 3 ft (1 m) in diameter, 40-ft (12-m) long and belled at the bottom so that they interconnected to form a somewhat continuous curtain. The drain had an 8-in. (20-cm) perforated pipe in the center for the full depth of the vertical drain and was back-

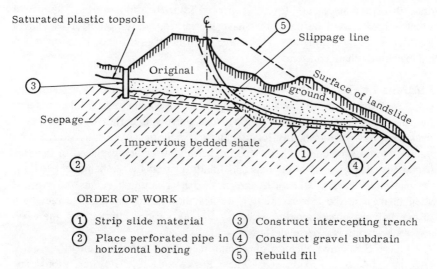

ORDER OF WORK

① Strip slide material   ③ Construct intercepting trench
② Place perforated pipe in ④ Construct gravel subdrain
   horizontal boring   ⑤ Rebuild fill

FIGURE 5.5. Corrective measures for Castaic-Alamos Creek slide. (After Dennis and Allan, 1941)

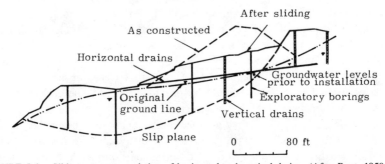

FIGURE 5.6.   Slide treatment consisting of horizontal and vertical drains. (After Root, 1958; 1 ft = 0.305 m)

filled with pervious material. The horizontal drains were then drilled to intersect and drain the belled portion of the vertical well.

Vegetation.   Slope movements generally disturb the vegetative cover, including both trees and grass. The reforestation of the slope is an important task in corrective treatment. It is carried out during the last stage, invariably after at least partial stabilization of the slide. Forestation is most beneficial for shallow slides. Slides with deep-lying failure surfaces cannot be detained by vegetation, although in this case too, vegetation can lower the infiltration of surface water into the slope and thus contribute indirectly to the stabilization of the slide.

It is generally accepted that forest growth has two functions: drying out of the surface layers and consolidating them by a network of roots. As trees draw the water necessary for their growth from the slope surface, the most suitable species will be those that have the largest consumption of water and the highest evaporation rate. Therefore, it is more advantageous to plant deciduous trees than conifers.

Buttress or Retaining Walls.   Figure 5.7 shows the use of a stabilizing berm for correcting the slide in the shale embankment on I-74 in southern Indiana (Haugen and DiMillio, 1974). The borrow material used in the embankment was predominantly local shale materials that were interbedded with limestone and sandstone. These shale materials, after being placed in the embankment, deteriorated with time and finally caused the embankment to fail. On one slide, careful studies of the in situ shear strength versus the original strength used in the preconstruction studies showed an approximate reduction of one half in shear strength. After considering various alternatives, the earth and rock stabilizing berm design was finally selected.

Figure 5.8 shows the use of a reinforced earth buttress wall to correct a large landslide on a section of I-40 near Rockwood, TN, (Royster, 1966). The slope-forming materials were essentially a thick surface deposit of colluvium underlain by residual clays and clay shales. The groundwater table was sea-

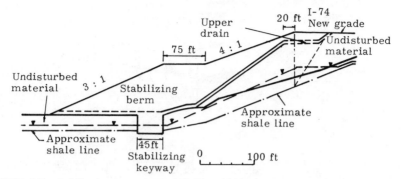

FIGURE 5.7.   Stabilization berm used to correct landslide in shale on I-74 in Indiana. (After Haugen and DiMillio, 1974; 1 ft = 0.305 m)

sonally variable but was generally found to be above the colluvium and residuum interface. This particular slide occurred within an embankment placed as a sidehill fill directly on a colluvium-filled drainage ravine. Because of blocked subsurface drainage and weakened foundation soils, the fill failed some 4 years after construction. Final design plans called for careful excavation of the failed portion of the fill to a firm unweathered shale base, installation of a highly permeable drainage course below the wall area, placing of the reinforced earth wall, and final backfill operations behind the reinforced earth mass.

Figure 5.9 shows the use of a retaining wall to correct a cut slope failure on I-94 in Minneapolis-St. Paul, MN (Shannon and Wilson, Inc., 1968). The use of a retaining wall is often occasioned by the lack of space necessary for the development of the slope to a full length. As retaining walls are subject to an unfavorable system of loading, a large wall width is necessary to increase stability. Although the methods of stability analysis can be applied to deter-

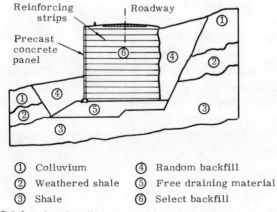

FIGURE 5.8.   Reinforced earth wall to correct slide on I-40 in Tennessee. (After Royster, 1966)

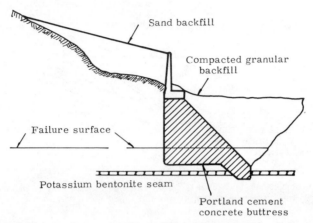

FIGURE 5.9.   Retaining wall to correct slide on I-94 in Minnesota. (After Shannon and Wilson, 1968)

mine the factor of safety of a slope with failure surfaces below the wall, the design of the wall will require special considerations to ensure that the wall itself is stable.

In particularly serious cases, a retaining wall may not be sufficient, and it is necessary to construct a tunnel as shown in Fig. 5.10 for the slide on the Spaichingen-Nusplingen railway line in Germany (Zaruba and Mencl, 1969).

Pile Systems. Recorded attempts to use driven steel or wooden piles of nominal diameter to retard or prevent landslides have seldom been successful. Unless the slide is shallow, such piles are incapable of providing adequate shear resistance. Shallow slides can be controlled by piling because the piles

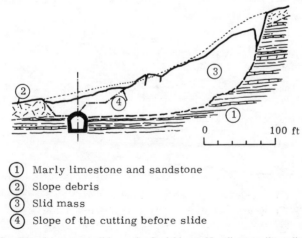

① Marly limestone and sandstone
② Slope debris
③ Slid mass
④ Slope of the cutting before slide

FIGURE 5.10.   Tunnel to correct slide on the Spaichingen-Nusplingen railway line in Germany. (After Zaruba and Mencl, 1969; 1 ft = 0.305 m)

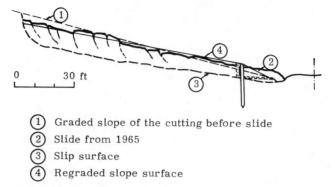

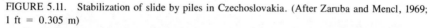

① Graded slope of the cutting before slide
② Slide from 1965
③ Slip surface
④ Regraded slope surface

FIGURE 5.11.   Stabilization of slide by piles in Czechoslovakia. (After Zaruba and Mencl, 1969; 1 ft = 0.305 m)

can be driven to an adequate depth. Otherwise, they may tilt from the vertical position, thus disturbing the adjacent material and the material underneath the piles and causing the development of a slip surface below the pile tip.

Figure 5.11 shows the use of piles to correct a shallow slide in a railway cutting at East Slovakie, Czechoslovakia (Zaruba and Mencl, 1969). The cutting had a slope of $4h:1v$ and was made in a fissured marly clay subjected to slaking. During the rainy spring of 1965, a small sheet slide developed at the toe of the slope, which extended to a length of 165 ft (50 m) and reached up to the top of the slope. As the site was not accessible and the removal of large volume of soil was difficult, piles were employed to prevent the further spreading of the slide. Forty-two piles, 20-ft (6-m) long, were driven into the prepared bore holes to a depth of 13 ft (4 m). Reinforced concrete slabs were supported against the piles to prevent movement of the soil between and around the piles. The pile spacing was 3 to 5 ft (1 to 1.5 m). A sand drain was constructed along the slab, discharging water to a ditch. After the treatment, the slope was flattened to $5h:1v$.

Figure 5.12 shows a cylinder pile wall system for stabilizing the deep-seated slope failure in I-94 in Minneapolis-St. Paul, MN, (Shannon and Wilson, Inc. 1968). The pile wall was placed as a restraining system, in which the forces tending to cause movement were carefully predicted. The case-in-place piles were designed as cantilevers to resist the full earth thrust imposed by the soil.

Anchor Systems.   One type of anchor system is the tie-back wall, which carries the backfill forces on the wall by a "tie" system to transfer the imposed load to an area behind the slide mass where satisfactory resistance can be established. The ties may consist of pre- or post-tensioned cables, rods, or wires and some form of deadmen or other method to develop adequate passive earth pressure. Figure 5.13 shows a section of tie-back wall to correct the slide condition on New York Avenue in Washington, DC (O'Colman and Trigo, 1970).

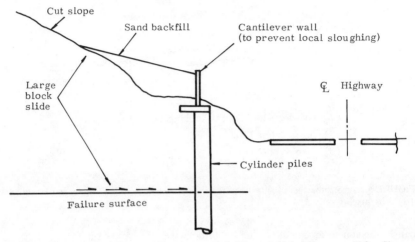

FIGURE 5.12.   Cylinder pile to stabilize deep-seated slide on I-94 in Minnesota. (After Shannon and Wilson, 1968)

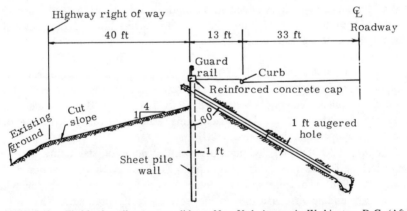

FIGURE 5.13.   Tied-back wall to correct slide on New York Avenue in Washington, D.C. (After O'Colman and Trigo, 1970; 1 ft = 0.305 m)

Hardening of Soils.   If the water in the slope cannot be drained by subsurface drainage methods, methods used by foundation engineers and known as hardening of soils may be considered. These methods can be divided into chemical treatments, electro-osmosis, and thermal treatments.

Chemical treatments, which have been used with varying degrees of success, are lime or lime soil mixtures, cement grout, and ion exchange. One successful treatment, in which quicklime was placed in predrilled 0.5-ft (0.2-m) diameter holes on 5-ft (1.5-m) centers throughout an extensive slide area, was reported by Handy and Williams (1967). The lime migrated a distance of 1 ft (0.3 m) from the drilled holes in one year. Cement grout has been used in England for the stabilization of embankments and cuttings (Zaruba and Mencl, 1969). Experience shows that this method yields fine results with rather shallow landslides in stiff materials such as clayshales, claystones, and stiff clays, which break into blocks separated by distinct fissures. The method is actually a mechanical stabilization of the slope by filling the fissures with cement grout rather than changing the consistency of soil mass, as the cement mortar cannot enter into the soil mass. Cement grout was used for a 300-ft (90-m) benched cut slope on I-40 along the Pigeon River in North Carolina (Schuster and Krizek, 1978). Large volumes of cement grout were injected into the voids of broken rubble and talus debris to stabilize the slope. The ion exchange technique, which consists of treating the clay minerals along the plane of potential movement with a concentrated chemical, was reported by Smith and Forsyth (1971).

The electro-osmosis technique has the same final effect as subsurface drainage, but differs in that water is drained by an electric field rather than by gravity. The loss of pore water causes consolidation of the soil and a subsequent increase in shear strength. Casagrande, Loughney, and Matich (1961) described the use of this method to stabilize a cut slope during the construction of a bridge foundation.

The use of thermal treatments for preventing landslides was first reported by Hill (1934). Since 1955, the Russians have experimented and reported on the success of thermal treatment on plastic loessial soils. The high temperature treatments cause a permanent drying of the embankments and cut slopes. Beles and Stanculescu (1958) described the use of thermal methods to reduce the in situ water contents of heavy clay soils in Romania. Applications to highway landslides and unstable railroad fills were also cited.

Part II
Simplified Methods of
Stability Analysis

6
Simplified Methods for Plane Failure

6.1 INFINITE SLOPES

The purpose of this chapter is to present some simple equations for determining the safety factor of slopes with plane failure surfaces. Only three simple cases will be considered: one involving an infinite slope with a failure plane parallel to the slope surface, one involving a triangular cross section with a single failure plane, and the other involving a trapezoidal cross section with two failure planes (Huang, 1977a, 1978b). For three failure planes, the computerized method presented in Chap. 8 should be used.

Figure 6.1 shows an infinite slope underlaid by a rock surface with an inclination, α. The slope is considered as infinite because it has a length much greater than the depth, d. If a free body of width a is taken, the forces on the two vertical sides are the same because every plane is considered the same in an infinite slope. As the side forces neutralize each other, the only forces to be considered are the weight, W, and the seismic force, C_sW. In contrast to the cylindrical failure, the effect of seismic force on the force

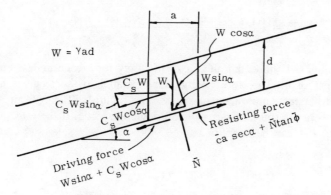

FIGURE 6.1. Analysis of infinite slope.

normal to the failure plane and thus on the shear resistance is considered. In all the derivations which follow, only the effective stress analysis will be presented. The equations can also be applied to a total stress analysis by simply replacing the effective strength parameters by the total strength parameters.

The factor of safety is defined as a ratio of the resisting force due to the shear strength of soil along the failure surface to the driving force due to the weight of the sliding mass. The resisting force is composed of two parts: one due to cohesion and equal to $\bar{c}a \sec \alpha$ and the other due to friction and equal to $\bar{N} \tan \bar{\phi}$, where $\bar{N}$ is the effective force normal to the failure plane. With a pore pressure ratio, r_u,

$$\bar{N} = W\,[(1 - r_u)\,\cos\,\alpha - C_s\,\sin\,\alpha] \qquad (6.1)$$

The driving force is always equal to the component of weight and seismic force parallel to the failure surface, or $W \sin \alpha + C_s W \cos \alpha$, regardless of whether seepage exists or not. Therefore, the factor of safety, F, can be written as

$$F = \frac{\bar{c}a\,\sec\,\alpha + W\,[\,(1 - r_u)\,\cos\,\alpha - C_s\,\sin\,\alpha]\,\tan\,\bar{\phi}}{W(\sin\,\alpha + C_s\,\cos\,\alpha)} \qquad (6.2)$$

Replacing W by $\gamma a d$, where γ is the unit weight of the sliding mass

$$F = \frac{(\bar{c}/\gamma d)\,\sec\,\alpha + [(1 - r_u)\,\cos\,\alpha - C_s\,\sin\,\alpha]\,\tan\,\bar{\phi}}{\sin\,\alpha + C_s\,\cos\,\alpha} \qquad (6.3)$$

Equation 6.3 is applicable to an infinite slope possessing both cohesion and angle of internal friction. The factor of safety decreases with the increase in d, so the most critical plane is along the rock surface.

If there is no cohesion ($\bar{c} = 0$), from Eq. 6.3

$$F = \frac{[(1 - r_u)\,\cos\,\alpha - C_s\,\sin\,\alpha]\,\tan\,\bar{\phi}}{\sin\,\alpha + C_s\,\cos\,\alpha} \qquad (6.4)$$

Equation 6.4 shows that the factor of safety for a cohesionless material is independent of d. Therefore, every plane parallel to the slope is a critical plane and has the same factor of safety. The failure will start from the slope surface where the soil particles will roll down the slope. If there is no seismic force, Eq. 6.4 can be simplified to

$$F = (1 - r_u)\,\frac{\tan\,\bar{\phi}}{\tan\,\alpha} \qquad (6.5)$$

6.2 TRIANGULAR CROSS SECTION

Figure 6.2 shows a triangular fill on a sloping surface. It is assumed that the failure plane is along the bottom of the fill. Good examples for this type of failure are the spoil banks created by surface mining, where the original ground surface is not properly scalped and a weak layer exists at the bottom of the fill. In addition to the vertical weight, a horizontal seismic force equal to $C_s W$ is also applied.

Similar to Eq. 6.2 for infinite slope except that the length of failure plane is $H \csc \alpha$ instead of $a \sec \alpha$, the factor of safety is

$$F = \frac{\bar{c}H \csc \alpha + W[(1 - r_u) \cos \alpha - C_s \sin \alpha] \tan \bar{\phi}}{W \sin \alpha + C_s W \cos \alpha} \quad (6.6)$$

and

$$W = \tfrac{1}{2} \gamma H^2 \csc \beta \csc \alpha \sin (\beta - \alpha) \quad (6.7)$$

in which γ is the unit weight of soil and β is the angle of outslope.

Substituting W from Eq. 6.7 into Eq. 6.6

$$F = \frac{2 \sin \beta \csc (\beta - \alpha) (\bar{c}/\gamma H) + [(1 - r_u) \cos \alpha - C_s \sin \alpha] \tan \bar{\phi}}{\sin \alpha + C_s \cos \alpha} \quad (6.8)$$

Note that $\bar{c}$, γ, and H can be lumped together as a single parameter for determining the factor of safety. If the fill width, W_f, is given instead of the height, H,

$$F = \frac{\bar{c}W_f \sin \beta/\sin (\beta - \alpha) + W[(1 - r_u) \cos \alpha - C_s \sin \alpha] \tan \bar{\phi}}{W \sin \alpha + C_s W \cos \alpha} \quad (6.9)$$

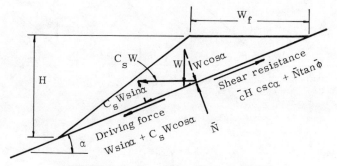

FIGURE 6.2. Plane failure of triangular fill.

in which

$$W = \tfrac{1}{2} W_f^2 \sin \alpha \sin \beta / \sin(\beta - \alpha) \qquad (6.10)$$

Substituting W from Eq. 6.10 into Eq. 6.9

$$F = \frac{2 \csc \alpha \ (\bar{c}/\gamma W_f) + [(1 - r_u) \cos \alpha - C_s \sin \alpha] \tan \bar{\phi}}{\sin \alpha + C_s \cos \alpha} \qquad (6.11)$$

Equation 6.11 shows that the factor of safety is independent of the angle of outslope, β, when the fill width, W_f, is used as a criterion for design because when β changes, both the resisting force and the driving force change in the same proportion.

6.3 TRAPEZOIDAL CROSS SECTION

Figure 6.3 shows the forces acting on a trapezoidal fill. Examples of this type are the hollow fills created by surface mining. When mining on a relatively steep slope, the spoil material is not permitted to be pushed down and placed on the hill. Instead, it must be used for backfilling the stripped bench with the remains to be disposed in a hollow where the natural ground is relatively flat. The actual ground surface below the hollow fill may be quite irregular but, for the purpose of analysis, can be approximated by a horizontal line and a line inclined at an angle, α, with the horizontal. The factor of safety with respect to failure along the natural ground surface can be determined by dividing the fill into two sliding blocks. Assuming that the force, P, acting between the two blocks is horizontal, and that the shear force has a factor of safety, F, four equations, two from each block, can be established from statics to solve for the four unknowns, P, F, $\bar{N}_1$, and $\bar{N}_2$, where $\bar{N}_1$ and $\bar{N}_2$ are the effective forces normal to the failure planes. It has been found that the as-

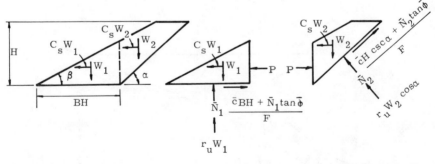

(a) TWO BLOCKS   (b) LOWER BLOCK   (c) UPPER BLOCK

FIGURE 6.3.   Plane failure of trapezoidal fill.

sumption of a horizontal P, or neglecting the friction between the two blocks, always results in a smaller factor of safety and is therefore on the safe side.

From the equilibrium of forces on the lower block, as shown in Fig. 6.3b,

$$\bar{N}_1 = (1 - r_u)W_1 \tag{6.12}$$

$$P + C_s W_1 = (\bar{c}BH + \bar{N}_1 \tan \bar{\phi})/F \tag{6.13}$$

in which W_1 is the weight of the lower block and B is the ratio between the base width and height. Substituting Eq. 6.12 into Eq. 6.13

$$P = [\bar{c}BH + (1 - r_u)W_1 \tan \bar{\phi}]/F - C_s W_1 \tag{6.14}$$

The equilibrium of forces on the upper block gives

$$\bar{N}_2 = P \sin \alpha + (1 - r_u)W_2 \cos \alpha - C_s W_2 \sin \alpha \tag{6.15}$$

$$W_2 \sin \alpha + C_s W_2 \cos \alpha = P \cos \alpha + (\bar{c}H \csc \alpha + \bar{N}_2 \tan \bar{\phi})/F \tag{6.16}$$

in which W_2 is the weight of upper block. Substituting Eqs. 6.14 and 6.15 into Eq. 6.16, a quadratic equation can be obtained which can be solved for the factor of safety, F.

$$a_1 F^2 + a_2 F + a_3 = 0 \tag{6.17}$$

where

$$a_1 = a_4 \sin \alpha + C_s (a_4 + a_5) \cos \alpha \tag{6.18}$$

$$a_2 = - \left\{ \frac{\bar{c}}{\gamma H} (B \cos \alpha + \csc \alpha) + [(1 - r_u) \cos \alpha - C_s \sin \alpha] \right. \\ \left. (a_4 + a_5) \tan \bar{\phi} \right\} \tag{6.19}$$

$$a_3 = -B \sin \alpha \tan \bar{\phi} \left[\frac{\bar{c}}{\gamma H} + (1 - r_u) \frac{a_5}{B} \tan \bar{\phi} \right] \tag{6.20}$$

For an irregular outslope with W_1 and W_2 given

$$a_4 = \frac{W_2}{\gamma H^2} \tag{6.21}$$

$$a_5 = \frac{W_1}{\gamma H^2} \tag{6.22}$$

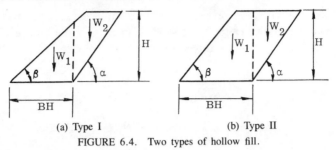

(a) Type I   (b) Type II

FIGURE 6.4. Two types of hollow fill.

For a uniform slope of type-I as shown in Fig. 6.4a

$$a_4 = \tfrac{1}{2} [\cot \alpha - (1 - B \tan \beta)^2 \cot \beta] \qquad (6.23)$$

$$a_5 = \tfrac{1}{2} B^2 \tan \beta \qquad (6.24)$$

For type-II fill as shown in Fig. 6.4b

$$a_4 = \tfrac{1}{2} \cot \alpha \qquad (6.25)$$

$$a_5 = B - \tfrac{1}{2} \cot \beta \qquad (6.26)$$

Note that in using Eqs. 6.21 and 6.22 for irregular slopes, the weights, W_1 and W_2, must be calculated from measurements on the cross section. If the slope is uniform, as shown in Fig. 6.4, either Eqs. 6.23 and 6.24 or Eqs. 6.25 and 6.26 should be used, depending on whether the fill is type-I or type-II. Type-I fill, in which $B \leq \cot \beta$, has a triangular lower block; while type-II fill, in which $B > \cot \beta$, has a trapezoidal lower block.

It should be noted that the purpose of using the simplified cross section is to obtain equations which can be calculated by hand or with a pocket calculator. A computerized method, in which P and the failure plane at the toe are not horizontal or the natural slope is approximated by three straight lines, will be discussed in Chap. 8.

6.4 ILLUSTRATIVE EXAMPLES

Two examples, one for a spoil bank and the other for a hollow fill, will be illustrated. To determine the factor of safety for plane failure, it is necessary to know the shear strength, $\bar{c}$ and $\bar{\phi}$, along the failure surface, the unit weight, γ, and the pore pressure ratio, r_u. In these two examples, it is assumed that $\bar{c} = 160$ psf (7.7 kPa), $\bar{\phi} = 24°$, $\gamma = 125$ pcf (19.6 kN/m³), and $r_u = 0.05$, and that the most critical failure planes are along the bottom of the fill. Note that a pore pressure ratio of 0.05 implies that 10 percent of the fill is under water.

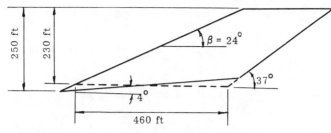

FIGURE 6.5.   Illustrative example of hollow fill. (1 ft = 0.305 m)

Example 1:   Determine the static factor of safety of a spoil bank with a height, H, of 40 ft (12.2 m), an outslope, β, of 36°, and a natural slope, α, of 20°.

Solution: Assuming $C_s = 0$, from Eq. 6.8, $F = \{2 \sin 36° \csc (36° - 20°) [160/(125 \times 40)] + (1 - 0.05) \cos 20° \tan 24°\}/\sin 20° = 1.56$.

Example 2:   Figure 6.5 shows a hollow fill. The natural slope at the bottom of the fill can be approximated by two straight lines. The straight line at the toe has an angle of 4° with the horizontal. The other straight line makes an angle of 37° with horizontal. Determine the seismic factor of safety with a seismic coefficient of 0.1.

Solution: In order to use the simplified method for plane failure, it is necessary to approximate the original ground surface by a horizontal line and an inclined line. One simple way is to draw a horizontal line, as indicated by the dashed line in the figure, such that the weight of the resisting, or lower, block is removed from the toe to the inside, keeping the total weight unchanged. Although the weight of the driving, or upper, block is increased, this increase will be compensated by the increase in factor of safety due to the use of a horizontal slope at the toe. Consequently, the factor of safety will still remain the same. This approximation reduces the height of fill from 250 ft (76.2m) to 230 ft (70.1 m).

The fill is type-I because it has a triangular cross section at the bottom. The base width is 460 ft (140.2 m), or $B = 460/230 = 2$. With $\beta = 24°$, $\alpha = 37°$, and $B = 2$, from Eqs. 6.23 and 6.24

$$a_4 = \tfrac{1}{2} [\cot 37° - (1 - 2 \tan 24°)^2 \cot 24°] = 0.650$$

$$a_5 = \tfrac{1}{2} \times (2)^2 \tan 24° = 0.890$$

With $\bar{c} = 160$ psf (7.7 kPa), $\bar{\phi} = 24°$, $\gamma = 125$ pcf (19.6 kPa), $r_u = 0.05$ and $C_s = 0.1$, from Eqs. 6.18 to 6.20.

$$a_1 = 0.650 \sin 37° + 0.1(0.650 + 0.890) \cos 37° = 0.514$$

$$a_2 = - \left\{ \frac{160}{125 \times 230} (2 \cos 37° + \csc 37°) + [(1 - 0.05) \cos 37° \right.$$

$$\left. - 0.1 \sin 37°] (0.650 + 0.890) \tan 24° \right\} = - 0.497$$

$$a_3 = - 2 \sin 37° \tan 24° \left[\frac{160}{125 \times 230} + (1 - 0.05) \left(\frac{0.890}{2} \right) \tan 24° \right]$$

$$= - 0.104$$

so the quadratic equation becomes

$$0.514F^2 - 0.497F - 0.104 = 0$$

$$F = \frac{0.497 \pm \sqrt{(0.497)^2 + 4 \times 0.514 \times 0.104}}{2 \times 0.514} = 1.144 \text{ or } -0.177$$

Disregarding the negative value the seismic factor of safety is 1.144.

7
Simplified Methods for Cylindrical Failure

7.1 EXISTING STABILITY CHARTS

Since Taylor (1937) first published his stability charts, various charts have been successively presented by Bishop and Morgenstern (1960), Morgenstern (1963), and Spencer (1967). The application of these charts were reviewed by Hunter and Schuster (1971).

Although computer programs are now available for the stability analysis of slopes, these programs generally do not serve the same purpose as stability charts. For major projects in which the shear strength parameters of various soils are carefully determined and the geometry of the slope is complex, a computer program will provide the most economical and probably the most useful solution, even though stability charts can still be used as a guide for preliminary designs or as a rough check for the computer solution. For minor projects in which the shear strength parameters are estimated from the results of standard penetration tests, Dutch-cone tests, or index properties, the application of sophisticated computer programs may not be worthwhile, and stability charts may be used instead.

Although the purpose of this chapter is to present a number of stability charts developed recently by the author, the existing charts developed by others will also be presented. These existing charts are presented in this section, while the remaining sections will be devoted exclusively to the charts developed by the author.

Taylor's Charts. Figure 7.1 shows the stability chart which can be used for the $\phi = 0$ analysis of a simple slope (Taylor, 1948). The slope has an angle, β, a height, H, and a ledge at a distance of DH below the toe, where D is a depth ratio. The chart can be used to determine not only the developed cohesion, c_d, as shown by the solid curves, but also nH, which is the distance

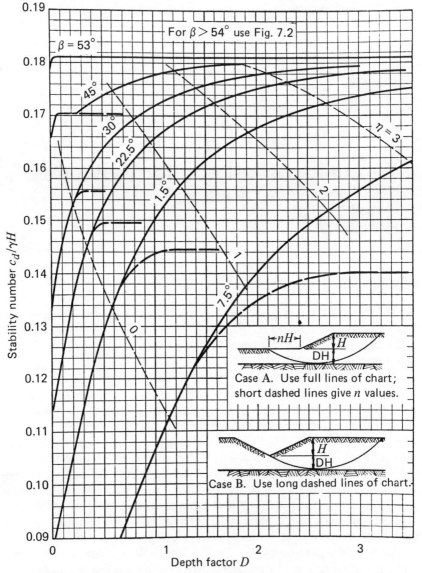

FIGURE 7.1. Stability chart for soils with zero friction angle. (After Taylor, 1948)

from the toe to the failure circle, as indicated by the short dashed curve. If there are loadings outside the toe which prevent the circle from passing below the toe, the long dashed curve should be used to determine the developed cohesion. Note that the solid and the long dashed curves converge when n approaches 0. The circle represented by the curves on the left of $n = 0$ does not pass below the toe, so the loading outside the toe has no effect on the developed cohesion.

Example 1: Given $H = 40$ ft (12.2 m), $DH = 60$ ft (18.3 m), $\beta = 22.5°$, a soil cohesion, c, of 1200 psf (57.5 kPa), and a total unit weight, γ, of 120 pcf (18.9 kN/m³). Determine the factor of safety and the distance from the toe to the point where the most critical circle appears on the ground surface. What is the factor of safety if there are heavy loadings outside the toe?

Solution: For $D = 60/40 = 1.5$ and $\beta = 22.5°$, from the solid curve in Fig. 7.1, $c_d/\gamma H = 0.1715$, or $c_d = 0.1715 \times 120 \times 40 = 823.2$ psf (39.4 kPa). The factor of safety is $c/c_d = 1200/823.2 = 1.46$. From the short dashed curve, $n = 1.85$, or the distance between the toe and the failure circle $= nH = 1.85 \times 40 = 74$ ft (22.6 m). By extending the horizontal portion of the long dashed curve, $c_d/\gamma H = 0.1495$, or $c_d = 0.1495 \times 120 \times 40 = 717.6$ psf (34.4 kPa). The factor of safety when there is heavy loading outside the toe is $1200/717.6 = 1.67$.

Figure 7.2 shows the stability chart of a simple slope when the soil in the slope exhibits both cohesion and angle of internal friction (Taylor, 1948). For a given developed friction angle, ϕ_d, the developed cohesion, c_d, is determined by the friction circle method. When the friction angle is not zero, the most critical circle is a shallow circle. If the ledge lies at a considerable depth below the toe, the location of the ledge, as indicated by the depth factor D, should have no effect on the developed cohesion. This case is shown by the solid curves for circles passing through the toe and by the long dashed curves for circles passing below the toe. However, if $D = 0$, the developed cohesion is much smaller, as indicated by the short dashed curves. The figure can be used to determine the factor of safety with respect to cohesion, F_c, by assuming that the angle of internal friction is fully developed, or the factor of safety with respect to internal friction, F_ϕ, by assuming that the cohesion is fully developed. To find the factor of safety with respect to shear strength, F, as defined by Eq. 1.1, a trial and error method must be used.

Example 2: Given $H = 40$ ft (12.2 m) and $\beta = 30°$. The ledge is far away from surface. The soil has a cohesion, c, of 800 psf (38.3 kPa), a friction angle, ϕ, of 10°, and a total unit weight of 100 pcf (15.7 kN/m³). Determine F_c, F_ϕ, and F.

Solution: Assume that the angle of internal friction is fully developed, or $\phi_d = 10°$. From Fig. 7.2, for $\beta = 30°$, $c_d/\gamma H = 0.075$ or $c_d = 0.075 \times 100 \times 40 = 300$ psf (14.4 kPa), so $F_c = c/c_d = 800/300 = 2.67$.

Next, assume that the cohesion is fully developed, or $c_d/\gamma H = 800/(100 \times 40) = 0.2$. It can be seen from Fig. 7.2 that when $c_d/\gamma H = 0.2$ and $\beta = 30°$, the developed friction angle is less than zero, or the factor of safety with respect to internal friction is infinity. This occurs when the resisting moment due to cohesion is greater than the driving moment.

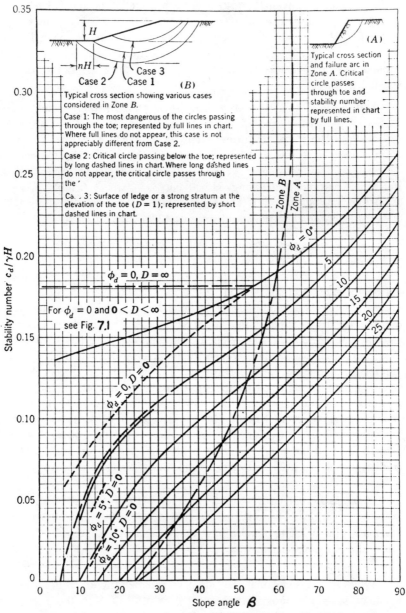

The chart contains the following annotations:

Case 2   Case 3   Case 1   (B)

Typical cross section showing various cases considered in Zone B.

Case 1: The most dangerous of the circles passing through the toe; represented by full lines in chart. Where full lines do not appear, this case is not appreciably different from Case 2.

Case 2: Critical circle passing below the toe; represented by long dashed lines in chart. Where long dashed lines do not appear, the critical circle passes through the '

Ca. . 3: Surface of ledge or a strong stratum at the elevation of the toe ($D = 1$); represented by short dashed lines in chart.

(A)

Typical cross section and failure arc in Zone A. Critical circle passes through toe and stability number represented in chart by full lines.

Zone B   Zone A

$\phi_d = 0$

5

10

15

20

25

$\phi_d = 0, D = \infty$

For $\phi_d = 0$ and $0 < D < \infty$ see Fig. 7.1

$\phi_d = 0, D = 0$

$\phi_d = 5°, D = 0$

$\phi_d = 10°, D = 0$

Stability number $c_d/\gamma H$

Slope angle β

FIGURE 7.2.   Stability chart for soils with friction angle. (After Taylor, 1948)

To determine the factor of safety with respect to shear strength, the same factor of safety should be applied to both cohesion and internal friction. A value of F_c is assumed and a value of F_ϕ, which is equal to $\tan \phi/\tan \phi_d$, is determined from the chart. By trial and error, the factor of safety with respect to shear strength is obtained when $F_c = F_\phi$. This can be accomplished by plotting F_c versus F_ϕ and finding its intersection with a 45° line.

One point on the $F_c - F_\phi$ curve was determined previously as $F_c = 2.67$ and $F_\phi = 1$. It is necessary to have two more points in order to plot the curve. First, assume $F_c = c/c_d = 2$ or $c_d = 800/2 = 400$ psf (19.2 kPa). For $c_d/\gamma H = 400/(100 \times 40) = 0.1$, from Fig. 7.2, $\phi_d = 7°$, or $F_\phi = \tan 10°/\tan 7° = 1.44$. Next, assume $F_c = 1.8$, or $c_d = 800/1.8 = 444$ psf (21.3 kPa). For $c_d/\gamma H = 444/(100 \times 40) = 0.111$, from Fig. 7.2, $\phi_d = 5°$ or $F_\phi = \tan 10°/\tan 5° = 2.02$. Figure 7.3 shows the plot of the three points. The factor of safety with respect to shear strength is 1.82.

Bishop and Morgenstern's Charts.   Figure 7.4 shows the stability charts for effective stress analysis when $\bar{c}/\gamma H = 0$ and 0.025, while Fig. 7.5 shows those for $\bar{c}/\gamma H = 0.05$. The factor of safety is based on the simplified Bishop method (Bishop, 1955) and can be expressed as

$$F = m - r_u n \tag{7.1}$$

in which m and n are the stability coefficients determined from the charts. The values of m and n depend on the depth ratio, D. The charts show three different depth ratios, i.e., 0, 0.25, and 0.5. If the ledge is far from the surface, it is necessary to determine which depth ratio is most critical. This determination can be facilitated by using the lines of equal pore pressure ratio, r_{ue}, on the charts defined as

$$r_{ue} = \frac{m_2 - m_1}{n_2 - n_1} \tag{7.2}$$

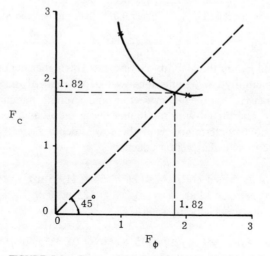

FIGURE 7.3.   Factor of safety with respect to strength.

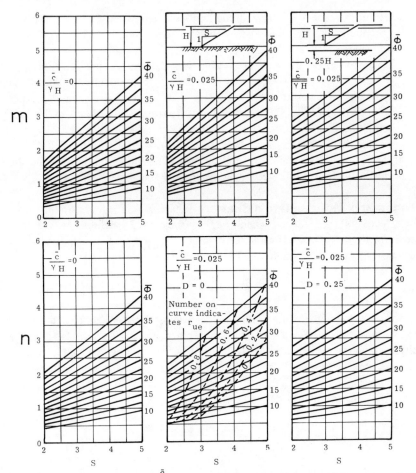

FIGURE 7.4. Stability chart for $\frac{\bar{c}}{\gamma H} = 0$ and 0.025. (After Bishop and Morgenstern, 1960)

in which m_2 and n_2 are the stability coefficients for higher depth ratio, and m_1 and n_1 are the stability coefficients for lower depth ratio. If the given value of r_u is higher than r_{ue} for the given section and strength parameters, then the factor of safety determined with the higher depth ratio has a lower value than the factor of safety determined with the lower depth factor. The following example problem will clarify this concept.

Example 3: Given cot $\beta = 4$, $H = 64$ ft (19.5 m), $DH = 40$ ft (12.2 m), $\bar{c} = 200$ psf (9.6 kPa), $\bar{\phi} = 30°$, $\gamma = 125$ pcf (19.6 kN/m³), and $r_u = 0.5$. Determine the factor of safety.

Solution: For $\bar{c}/\gamma H = 200/(125 \times 64) = 0.025$, cot $\beta = 4$ and $\bar{\phi} = 30°$, from Fig. 7.4, $r_{ue} = 0.43$ when $D = 0$. Because the given r_u of 0.5

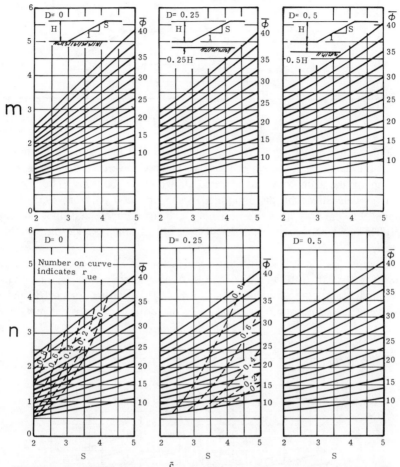

FIGURE 7.5. Stability chart for $\dfrac{\bar{c}}{\gamma H} = 0.05$. (After Bishop and Morgenstern, 1960)

is greater than r_{ue} of 0.43, $D = 0.25$ is more critical. When $D = 0.25$, from Fig. 7.4, $m = 2.95$ and $n = 2.82$, or $F = 2.95 - 0.5 \times 2.82 = 1.54$. If $\bar{c}/\gamma H$ is not exactly equal to 0, 0.025, or 0.05, an interpolation procedure will be needed.

In the above example, if the lines of equal pore pressure ratio are not used, it is necessary to determine the factors of safety for $D = 0$ and $D = 0.25$ and pick up the smaller one. When $D = 0$, from Fig. 7.4, $m = 2.89$ and $n = 2.64$, or $F = 2.89 - 0.5 \times 2.64 = 1.57$, which is slightly greater than the 1.54 for $D = 0.25$.

Morgenstern's Charts. Figure 7.6 shows the cases considered by Morgenstern (1963) in developing his stability charts. An earth dam is placed on

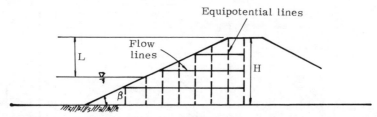

FIGURE 7.6. Slope subject to rapid drawdown.

an impervious surface. The original water level is at the same elevation as the top of the dam. Then the water level is suddenly lowered a distance L below the top of the dam to simulate rapid drawdown conditions. The factor of safety is determined by the simplified Bishop method (Bishop, 1955) by assuming that the flow lines are horizontal and the equipotential lines are vertical after rapid drawdown, and that the weight of soil is twice the weight of water.

Figures 7.7, 7.8, and 7.9 show the factors of safety under rapid drawdown for $\bar{c}/\gamma H$ of 0.0125, 0.025, and 0.05, respectively. The factor of safety is plotted against the drawdown ratio, L/H, for various S, or cot β, and $\bar{\phi}$. When the drawdown ratio is equal to 1, the critical circle is tangent to the

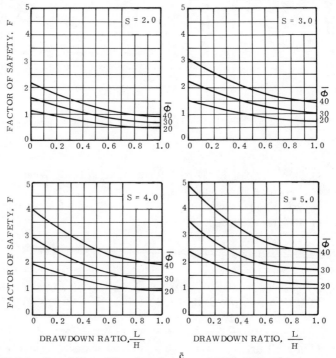

FIGURE 7.7. Drawdown stability chart for $\dfrac{\bar{c}}{\gamma H} = 0.0125$. (After Morgenstern, 1963).

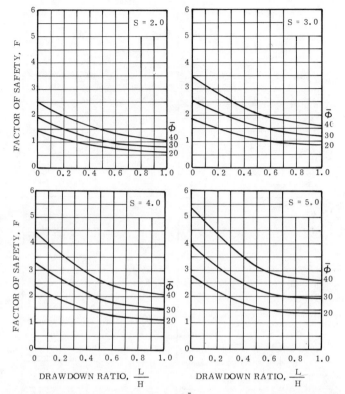

FIGURE 7.8.   Drawdown stability chart for $\frac{\bar{c}}{\gamma H}$ = 0.025. (After Morgenstern, 1963)

base of the dam. When it is less than 1, several circles must be tried to determine the one with the lowest factor of safety.

Example 4: Given cot β = 3, H = 65 ft (19.8 m), $\bar{c}$ = 200 psf (9.6 kPa), $\bar{\phi}$ = 30°, and γ = 124.8 pcf (19.6 kN/m³). Determine the factors of safety when L/H = 1 and L/H = 0.5.

Solution: For $\bar{c}/\gamma H$ =. 200/(124.8 × 65) = 0.025, cot β = 3, and $\bar{\phi}$ = 30°, from Fig. 7.8, the factor of safety is 1.20 when L/H = 1.

When L/H = 0.5, first consider the circle tangent to the base of the dam with an equivalent height, H_e, equal to H. For $\bar{c}/\gamma H_e$ = 0.025, L/H_e = 0.5, from Figure 7.8, F = 1.52. Next consider the circle tangent to the midheight of the dam, or H_e = $H/2$. For $\bar{c}/\gamma H_e$ = 0.05, L/H_e = 1, from Fig. 7.9, F = 1.48. Finally, consider the circle tangent to a level $H/4$ above the base of the dam with H_e = $3H/4$, $\bar{c}/\gamma H_e$ = 0.033, and L/H_e = 0.67. The factor of safety can be determined by a straight-line interpolation of $\bar{c}/\gamma H_e$ between 0.025 and 0.05. From Fig. 7.8 with $\bar{c}/\gamma H_e$ = 0.025, F = 1.37. From Fig. 7.9 with $\bar{c}/\gamma H_e$ = 0.05, F = 1.66. By interpolation F

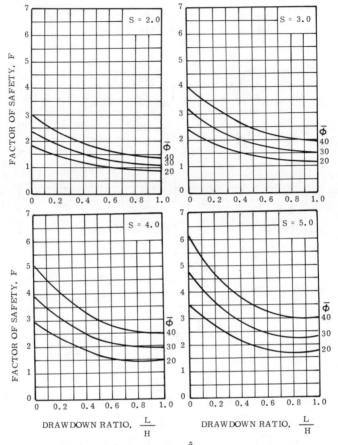

FIGURE 7.9. Drawdown stability chart for $\dfrac{\bar{c}}{\gamma H} = 0.05$. (After Morgenstern, 1963)

= 1.37 + 0.29/3 = 1.47. Although a slightly lower value could perhaps be found, further refinements are unwarranted. This example demonstrates that for partial drawdown, the critical circle may often lie above the base of the dam and it is important to investigate several levels of tangency.

Spencer's Charts. Figure 7.10 shows the stability charts for determining the required slope angle when the factor of safety is given. If the angle of slope is given, the factor of safety can be determined by a method of successive approximations, or trial and error. The Spencer's method assumes parallel interslice forces and satisfies both force and moment equilibrium. It checks well with the simplified Bishop method, which only satisfies the moment equilibrium, because the factor of safety based on moment equilibrium is insensitive to the direction of interslice forces. The charts use three different

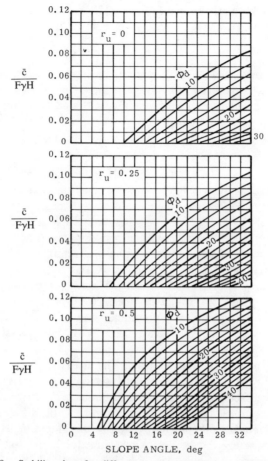

FIGURE 7.10. Stability chart for different pore pressure ratios. (After Spencer, 1967)

pore pressure ratios, that is, 0, 0.25, and 0.5, and assume that the ledge or firm stratum is at a great depth below the surface. In using the charts, it is necessary to find the developed friction angle which is defined as

$$\phi_d = \tan^{-1} (\tan \bar{\phi}/F) \qquad (7.3)$$

Spencer (1967) also developed charts for locating the critical surface, which are not reproduced here. If the ledge is very close to the surface, the design based on Fig. 7.10 is somewhat conservative.

Example 5: Given $H = 64$ ft (19.5 m), $\bar{c} = 200$ psf (9.6 kPa), $\bar{\phi} = 30°$, $\gamma = 125$ pcf (19.6 kN/m³), $r_u = 0.5$, and $F = 1.5$. Determine the slope angle, β.

Solution: With $\bar{c}/F\gamma H = 200/(1.5 \times 125 \times 64) = 0.0167$, $\phi_d = \tan^{-1}$ (tan $30°/1.5$) $= 21°$, and $r_u = 0.5$, from Fig. 7.10, $\beta = 14.5°$ or cot $\beta = 3.9$.

Comparisons of Charts. All the charts presented above involve a homogeneous slope with a given cohesion and angle of internal friction. Some charts assume that a ledge or stiff stratum is located at a great depth while others assume at a given depth ratio, D, ranging from 0 to 0.5. The advantages and limitations of the charts are discussed below.

1. Taylor's charts. The chart in Fig. 7.1 is the only one applicable to the total stress analysis with $\phi = 0$. For total stress analysis with $\phi \neq 0$, the chart in Fig. 7.2 can be used for $D = 0$ or $D = \infty$. The chart can show whether the critical circle passes through or below the toe. If it is below the toe, the case will lie between $D = 0$ and $D = \infty$. Figure 7.2 is the only chart for determining the factors of safety with respect to cohesion and internal friction, as well as shear strength. If the factor of safety is given, the slope angle can be determined directly from the chart. If the slope is given, the factor of safety can only be determined by a trial-and-error procedure. The chart cannot be applied to effective stress analysis because the effect of pore pressure is not considered.

2. Bishop and Morgenstern's Charts. The charts in Figs. 7.4 and 7.5 are applicable to effective stress analysis only with $\bar{c}/\gamma H$ smaller than 0.05. The factor of safety can be determined for different depth ratios, D, and the most critical depth ratio can be easily located by using the lines of equal pore pressure ratio. If the slope angle is given, the factor of safety can be determined directly. If the factor of safety is given, the slope angle can only be determined by a trial-and-error procedure. The charts can be applied to the case of full rapid drawdown, i.e., water level lowered from the top to the toe of the dam, by assuming $r_u = 0.5$, or to the case of total stress analysis by assuming $r_u = 0$.

3. Morgenstern's Charts. The charts in Figs. 7.7 to 7.9 are the only ones which can be used for the effective stress analysis of earth dams during partial drawdown with L/H ranging from 0 to 1. They can be used only when the ledge or stiff stratum is at the toe with $D = 0$, and only when the weight of soil is assumed to be about 125 pcf (19.6 kN/m³). Other features are similar to those of Bishop and Morgenstern's.

4. Spencer's Charts. The charts in Fig. 7.10 are applicable to effective stress analysis when a ledge or stiff stratum is at a great depth. The charts cover a larger range of $\bar{c}/\gamma H$, compared to Bishop and Morgenstern's charts, and can be used for total stress analysis or full rapid drawdown by assuming $r_u = 0$ or 0.5, respectively. Similar to Tay-

lor's chart in Fig. 7.2, the slope angle can be determined directly if the factor of safety is given. However, if the slope angle is given, the factor of safety can only be determined by a trial-and-error procedure.

It will be interesting to compare the results obtained by different methods as illustrated by Examples 6 and 7.

Example 6: Given $H = 100$ ft (30.5 m), $D = \infty$, $c = 1000$ psf (47.9 kPa), $\phi = 20°$, and $\gamma = 125$ pcf (19.6 kN/m³). Determine the slope angle, β, by total stress analysis for a factor of safety of 1.5.

Solution: This problem can be solved by Taylor's or Spencer's charts. If Taylor's chart in Fig. 7.2 is used, $c_d = c/F = 1000/1.5 = 667$ psf (31.9 kPa), $c_d/\gamma H = 667/(125 \times 100) = 0.0533$, and $\phi_d = \tan^{-1}(\tan \phi/F) = \tan^{-1}(\tan 20°/1.5) = 13.6°$, so $\beta = 29°$. If Spencer's chart in Fig. 7.10 with $r_u = 0$ is used, $\bar{c}/F\gamma H = 0.0533$ and $\phi_d = 13.6°$, so $\beta = 29°$. It can be seen that both charts yield the same result. Bishop and Morgenstern's charts cannot be used because the value of $c/\gamma H$ is outside the range of the charts.

Example 7: Given $H = 48$ ft (14.6 m), $D = 0$, $\cot \beta = 3$, $\bar{c} = 300$ psf (14.4 kPa), $\bar{\phi} = 30°$, and $\gamma = 125$ pcf (19.6 kN/m³). Determine the factor of safety under full rapid drawdown.

Solution: This problem can be solved by Bishop and Morgenstern's or Morgenstern's charts. If Bishop and Morgenstern's charts are used, $\bar{c}/\gamma H = 300/(125 \times 48) = 0.05$; from Fig. 7.5, $m = 2.57$ and $n = 2.17$; for $r_u = 0.5$, $F = 2.57 - 0.5 \times 2.17 = 1.49$. If Morgenstern's charts are used, from Fig. 7.9 with $L/H = 1$, $F = 1.49$. It can be seen that both charts yield the same factor of safety.

Although Spencer's charts assumed that the ledge is at a great depth, or $D = \infty$, the result is not too much different if they are applied to $D = 0$. For a safety factor of 1.49, $\bar{c}/F\gamma H = 300/(1.49 \times 125 \times 48) = 0.0336$, and $\phi_d = \tan^{-1}(\tan \bar{\phi}/F) = \tan^{-1}(\tan 30°/1.49) = 21.2°$; from Fig. 7.10 with $r_u = 0.5$, $\beta = 18.4°$, or $\cot \beta = 3$.

7.2 TRIANGULAR FILLS ON ROCK SLOPES (Huang, 1977a, 1978b)

Figure 7.11 shows a triangular fill deposited on a rock slope. The fill has a height, H, an angle of outslope, β, and a degree of natural slope, α. The natural ground is assumed to be much stiffer than the fill, so the failure surface will lie entirely within the fill.

In total stress analysis when the angle of internal friction is assumed to be 0, the most critical circle is tangent to the rock. The average shear stress developed along the failure surface can be easily determined by equating the

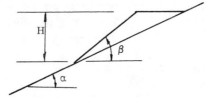

FIGURE 7.11. Triangular fill on rock slope.

moment about the center of the circle due to the weight of sliding mass to that due to the average shear stress distributed uniformly over the failure arc. This developed shear stress is proportional to the unit weight and the height of fill and can be expressed as

$$\tau = \frac{\gamma H}{N_s} \tag{7.4}$$

in which τ is the developed shear stress and N_s is the stability number, which is a function of α and β. The factor of safety is defined as

$$F = \frac{s}{\tau} \tag{7.5}$$

in which s is the shear strength, which is equal to the cohesion, c, in a total stress analysis. Substituting Eq. 7.4 into Eq. 7.5

$$F = \frac{cN_s}{\gamma H} \tag{7.6}$$

By comparing Eq. 7.6 with Eq. 2.3 and noting ϕ as zero, the stability number, N_s, can be expressed as

$$N_s = \frac{\gamma H \sum\limits_{i=1}^{n} L_i}{\sum\limits_{i=1}^{n} (W_i \sin \theta_i)} \tag{7.7}$$

Values of N_s for various combinations of α and β can be computed by the REAME computer program (Huang, 1981b) and are plotted in the top part of Fig. 7.12. This plot can be achieved by assuming $H = 10$ ft (3.1 m), $\gamma = 100$ pcf (15.7 kN/m³), and $c = 1000$ psf (47.9 kPa), so the factor of safety obtained from the computer program is actually the stability number, as can be seen by substituting the preceding values into Eq. 7.6. The program also gives the center of the most critical circle and the corresponding radius. The most critical circle is always tangent to the rock.

In the effective stress analysis, the angle of internal friction is not zero

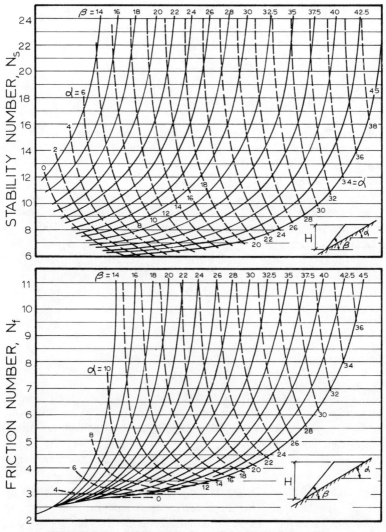

FIGURE 7.12. Stability chart for spoil banks and hollow fills.

and the shear strength can be expressed as

$$s = \bar{c} + \frac{(1 - r_u)\gamma H \tan \bar{\phi}}{N_f} \qquad (7.8)$$

in which N_f is the friction number, which is also a function of α and β. Substituting Eqs. 7.4 and 7.8 into Eq. 7.5

$$F = N_s\left[\frac{\bar{c}}{\gamma H} + \frac{(1 - r_u) \tan \bar{\phi}}{N_f}\right] \qquad (7.9)$$

By substituting N_s from Eq. 7.7 into Eq. 7.9 and comparing with Eq. 2.9

$$N_f = \frac{\gamma H \sum\limits_{i=1}^{n} L_i}{\sum\limits_{i=1}^{n} (W_i \cos \theta_i)} \tag{7.10}$$

By selecting the circle used for determining N_s, the friction number, N_f, for various combinations of α and β can be computed from Eq. 7.10 and is plotted in the bottom part of Fig. 7.12.

The factor of safety determined from Eq. 7.9 in conjunction with Fig. 7.12 may not be the minimum because the most critical circle for $\phi = 0$, as used in developing Fig. 7.12, is different from that for $\phi \neq 0$, especially when the soil has a high internal friction relative to cohesion. Thus a correction factor is needed.

Figure 7.13 gives the correction factor, c_f, for the factor of safety computed by Eq. 7.9. The correction factor depends on the angle of outslope, β, the degree of natural slope, α, and the percent of cohesion resistance, P_c, which is defined as

$$P_c = \frac{(\bar{c}/\gamma H)}{(\bar{c}/\gamma H) + (1 - r_u) (\tan \bar{\phi}/N_f)} \tag{7.11}$$

These curves were obtained from a series of analyses by comparing the factor of safety from Eq. 7.9 with that from the REAME computer program. The corrected factor of safety is the product of the correction factor and the factor of safety from Eq. 7.9. Although Fig. 7.13 only gives the correction factor for three values of β, that is, $37°$, $27°$, and $17°$, the correction factor for other values of β can be obtained by a straight-line interpolation.

Although Fig. 7.12 is based on the triangular cross section shown in Fig. 7.11, it can also be applied to the effective stress analysis of the slope shown in Fig. 7.14, where the rock slope is quite irregular. In such a case, α is the degree of natural slope at the toe. Because of the small cohesion used in the effective stress analysis, the failure surface will be a shallow circle close to the surface of the slope near the toe and is independent of all slopes behind the toe.

As an illustration of the application of the method, consider the hollow fill shown in Fig. 7.11 with $\alpha = 8.7°$, $\beta = 27.4°$, and $H = 140$ ft (42.7 m). Assuming that the fill has an effective cohesion of 200 psf (9.6 kPa), an effective friction angle of $30°$, a total unit weight of 125 pcf (19.6 kN/m³), and that there is no seepage, or $r_u = 0$, determine the factor of safety of the fill.

From Fig. 7.12, $N_s = 9.5$ and $N_f = 3.6$. From Eq. 7.9, $F = 9.5 \times$ [200/(125 × 140) + (1 − 0) (tan 30°)/3.6] = 9.5 × (0.0114 + 0.1603) = 9.5

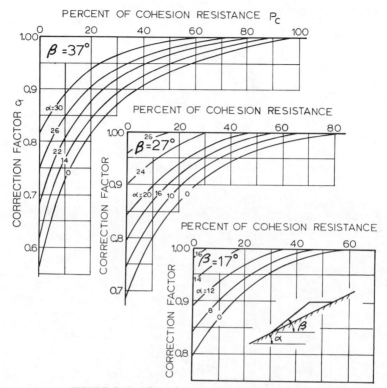

FIGURE 7.13. Chart for correcting factor of safety.

$\times$ 0.1717 = 1.631. From Eq. 7.11, P_c = 0.0114/0.1717 = 0.066 = 6.6 percent. From Fig. 7.13, c_f = 0.77. The corrected factor of safety is 0.77 $\times$ 1.631 = 1.256.

7.3 TRAPEZOIDAL FILLS ON ROCK SLOPES (Huang, 1978a, 1978b)

Figure 7.15 shows the cross section of a trapezoidal fill with a height, H, an outslope S:1, an angle of natural slope, α, and a base width BH, in which B is a ratio between base width and height. The triangular fill is a special case of trapezoidal fill when B = 0. The natural ground is assumed to be much stiffer than the fill, so the failure circle will be entirely within the fill.

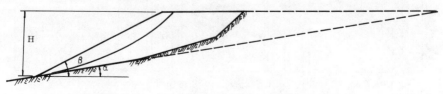

FIGURE 7.14. Approximation of hollow fill by triangular fill.

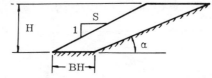

FIGURE 7.15.   Trapezoidal fill on rock slope.

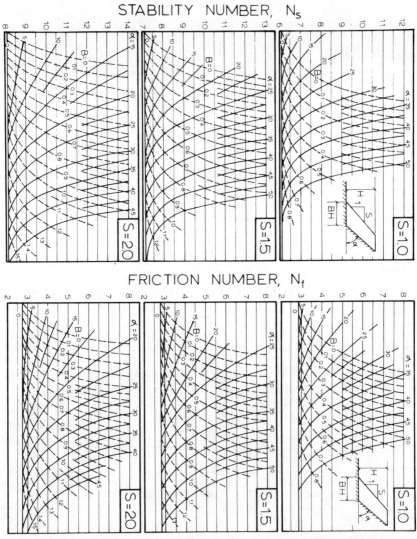

FIGURE 7.16.   Stability chart for trapezoidal fill on rock slope, S=1.0, 1.5, and 2.0.

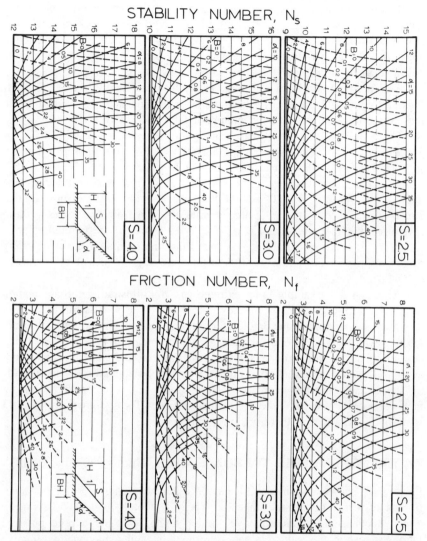

FIGURE 7.17. Stability chart for trapezoidal fill on rock slope, S=2.5, 3.0 and 4.0.

By using the same procedure as in triangular fills, values of N_s and N_f for various combinations of S, B, and α are presented in Figs. 7.16 and 7.17. The friction number, N_f, is based on the circle with a center and radius the same as the most critical circle for $\bar{\phi} = 0$. In other words, the same circle is used for determining N_f and N_s. If α or $\bar{c}$ is very small, the most critical circle for $\bar{\phi} = 0$ may be quite different from that for $\bar{\phi} \neq 0$, so a trial-and-error procedure, as described by the following example, should be used to determine the minimum factor of safety.

Figure 7.18 shows the extreme case of an embankment on a horizontal ledge. The embankment is 50 ft (15.3 m) high with a side slope of 1.5:1. The soil has an effective cohesion of 300 psf (14.4 kPa), an effective friction angle of 30°, and a total unit weight of 120 pcf (18.8 kN/m³). It is now desirable to determine the factor of safety when there is no pore pressure.

This problem can be solved most easily by using Figs. 7.12 and 7.13. If Figs. 7.16 and 7.17 are used, a trial-and-error method must be used to locate the most critical circle with the minimum factor of safety.

Given $S = 1.5$ and $\alpha = 0$ (B can be any value), from Fig. 7.16, $N_s = 7.1$ and $N_f = 2.8$. Given $\bar{c}/(\gamma H) = 300/(120 \times 50) = 0.05$, $\tan \bar{\phi} = \tan 30°$ $= 0.577$, and $r_u = 0$; from Eq. 7.9, $F = 7.1 \times (0.05 + 0.577/2.8) = 1.82$, which is based on the most critical circle for $\bar{\phi} = 0$, as indicated by the dashed arc. When $\bar{c} = 300$ psf (14.4 kPa) and $\bar{\phi} = 30°$, the most critical circle, as indicated by the solid arc, is quite different from that for $\bar{\phi} = 0$, so several combinations of α and B must be tried to obtain the minimum factor of safety. After several trials, it was found that the minimum factor of safety occurs when $\alpha = 50°$ and $B = 0.9$. In this case, $N_s = 9.0$ and $N_f = 5.4$, or $F = 1.41$.

Although this example is an extreme case for illustrating the application of the charts, it does demonstrate how a trial-and-error process can be used to determine the minimum factor of safety. If the natural ground surface is not too far away from the slope surface, the most critical circle is most probably tangent to the natural ground surface, and the trial-and-error process is not needed. The same trial-and-error process can also be applied to triangular fills by simply changing the degree of natural slope, α, to locate the minimum factor of safety. However, the use of the correction factor shown in Fig. 7.13 is much quicker and is therefore recommended.

7.4 TRIANGULAR FILLS ON SOIL SLOPES (Huang, 1977b, 1978b)

Figure 7.19 shows a triangular fill having a height, H, an angle of outslope, β, and a degree of natural slope, α. The fill is built by soil 2 having an effective cohesion $\bar{c}_2$, an effective friction angle, $\bar{\phi}_2$, and a pore pressure ratio, r_{u2}, while the natural slope is formed by soil 1 having an effective cohesion, $\bar{c}_1$, an effective friction angle, $\bar{\phi}_1$, and a pore pressure ratio, r_{u1}. The method described here can also be applied to the total stress analysis by simply replacing the effective strength parameters by the total strength parameters. It is also assumed that soils 1 and 2 have the same unit weight, γ. If the two unit weights are different, an average value should be used.

In Fig. 7.19, the most critical circle, or the circle with a minimum factor of safety, is assumed tangent to a line at a depth, DH, below the surface of the natural slope, in which D is the depth ratio, and H is the height of fill. When $D = 0$, the circle is tangent to the natural slope and Figs. 7.12 and 7.13 apply. The maximum permissible D is governed by the depth of stiff stratum. For a given depth ratio, a critical circle with the lowest factor of

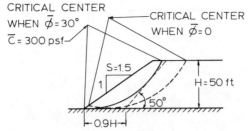

FIGURE 7.18.   Simple slope on horizontal ledge. (1 ft. = 0.305 m, 1 psf = 47.9 Pa)

safety can be found. By comparing the lowest factors of safety for various depth ratios, the most critical circle with the minimum factor of safety can be obtained.

   To determine the stability and friction numbers, it is necessary to know the location of the critical circle, which depends not only on the geometry of the slope (i.e., α, β, and D) but also on the soil parameters (i.e., $\bar{c}_1$, $\bar{\phi}_1$, r_{u1}, $\bar{c}_2$, $\bar{\phi}_2$, r_{u2}, and γ). It is practically impossible to present stability charts in terms of all the preceding parameters. Fortunately, it was found that geometry has far more effect than the soil parameters. As a result, a homogeneous cohesive soil with $\bar{\phi} = 0$ can be assumed to determine the location of the most critical circle, and the stability and friction numbers can be presented in terms of α, β, and D.

   The factor of safety by the normal method can be expressed as

$$F = \frac{\sum\limits_{i=1}^{n} \left[\bar{c}_1 L_1 + \bar{c}_2 L_2 + (1 - r_{u1}) W_{i1} \cos \theta_i \tan \bar{\phi}_1 + (1 - r_{u2}) W_{i2} \cos \theta_i \tan \bar{\phi}_2 \right]}{\sum\limits_{i=1}^{n} W_i \sin \theta_i} \qquad (7.12)$$

in which L_1 and L_2 are the lengths of the failure arcs in soils 1 and 2, respectively; and W_{i1} and W_{i2} are the weights of slice i above the arc in soils 1 and

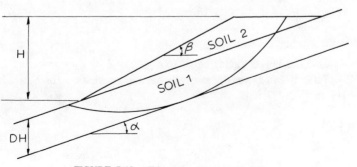

FIGURE 7.19.   Triangular fill on soil slope.

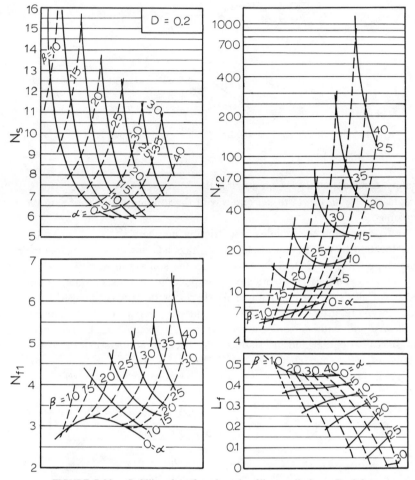

FIGURE 7.20.   Stability chart for triangular fill on soil slope, $D=0.2$.

2, respectively. Eq. 7.12 can be written as

$$F = N_s \left[\frac{\bar{c}_1}{\gamma H} (1 - L_f) + (\bar{c}_2/\gamma H)L_f + (1 - r_{u1}) \frac{\tan \bar{\phi}_1}{N_{f1}} + (1 - r_{u2}) \frac{\tan \bar{\phi}_2}{N_{f2}} \right]$$
(7.13)

in which L_f is the length factor equal to $L_2/(L_1 + L_2)$. By comparing Eq. 7.13 with Eq. 7.12, it can be easily proved that the expression for N_s is the same as shown in Eq. 7.7 and

$$N_{f1} = \frac{\gamma H \sum_{i=1}^{n} L_i}{\sum_{i=1}^{n} (W_{i1} \cos \theta_i)}$$
(7.14)

$$N_{f2} = \frac{\gamma H \sum\limits_{i=1}^{n} L_i}{\sum\limits_{i=1}^{n} (W_{i2} \cos \theta_i)} \qquad (7.15)$$

Values of N_s, N_{f1}, N_{f2}, and L_f for various values of α and β are presented in Figs. 7.20 through 7.23 for various values of D. The degree of natural slope, α, ranges from 0 to 30°, the angle of outslope, β, starts from 5° above α and goes to 40°, and the depth ratio, D, ranges from 0.2 to 0.8 (when $D = 0$, refer to Fig. 7.12). These ranges should cover most of the cases encountered in practice. When α is very small and $\bar{\phi} = 0$, the depth ratio may be very large and fall outside the range of the charts. In such a special case, the

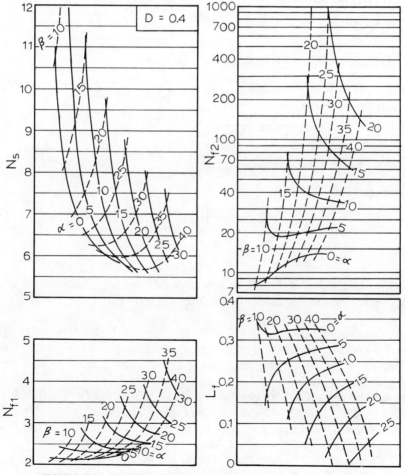

FIGURE 7.21. Stability chart for triangular fill on soil slope, D=0.4.

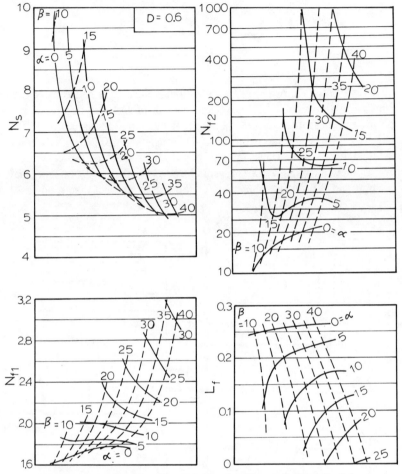

FIGURE 7.22.   Stability chart for triangular fill on soil slope, D=0.6.

stability chart shown in Fig. 7.36 may be used. The application of the stability charts can be illustrated by the following example.

Example 8: Given $\alpha = 15°$, $\beta = 35°$, $H = 20$ ft (6 m), $\gamma = 125$ pcf (19.6 kN/m³), $\bar{c}_1 = 100$ psf (4.8 kPa), $\bar{\phi}_1 = 20°$, $\bar{c}_2 = 150$ psf (7.2 kPa), and $\bar{\phi}_2 = 30°$, determine the factor of safety when there is no seepage.

Solution: $\bar{c}_1/(\gamma H) = 100/(125 \times 20) = 0.04$, $\bar{c}_2/(\gamma H) = 150/(125 \times 20) = 0.06$, $\tan 20° = 0.364$, and $\tan 30° = 0.577$.

When $D = 0$, from Fig. 7.12, $N_s = 8.9$, $N_f = 4.1$; from Eq. 7.9, $F = 8.9 \times (0.06 + 0.577/4.1) = 1.79$.

Where $D = 0.2$, from Fig. 7.20, $N_s = 6.8$, $N_{f1} = 3.5$, $N_{f2} = 26$, and $L_f = 0.29$; from Eq. 7.13, $F = 6.8(0.04 \times 0.71 + 0.06 \times 0.29 + 0.364/3.5 + 0.577/26) = 1.17$.

When $D = 0.4$, from Fig. 7.21, $N_s = 6.0$, $N_{f1} = 2.6$, $N_{f2} = 72$, and $L_1 = 0.18$; from Eq. 7.13, $F = 6.0(0.04 \times 0.82 + 0.06 \times 0.18 + 0.364/2.6 + 0.577/72) = 1.15$.

When $D = 0.6$, from Fig. 7.22, $N_s = 5.4$, $N_{f1} = 2.1$, $N_{f2} = 140$, and $L_1 = 0.12$; from Eq. 7.13, $F = 5.4(0.04 \times 0.88 + 0.06 \times 0.12 + 0.364/2.1 + 0.577/140) = 1.19$.

The minimum factor of safety is 1.15 and occurs when $D = 0.4$.

7.5 EFFECTIVE STRESS ANALYSIS OF HOMOGENEOUS DAMS (Huang, 1975)

Figure 7.24 shows the stability chart for a homogeneous slope, such as an earth dam. This is the only chart developed by the author where the factor of safety is based on the simplified Bishop method; all others are based on the

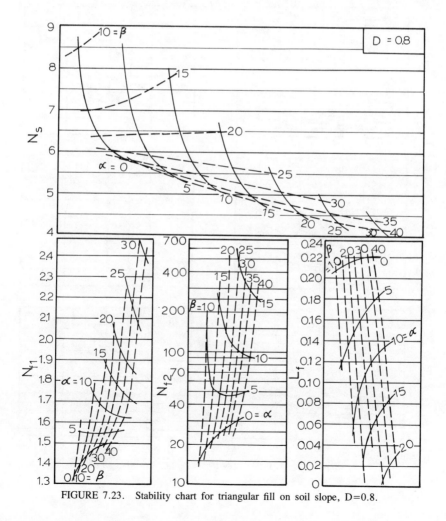

FIGURE 7.23. Stability chart for triangular fill on soil slope, D=0.8.

normal method. The chart is applicable to a homogeneous dam on a deep foundation with the same shear strength for the dam and the foundation. The assumption of a homogeneous slope is not a serious limitation because the effective strength parameters, $\bar{c}$ and $\bar{\phi}$, of most soils do not change significantly, so average values of cohesion and angle of internal friction can be reasonably assumed. The chart can be applied to inclined ground by slightly

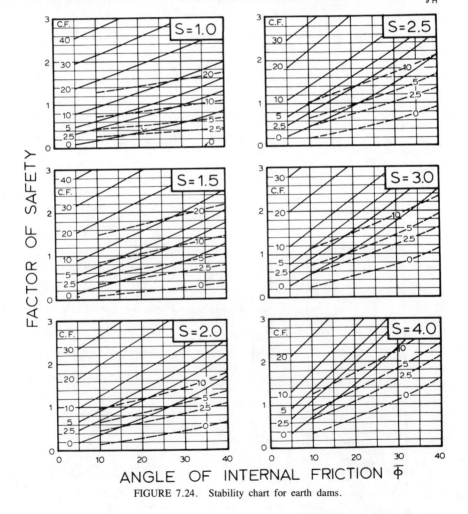

FIGURE 7.24. Stability chart for earth dams.

adjusting the configuration of the slope at the toe. This small change in configuration should have very little effect on the computed factor of safety.

The left upper corner of Fig. 7.24 shows an earth embankment with a height, H, and an outslope of S:1. In the effective stress analysis, the soil has a small cohesion relative to the angle of internal friction, so the most critical failure surface may be a toe circle or a circle slightly below the toe. As long as the bedrock is at a considerable distance from the surface, its location has no effect on the factor of safety.

In Fig. 7.24, the solid curves indicate zero pore pressure and the dashed curves indicate a pore pressure ratio of 0.5. The factor of safety for other pore pressure ratios can be obtained by a straight-line interpolation between the solid and dashed curves. The number on each curve is the cohesion factor in percent, which is equal to $100\bar{c}/(\gamma H)$.

For a given effective friction angle and a given cohesion factor, the factor of safety for a given slope can be determined directly from the chart. This chart cannot be applied to total stress analysis when $\phi = 0$ because when $\phi = 0$ the most critical circle will be a deep circle tangent to the bedrock. As the depth to the bedrock is not given, the factor of safety cannot be determined. This is why all curves stop at $\bar{\phi} = 5°$ and should not be extended to $\bar{\phi} = 0$.

Figure 7.25 shows a practical example for the application of the stability chart. This dam, which provides water supply to Springfield, is located in Washington County, KY. The dam was built quite a while ago. Several years ago a failure took place. The material on the downstream face slid away. The location of the failure surface is very close to the theoretical circle shown in the figure. This provides a good opportunity to back calculate the shear strength of the soil in the field. The failure can be considered as a full-scale model test. When the dam failed, the factor of safety should have decreased to 1.

The original downstream slope is not uniform, being flatter at the toe than at the top. However, it can be changed to a uniform slope by approximating the slope at the toe with a horizontal line and an inclined line, so that the cut is equal to the fill. The downstream slope is 1.75:1 and the height is 37 ft (11.3 m). The phreatic surface is determined by the theoretical method, as described in Sect. 4.2. By using the REAME computer program and assuming an effective cohesion of 200 psf (9.6 kPa) and an effective friction angle of 25°, a factor of safety of 0.97 was obtained. This indicates that the assumed shear strength is reasonable because it yields a factor of safety close to 1. Therefore, this shear strength can be used for the redesign of the dam. Unfortunately, there was an office building not far from the dam, so the downstream slope could not be flattened. Finally, a rock berm at a slope of 1:1 was constructed to increase the factor of safety.

In using the simplified method, it is necessary to estimate the percent of fill under water. In this example, it is estimated that 75 percent of the area is below and 25 percent above the water table.

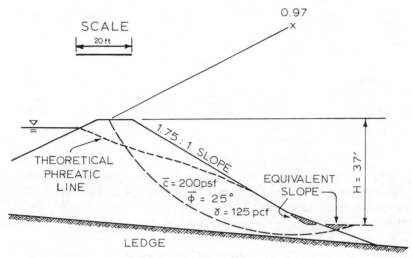

FIGURE 7.25. Stability analysis of Springfield Dam in Kentucky. (1 ft = 0.305 m, 1 psf = 47.9 Pa, 1 pcf = 157.1 N/m³)

Figure 7.26 illustrates the application of stability charts for determining the factor of safety of the dam. For an outslope of 1.5:1, a cohesion factor of 4.32, and a friction angle of 25°, as indicated by the chart on the left, the factor of safety is 1.3 from the solid curves, where no fill is under water, and 0.7 from the dashed curves, where the entire fill is under water. If 75 percent of the fill is under water, the factor of safety is

$$F = 0.75 \times 0.7 + 0.25 \times 1.3 = 0.85$$

This interpolation is based on the assumption that the soil weighs twice as much as the water, so a pore pressure ratio of 0.5 implies that the entire fill is under water. If this is not the case, an interpolation based on the pore pressure ratio is required.

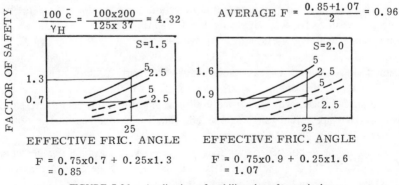

FIGURE 7.26. Application of stability chart for earth dam.

For an outslope of 2:1, as indicated by the chart on the right, the factor of safety is 1.6 from the solid curves and 0.9 from the dashed curves, so the factor of safety is

$$F = 0.75 \times 0.9 + 0.25 \times 1.6 = 1.07$$

The actual outslope is 1.75:1, which is the average of 1.5:1 and 2:1; so the average factor of safety is

$$F = \frac{0.85 + 1.07}{2} = 0.96$$

which checks with the 0.97 obtained from the REAME computer program.

7.6 EFFECTIVE STRESS ANALYSIS OF NONHOMOGENEOUS DAMS (Huang 1979, Huang 1980)

In order to determine more accurately the average cohesion, friction angle, and pore pressure of a nonhomogeneous dam along the failure surface, it is necessary to know the exact location of the critical circle. Therefore, a number of potential circles must be tried, and the factor of safety for each circle determined. The minimum factor of safety can then be found.

The stability charts were developed on the basis of a homogeneous dam but can be applied to a nonhomogeneous dam with proper calculations. Because the material has a small effective cohesion, as is usually the case in an effective stress analysis, the most critical failure surface is a shallow circle. Figure 7.27 shows three cases of shallow circle for a dam with a height, H, and a slope of S:1. Case 1, in which the circle passes through the top edge and the toe of the dam, is applicable when the ledge, or stiff stratum, is located at the bottom of the dam. It also applies when the circle intersects the surface of the slope, in which case H is the vertical distance between the two intersecting points. Case 2, in which the circle passes through the toe but intersects the top at a distance of $0.1SH$ from the edge, is applicable when the ledge is very close to the bottom of the dam, say less than $0.1H$. Case 3, in which the two end points of the circle lie at a distance of $0.1SH$ from the edge and the toe, is applicable when the ledge is located at some distance from the bottom of the dam. However, in most cases, the difference is not very significant, so only one case, based on the location of the ledge, need be tried.

In applying the stability charts, it is necessary to plot a cross section of the dam on a graph paper. In case 1, a bisector perpendicular to the surface of the slope is established, and several circles with centers along the bisector may be drawn. Each center is defined by YH, which is the vertical distance between the center of the circle and the top of the dam. In cases 2 and 3, a bisector perpendicular to the dashed line is drawn, and the factors of safety for several circles are determined and compared. If the ledge or stiff stratum

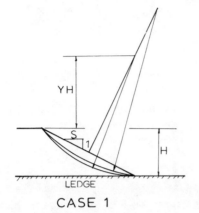

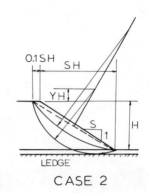

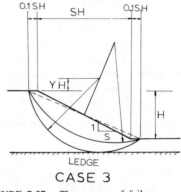

FIGURE 7.27. Three cases of failure.

is close to the bottom of the dam, the circle tangent to the ledge is usually the most critical.

Figures 7.28, 7.29, and 7.30 show the stability number, N_s, the friction number, N_f, and the earthquake number, N_e, for cases 1, 2, and 3, respectively. Note that both Y and S are dimensionless ratios between two distances. Since the slope angle, β, is related to S, the slope angle corresponding to each S is also shown. Figure 7.31 shows the friction number in a large scale.

The factor of safety can be determined by

$$F = \frac{(\bar{c}/\gamma H) + (1 - r_u)\,(\tan\,\bar{\phi}/N_f)}{(1/N_s) + (C_s/N_e)} \qquad (7.16)$$

If C_s equals zero, the equation is the same as Eq. 7.9. Expressions for N_s and N_f are the same as Eqs. 7.7 and 7.10. By substituting N_s and N_f into Eq. 7.16

and comparing with Eq. 2.11

$$N_e = \frac{\gamma H R \sum\limits_{i=1}^{n} L_i}{\sum W_i a_i} \qquad (7.17)$$

To compute the factor of safety, it is necessary to determine three geometric parameters (N_s, N_f, and N_e) from the stability charts and four soil parameters (r_u, $\bar{c}$, $\bar{\phi}$, and γ). The pore pressure ratio can be determined from the location of phreatic surface, as shown in Eq. 4.9 or 4.10. If there is only one soil, the soil parameters, $\bar{c}$, $\bar{\phi}$, and γ are given directly. If there are

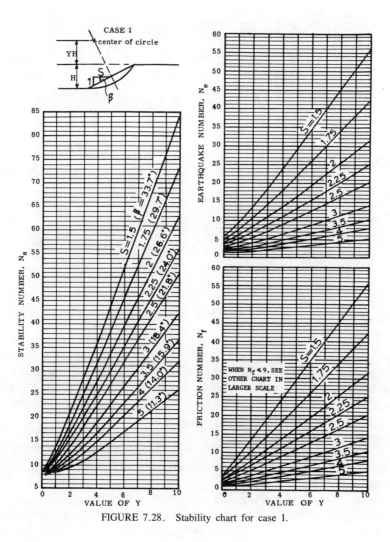

FIGURE 7.28.   Stability chart for case 1.

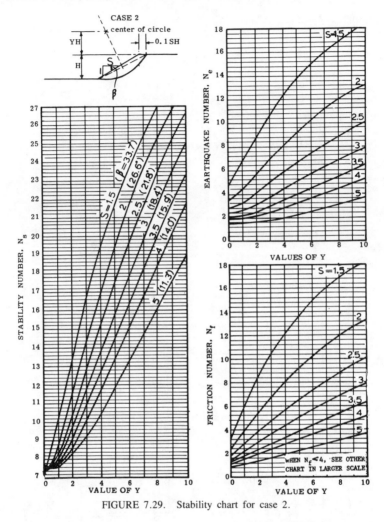

FIGURE 7.29.   Stability chart for case 2.

several different soils, the average values of $\bar{c}$, $\bar{\phi}$, and γ must be determined. To determine the average soil parameters, it is necessary to measure the length of failure arc in each soil and the area occupied by each soil. The work is tedious but can be simplified by Table 7.1 which relates the chord length, L_c, to the arc length, L, and the area between the chord and the arc, A. The length and area are expressed as a dimensionless ratio in terms of radius R and R^2, respectively. For a given ratio of L_c/R, the upper number is L/R and the lower number is A/R^2. The application of this table will be illustrated by the following example.

Figure 7.32 shows a refuse dam composed of both coarse and fine refuse and resting on a layer of natural soil. The total unit weight, the effective cohesion, and the effective friction angle of these three materials are tabulated

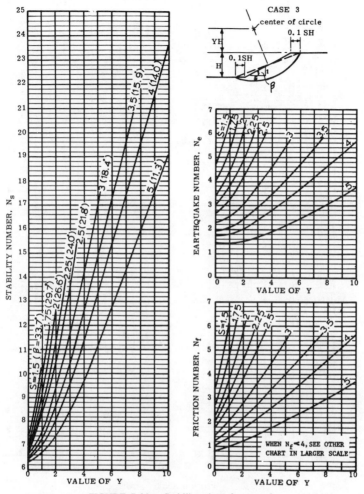

FIGURE 7.30.   Stability chart for case 3.

on the top of the figure. Because the fine refuse is in a loose and saturated condition, which may be subjected to liquefaction, an evaluation of both static and seismic factors of safety is needed. In the seismic analysis, it is assumed that a seismic coefficient of 0.1 will be used and that the fine refuse has an effective friction angle of only 16°, instead of the 31° obtained from the triaxial compression test. The assumption of a very low friction angle is to compensate the effect of pore pressure generated by the earthquake. The location of the phreatic surface is shown by the dotted line.

The refuse dam consists of two slopes. The upper slope has a height of 130 ft (39.6 m) and an outslope of 3.85:1. Because the stiff stratum is located at a considerable distance below the toe, case 3 should be applied. The lower

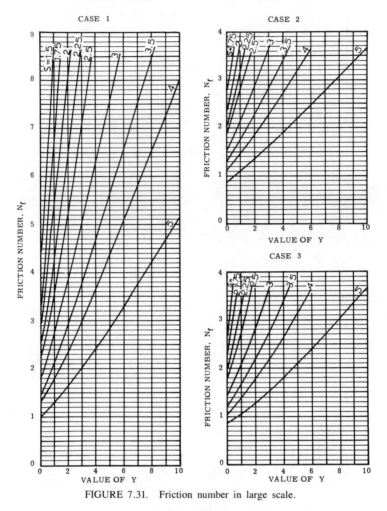

FIGURE 7.31.   Friction number in large scale.

slope has a height of 145 ft (44.2 m) and an outslope of 2.28:1. Because the stiff stratum is close to the toe, case 2 should be considered.

Figure 7.33 shows the factors of safety for different trial circles. For the upper slope, four circles of case 3 were tried. The minimum static factor of safety is 2.178 and occurs when the circle is tangent to the bottom of the natural soil. The minimum seismic factor of safety with a seismic coefficient of 0.1 is 0.977 and occurs when the circle is tangent to the bottom of fine refuse. The minimum factor of safety by the REAME computer program is 2.113 for the static case and 0.948 for the seismic case, based on the normal method. The corresponding factors of safety based on the simplified Bishop method are 2.758 and 1.270, respectively. It can be seen that the difference between the normal method and the simplified Bishop method is quite large, due to the very deep circle involved. However, compared to the simplified Bishop method, the use of stability charts is on the safe side.

Table 7.1 L/R and A/R^2 in terms of L_c/R.

L_c/R	0.00	0.01	0.02	0.03	0.04	0.05	0.06	0.07	0.08	0.09
0.0	0.000	0.010	0.020	0.030	0.040	0.050	0.060	0.070	0.080	0.090
	0.000	0.000	0.000	0.000	0.000	0.000	0.000	0.000	0.000	0.000
0.1	0.100	0.110	0.120	0.130	0.140	0.150	0.160	0.170	0.180	0.190
	0.000	0.000	0.000	0.000	0.000	0.000	0.000	0.000	0.000	0.001
0.2	0.200	0.210	0.220	0.231	0.241	0.251	0.261	0.271	0.281	0.291
	0.001	0.001	0.001	0.001	0.001	0.001	0.001	0.002	0.002	0.002
0.3	0.301	0.311	0.321	0.332	0.342	0.352	0.362	0.372	0.382	0.393
	0.002	0.003	0.003	0.003	0.003	0.004	0.004	0.004	0.005	0.005
0.4	0.403	0.413	0.423	0.433	0.444	0.454	0.464	0.474	0.485	0.495
	0.005	0.006	0.006	0.007	0.007	0.008	0.008	0.009	0.009	0.010
0.5	0.505	0.516	0.526	0.536	0.547	0.557	0.568	0.578	0.588	0.599
	0.011	0.011	0.012	0.013	0.013	0.014	0.015	0.016	0.017	0.018
0.6	0.609	0.620	0.630	0.641	0.651	0.662	0.673	0.683	0.694	0.704
	0.019	0.019	0.020	0.021	0.023	0.024	0.025	0.026	0.027	0.028
0.7	0.715	0.726	0.737	0.747	0.758	0.769	0.780	0.790	0.801	0.812
	0.030	0.031	0.032	0.034	0.035	0.037	0.038	0.040	0.042	0.043
0.8	0.823	0.834	0.845	0.856	0.867	0.878	0.889	0.900	0.911	0.922
	0.045	0.047	0.048	0.050	0.052	0.054	0.056	0.058	0.060	0.063
0.9	0.934	0.945	0.956	0.967	0.979	0.990	1.001	1.013	1.024	1.036
	0.065	0.067	0.070	0.072	0.074	0.077	0.080	0.082	0.085	0.088
1.0	1.047	1.059	1.070	1.082	1.094	1.105	1.117	1.129	1.141	1.153
	0.091	0.094	0.096	0.100	0.103	0.106	0.109	0.113	0.116	0.119
1.1	1.165	1.177	1.189	1.201	1.213	1.225	1.237	1.250	1.262	1.275
	0.123	0.127	0.130	0.134	0.138	0.142	0.146	0.150	0.155	0.159
1.2	1.287	1.300	1.312	1.325	1.337	1.350	1.363	1.376	1.389	1.402
	0.164	0.168	0.173	0.177	0.182	0.187	0.192	0.197	0.203	0.208
1.3	1.415	1.428	1.442	1.455	1.468	1.482	1.496	1.509	1.523	1.537
	0.214	0.219	0.225	0.231	0.237	0.243	0.249	0.256	0.262	0.269
1.4	1.551	1.565	1.579	1.593	1.608	1.622	1.637	1.651	1.666	1.681
	0.275	0.282	0.290	0.297	0.304	0.312	0.319	0.327	0.335	0.344
1.5	1.696	1.711	1.727	1.742	1.758	1.773	1.789	1.805	1.822	1.838
	0.352	0.361	0.369	0.378	0.388	0.397	0.407	0.416	0.426	0.437
1.6	1.855	1.871	1.888	1.905	1.923	1.940	1.958	1.976	1.995	2.013
	0.447	0.458	0.469	0.480	0.492	0.504	0.516	0.529	0.542	0.555
1.7	2.032	2.051	2.071	2.090	2.110	2.131	2.152	2.173	2.195	2.217
	0.568	0.582	0.596	0.611	0.626	0.642	0.658	0.674	0.692	0.709
1.8	2.240	2.263	2.287	2.311	2.336	2.362	2.389	2.417	2.445	2.475
	0.727	0.746	0.766	0.786	0.808	0.830	0.853	0.877	0.902	0.929
1.9	2.506	2.539	2.574	2.611	2.650	2.693	2.741	2.795	2.859	2.942
	0.957	0.986	1.018	1.052	1.089	1.130	1.175	1.227	1.290	1.371

Note: For each given value of L_c/R, the upper line is L/R and the lower line is A/R^2, where L_c = chord length, R = radius of circle, L = arc length, and A = area of segment.

For the lower slope, three circles of case 2 were tried. The minimum static factor of safety is 1.462 and the minimum seismic factor of safety is 1.190; both occur when the circle is tangent to the bottom of natural soil. The minimum factor of safety by REAME is 1.410 for the static case and 1.141 for the seismic case, based on the normal method. The corresponding factor of safety based on the simplified Bishop method are 1.579 and 1.214, respectively. Because of the relatively shallow circle, the discrepancy between the normal method and the simplified Bishop method is not as significant as that for the upper slope.

The most difficult task in analyzing a nonhomogeneous slope is the determination of average soil parameters. If these parameters for different soils in

PARAMETERS FOR ANALYSIS

SOIL NO.	MATERIAL	TOTAL UNIT WEIGHT(pcf)	EFFECTIVE STRENGTH	
			ϕ (deg)	$\bar{c}$(psf)
1	NATURAL SOIL	120	28	200
2	FINE REFUSE	78	31/16*	0
3	COARSE REFUSE	108	35	0

* USE 31 deg IN STATIC ANALYSIS AND 16 deg IN SEISMIC ANALYSIS

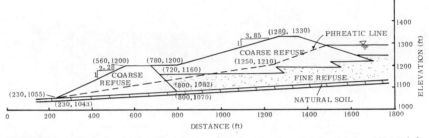

FIGURE 7.32.   Example of refuse dam. (1 ft = 0.305 m, 1 psf = 47.9 Pa, 1 pcf = 157.1 N/m³)

a dam do not vary significantly, average values can be estimated by visual inspection. Otherwise, Table 7.1 should be used in conjunction with a special form. The basic principle for determining the average parameters is that the average cohesion depends on the length of arc in each soil and is the weighted average with respect to the arc length, that the average unit weight is the weighted average with respect to the area occupied by each soil, and that the average friction angle is the weighted average with respect to the normal component of weight in each soil. Table 7.2 shows the form used for the upper slope. The measurements on the slope are shown in Fig. 7.34. In Table 7.2, enter case 3, $S = 3.85$, $H = 130$ ft (39.6 m), $YH = 95$ ft (29.0 m), $Y = 0.73$, and $R = 350$ ft (106.7 m). Each column of the table can be explained as follows:

Column 1: Soils are numbered from bottom to top, so soil 1 is the bottom-most soil.

Column 2: The chord length, L_c, is measured from the cross section. The phreatic surface has a chord length of 480 ft (146.3 m), soil 1 of 195 ft (59.4 m), soil 2 of 450 ft (137.2 m), and soil 3 of 615 ft (187.5 m). If the phreatic surface or soil boundaries are not a straight line, they should be replaced by an equivalent straight line. Note that the chord length for soil 3 is the distance between the two end points on the circle, as indicated by the dashed line.

Column 3: The radius, R, is measured from the cross section, which is 350 ft (106.7 m), and the ratio, L_c/R, is computed.

Column 4: The arc length-radius ratio, L/R, is obtained from Table 7.1. The value of L/R for soil 1 is not cumulative and is entered in col-

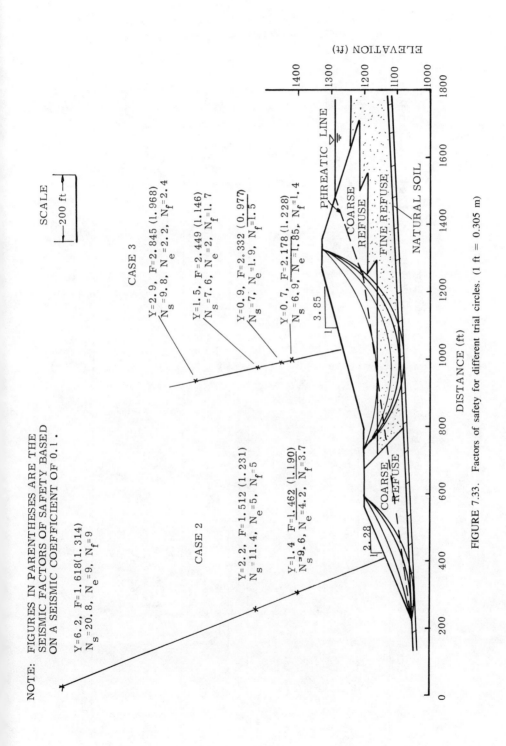

FIGURE 7.33. Factors of safety for different trial circles. (1 ft = 0.305 m)

NOTE: FIGURES IN PARENTHESES ARE THE SEISMIC FACTORS OF SAFETY BASED ON A SEISMIC COEFFICIENT OF 0.1.

$Y=6.2$, $F=1.618(1.314)$
$N_s=20.8$, $N_e=9$, $N_f=9$

CASE 2

$Y=2.2$, $F=1.512$ (1.231)
$N_s=11.4$, $N_e=5$, $N_f=5$

$Y=1.4$ $F=1.462$ (1.190)
$N_s=9.6$, $N_e=4.2$, $N_f=3.7$

CASE 3

$Y=2.9$, $F=2.845$ (1.968)
$N_s=9.8$, $N_e=2.2$, $N_f=2.4$

$Y=1.5$, $F=2.449$ (1.146)
$N_s=7.6$, $N_e=2$, $N_f=1.7$

$Y=0.9$, $F=2.332$ (0.977)
$N_s=7$, $N_e=1.9$, $N_f=1.5$

$Y=0.7$, $F=2.178$ (1.228)
$N_s=6.9$, $N_e=1.85$, $N_f=1.4$

SCALE
200 ft

ELEVATION (ft)

DISTANCE (ft)

PHREATIC LINE

COARSE REFUSE

FINE REFUSE

NATURAL SOIL

COARSE REFUSE

Table 7.2 Average Soil Parameters of Upper Slope.

Case 3 S 3.85 H 130 ft HY 95 ft Y 0.73 R 350 ft r_u 0.244 γ 99.5 pcf

ITEM (1)	L_c ft (2)	L_c/R (3)	L/R CUMULATIVE (4)	L/R INDIVIDUAL (5)	L/R FRACTION (6)	COHESION $\bar{c}$ psf (7)	A/R^2 CUMULATIVE (8)	A/R^2 INDIVIDUAL (9)	WEIGHT UNIT pcf (10)	WEIGHT TOTAL (11)	FRICTION $\bar{\phi}$ DEGREE (17)	FRICTION tan$\bar{\phi}$ (18)
Phreatic	480	1.37		1.509				0.256	62.4	15.974		
Soil 1	195	0.56		0.568	0.264	200		0.015	120.0	1.800	28	0.532
Soil 2	450	1.29	1.402	0.834	0.388	0	0.208	0.193	78.0	15.054	31	0.601
Soil 3	615	1.76	2.152	0.750	0.348	0	0.658	0.450	108.0	48.600	35	0.700
Total				2.152	1.000	52.8		0.658		65.454		0.577

FRICTION DISTRIBUTION BASED ON WEIGHT ABOVE FAILURE ARC

	FORMULA (12)	WEIGHT (13)	CORRECTION FACTOR (14)	CORRECTED WEIGHT (15)	FRACTION (16)
Soil 1	$1.800 + 195 \times (70 \times 78 + 100 \times 108) / (350)^2$	27.683	1.0	27.683	0.491
Soil 2	$65.454 - (27.683 + 10.046)$	27.725	1.2	23.104	0.410
Soil 3	$\frac{1}{2}(40 \times 40 + 163 \times 130) \times 108 / (350)^2$	10.046	1.8	5.581	0.099
Total		65.454		56.368	1.000

(1 ft=0.305 m. 1 psf=47.9 Pa. 1 pcf=157.1 N/m³)

umn 5. The value of L/R for other soils are cumulative because they include the arc length of all underlying soil layers.

Column 5: The individual L/R is the difference between the cumulative L/Rs. Note that the sum of individual L/Rs is equal to the cumulative L/R of soil 3.

Column 6: The fraction of arc length in each soil is obtained by dividing the individual L/R by the total L/R. Note that the sum of all fractions is unity.

Column 7: The average cohesion is obtained by multiplying the fraction in column 6 with the cohesion in column 7 and summing up the results, i.e., $c = 0.264 \times 200 + 0.388 \times 0 + 0.348 \times 0 = 52.8$.

Column 8: The values of A/R^2 are obtained from Table 7.1. The values for the phreatic surface and soil 1 are not cumulative and are entered in column 9.

Column 9: Similar to column 5, the individual A/R^2 is the difference between cumulative A/R^2s.

Column 10: The unit weight of water and each soil is entered here.

Column 11: The total weight in terms of R^2 is the product of A/R^2 in column 9 and the unit weight in column 10. A sum of 65.454 is the total weight for all three soils. The pore pressure ratio, r_u, can now be determined by dividing the weight of water by the weight of soils, or $r_u = 15.974/65.454 = 0.244$. The average unit weight can also be determined by dividing the total weight of soils by the total A/R^2 or $\gamma = 65.454/0.658 = 99.5$. Both r_u and γ are entered in the upper right corner of the table.

Column 12: Various methods can be used to compute the weight of soils above the failure arc through each soil. In this example, the weight above the arc through soil 1 is computed first by summing the weight of soil 1, which is 1.800, and soils 2 and 3, which consist of two trapezoids. Then the weight above the arc through soil 3 is computed as two triangles. Finally, the weight above soil 2 is the difference between the total weight, which is 65.454, and the weights above soils 1 and 3. Because all weights are in terms of R^2, they must be divided by $(350)^2$.

Column 13: The answer of the formula in column 12 is entered here. This is the weight in the vertical direction.

Column 14: The correction factor is $\sec \theta$, which is the reciprocal of $\cos \theta$, where θ is the angle of inclination of a given chord. The correction factor is used to obtain the weight normal to the failure surface. If a circle cuts through a given soil, the average $\sec \theta$ for that soil can be determined by passing a straight line through the two end points on the failure arc. The length measured along this straight line for a unit horizontal distance is the correction factor, which is 1.2 at the right and 1.15 at the left, as shown in Fig. 7.34, thus resulting in a weighted average of 1.2. If one of the two end points is not the

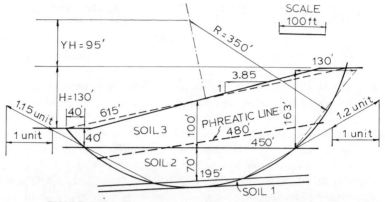

FIGURE 7.34.   Measurements on upper slope. (1 ft = 0.305 m)

lowest point on the failure arc, such as the arc in soil 1, the average of two lines, each passing through the lowest point and the end point should be used.

Column 15: The corrected weight is obtained by dividing the weight in column 13 by the correction factor in column 14.

Column 16: The fraction of weight above the arc through each soil is obtained by dividing the individual corrected weight by the total corrected weight. Note that the sum of all fractions is unity.

Column 17: The effective friction angle of each soil is entered here.

Column 18: The coefficient of friction, $\tan \bar\phi$, for each soil is computed. The average coefficient of friction is obtained by multiplying the fraction in column 16 with the coefficient of friction in column 18 and summing up the results, i.e., $\tan \bar\phi = 0.532 \times 0.491 + 0.601 \times 0.410 + 0.700 \times 0.099 = 0.577$. (Because the friction angle for soil 2 is reduced to 16° under earthquake, the average $\tan \bar\phi$ for determining the seismic factor of safety is $\tan \bar\phi = 0.532 \times 0.491 + 0.287 \times 0.410 + 0.700 \times 0.099 = 0.448$.)

Knowing the average values of r_u, γ, $\bar c$, and $\tan \bar\phi$, the factor of safety can be computed by Eq. 7.16.

Table 7.3 shows the form used for the lower slope. The major differences between Table 7.2 for the upper slope and Table 7.3 for the lower slope are that the former has three soils while the latter has only two, and also that a triangular area must be added to determine the total A/R^2 of the latter. This is shown in Fig. 7.35 by the shaded area.

7.7 TOTAL STRESS ANALYSIS OF SLOPES (Huang, 1975)

Figure 7.36 shows the stability chart for total stress analysis in which $\phi = 0$. The chart is based on a homogeneous simple slope and a circular failure

Table 7.3 Average Soil Parameters of Lower Slope.

Case 2   S 2.28   H 145 ft   HY 205 ft   R 360 ft r_u 0.117   γ 109.0 pcf

ITEM	L_c ft	L_c/R	L/R			COHESION $\bar{c}$ psf	A/R^2		WEIGHT	
			CUMULATIVE	INDIVIDUAL	FRACTION		CUMULATIVE	INDIVIDUAL	UNIT pcf	TOTAL
(1)	(2)	(3)	(4)	(5)	(6)	(7)	(8)	(9)	(10)	(11)
Phreatic	250	0.69		0.704				0.028	62.4	1.747
Soil 1	180	0.50		0.505	0.438	200		0.011	120.0	1.320
Soil 2	393	1.09	1.153	0.648	0.562	0	0.119	0.108	108.0	11.664
Additional A/R^2	½ × 33 × 145/(360)²							0.018	108.0	1.944
Total				1.153	1.000	87.6		0.137		14.928

FRICTION DISTRIBUTION BASED ON WEIGHT ABOVE FAILURE ARC

ITEM	FORMULA	WEIGHT	CORRECTION FACTOR	CORRECTED WEIGHT	FRICTION FRACTION	$\bar{\phi}$ DEGREE	$\tan\bar{\phi}$
	(12)	(13)	(14)	(15)	(16)	(17)	(18)
Soil 1	1.320 + ½ × 180 × 70 × 108/(360)²	6.570	1.0	6.570	0.505	28	0.532
Soil 2	14.298 − 6.570	8.358	1.3	6.429	0.495	35	0.700
Total		14.928		12.999	1.000		0.615

(1 ft=0.305 m. 1 psf=47.9 Pa. 1 pcf=157.1 N/m³)

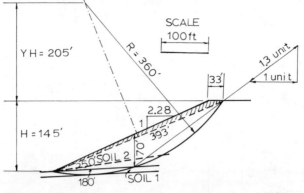

FIGURE 7.35.   Measurements on lower slope. (1 ft = 0.305 m)

surface, as shown in the upper left corner of Fig. 7.36. The embankment has a height, H, and an outslope S:1. A ledge is located at a depth DH below the toe, where D is the depth ratio. The center of the circle is at a horizontal distance XH and a vertical distance YH from the edge of the embankment. For a section of homogeneous soil with the critical circle passing below the toe, it can be easily proved that the center of the critical circle lies on a vertical line intersecting the slope at midheight, or $X = 0.5S$. This type of failure surface is called a midpoint circle, the results of which are shown by the solid curves in Fig. 7.36. If the depth ratio, D, is small, the failure surface may intersect the slope at or above the toe. This type of failure surface is called a toe or slope circle, the results of which are shown by the dashed curves. The factor of safety for each given D and Y is determined directly by taking the moment at the center of the circle. If the failure surface is not a midpoint circle, several circles with different values of X must be tried to determine the minimum factor of safety. It can be seen from the figure that in most cases the most critical circle occurs when $X = 0.5$ except for $D = 0$ and $S \leq 2.0$. After the factor of safety is determined, the stability number, N_s, can be computed by Eq. 7.18

Figure 7.36 shows the value of stability number, N_s, in terms of S, D, and Y. When $\phi = 0$, Eq. 7.9 becomes

$$F = \frac{cN_s}{\gamma H} \qquad (7.18)$$

It can be seen from the figure that the deeper the circle, the smaller the stability number and the smaller the factor of safety. Therefore, the critical circle is always tangent to the ledge. The chart is different from Taylor's (1937) in that the stability numbers for various circles are shown, and thus the application of the chart to nonhomogeneous soils is possible.

As an illustration of the application of the chart to nonhomogeneous slopes, consider the embankment shown in Fig. 7.37. The embankment is 20-ft (6.1-m) high and has an outslope of 3:1 and a cohesion of 1500 psf (71.9

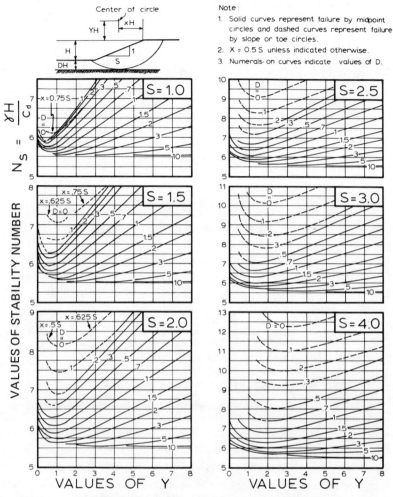

FIGURE 7.36.   Stability chart for total stress analysis.

kPa). The foundation consists of one 40-ft (12.2-m) soil layer having a cohesion of 800 psf (38.3 kPa) and one 20-ft (6.1-m) soil layer having a cohesion
of 100 psf (4.8 kPa), which is underlaid by a ledge. Although the unit weights
weights for different soils are generally not the same, an average unit weight
of 130 pcf (20.4 kN/m³) is assumed. Because the weakest layer lies directly
above the ledge, the critical circle will be tangent to the ledge, or $D = 60/20$
$= 3$.

If the slope is homogeneous with $S = 3$ and $D = 3$, from Fig. 7.36 the
most critical circle is a midpoint circle with a center located at $YH = 2 \times 20$
$= 40$ ft (12.2 m) above the top of the embankment and a minimum stability
number of 5.7 is obtained. The average cohesion can be determined by measuring the length of arc through each soil layer, or

$$c = \frac{24 \times 1500 + 2 \times 57 \times 800 + 142 \times 100}{24 + 2 \times 57 + 142} = 505 \text{ psf } (24.2 \text{ kPa})$$

From Eq. 7.18, the factor of safety is $505 \times 5.7/(130 \times 20) = 1.11$.

If a circle of larger radius with $Y = 6$ or $YH = 6 \times 20 = 120$ ft (36.6 m) is used, from Fig. 7.36, $N_s = 6.05$. The average cohesion is

$$c = \frac{28 \times 1500 + 2 \times 70 \times 800 + 180 \times 100}{28 + 2 \times 70 + 180} = 494 \text{ psf } (23.7 \text{ kPa})$$

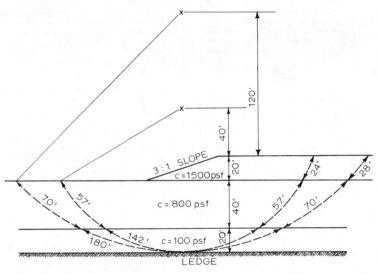

FIGURE 7.37. Failure circle in nonhomogeneous slope. (1 ft = 0.305 m, 1 psf = 47.9 Pa)

The factor of safety is $494 \times 6.05/(130 \times 20) = 1.15$.

It can be seen that the use of the critical center based on a homogeneous slope still yields a smaller factor of safety. This is usually the case, and only one circle need be tried. If the depth of the most critical circle is not apparent, several circles, each tangent to the bottom of a soil layer, should be investigated to locate the minimum factor of safety.

7.8 SUMMARY OF METHODS

Table 7.4 is a summary of the methods developed by the author. Methods 5 and 6 require the trial of a number of circles to locate the minimum factor of safety. Method 5 is quite cumbersome but has the advantage that the pore pressure ratio can be more accurately estimated from the phreatic surface; while the pore pressure ratio in methods 1 through 4 can only be estimated roughly by assuming the probable location of the most critical circle.

For a given problem, several methods can be used. Some methods may

Table 7.4   Simplified Methods of Stability Analysis for Cylindrical Failures

NO.	METHOD	APPLICABILITY
1	Triangular fills on rock slopes	For the total or effective stress analysis of a homogeneous fill on an inclined or horizontal rock surface
2	Trapezoidal fills on rock slopes	For the total or effective stress analysis of a homogeneous fill with the outer part on a horizontal and the inner part on an inclined rock surface
3	Triangular fills on soil slopes	For the total or effective stress analysis of a fill resting on a layer of soil underlain by rock, both of which may be horizontal or inclined
4	Effective stress analysis of homogeneous dams	For the effective stress analysis of a homogeneous dam on a thick layer of soil with the soil in the dam the same as that in the foundation
5	Effective stress analysis of non-homogeneous dams	For the effective stress analysis of a nonhomogeneous dam in which the location of the most critical circle and the corresponding pore pressure ratio can be determined
6	Total stress analysis of slopes	For the total stress analysis of a homogeneous or nonhomogeneous dam in which the location of the most critical circle can be determined

yield more accurate results but are more cumbersome, while others may be very crude but require less computational effort. The following example gives an interesting comparison of the various methods.

Figure 7.38 is an earth dam on an inclined soil foundation. The dimension of the fill and the location of the phreatic surface are as shown. Determine the factor of safety for both total and effective stress analyses.

In the total stress analysis, it is assumed that soil 1 in the foundation has a cohesion of 800 psf (38.3 kPa) and soil 2 in the dam has a cohesion of 2000 psf (95.8 kPa), and that both have an angle of internal friction equal to zero and a total unit weight equal to 125 pcf (19.6 kN/m³). As the dam is not constructed directly on rock, methods 1 and 2 cannot be used. Methods 4 and 5 also cannot be used because they are only applicable to the effective stress analysis. Consequently, only methods 3 and 6 can be applied.

Method 3: When $\phi = 0$, the most critical circle will be tangent to the rock, or $D = 30/50 = 0.6$. With $\alpha = 5°$ and $\beta = \tan^{-1}0.333 = 18.4°$, from Fig. 7.22, $N_s = 7.1$ and $L_f = 0.2$. From Eq. 7.13

$$F = 7.1 \times \left[\frac{800}{125 \times 50} \times (1 - 0.2) + \frac{2000}{125 \times 50} \times 0.2 \right] = 1.18$$

Method 6: By assuming that the center of circle lies on a vertical line through the midheight of the slope, a circle with $YH = 75$ ft (22.9 m) and

$DH = 22$ ft (6.7 m) is tried, as shown in Fig. 7.39. For $S = 3$ and $D = 22/50 = 0.44$, from Fig. 7.36, the minimum N_s is 7.2 and occurs at $Y = 1.5$

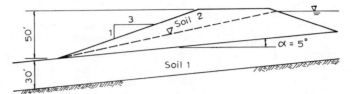

FIGURE 7.38. Earth dam on inclined soil foundation. (1 ft = 0.305 m)

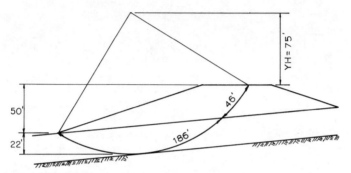

FIGURE 7.39. Location of most critical circle for total stress analysis. (1 ft = 0.305 m)

or $YH = 1.5 \times 50 = 75$ ft, which is exactly the center of the trial circle. The average cohesion is

$$c = \frac{186 \times 800 + 46 \times 2000}{186 + 46} = 1038 \text{ psf (49.7 kPa)}$$

From Eq. 7.18, the factor of safety is $F = 1038 \times 7.2/(125 \times 50) = 1.19$ which checks closely with method 3.

In the effective stress analysis, it is assumed that soil 1 has an effective cohesion of 100 psf (4.8 kPa) and an effective friction angle of 30°, while soil 2 has an effective cohesion of 200 psf (9.6 kPa) and an effective friction angle of 35°. Both soils have a total unit weight of 125 pcf (19.6 kN/m³). The methods applicable to this case are methods 3, 4, and 5.

Method 3: Assume $D = 0.2$, from Fig. 7.20, $N_s = 9.9$, $N_{f1} = 3.5$, $N_{f2} = 11.0$, and $L_f = 0.38$. Sketch a failure circle such that the length of arc in soil 2 is about 38 percent of the total length, as shown in Fig. 7.40a. It is estimated that about 60 percent of the weight above the arc in soil 1 is under water and that about 30 percent of that in soil 2 is under water, so $r_{u1} = 0.3$ and $r_{u2} = 0.15$. With $\bar{c}_1/(\gamma H) = 100/(125 \times 50) = 0.016$ and $\bar{c}_2/(\gamma H) = 0.032$, from Eq. 7.13, $F = 9.9(0.016 \times 0.62 + 0.032 \times 0.38 + 0.7 \times \tan 30°/3.5 + 0.85 \times \tan 35°/11.0) = 1.90$. Next, assume $D = 0.4$, from Fig.

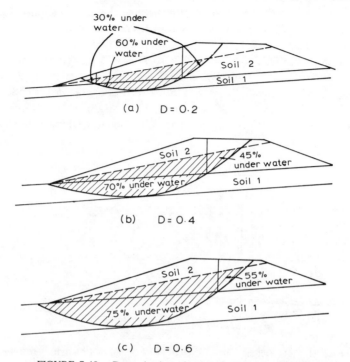

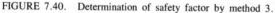

FIGURE 7.40. Determination of safety factor by method 3.

7.21, $N_s = 7.8$, $N_{f1} = 2.3$, $N_{f2} = 19.0$ and $L_f = 0.25$. From Fig. 7.40b, $r_{u1} = 0.35$ and $r_{u2} = 0.225$. $F = 7.8(0.016 \times 0.75 + 0.032 \times 0.25 + 0.65 \times \tan 30°/2.3 + 0.775 \times \tan 35°/19.0) = 1.65$. Finally, assume $D = 0.6$, from Fig. 7.22, $N_s = 7.0$, $N_{f1} = 1.84$, $N_{f2} = 28.0$ and $L_f = 0.2$. From Fig. 7.40c, $r_{u1} = 0.375$ and $r_{u2} = 0.275$. $F = 7.0(0.016 \times 0.8 + 0.032 \times 0.2 + 0.625 \times \tan 30°/1.84 + 0.725 \times \tan 35°/28.0) = 1.63$. It can be seen that the minimum factor of safety is 1.63 and occurs when the circle is tangent to the rock, or $D = 0.6$.

Method 4: As this method is only applicable to a homogeneous slope, average soil parameters must be determined. Based on Fig. 7.40b, it is assumed that 75 percent of the failure arc lies in soil 1, 80 percent of the weight is above the arc in soil 1, and 66 percent of the sliding mass is under water; therefore, the average cohesion $= 0.75 \times 100 + 0.25 \times 200 = 125$ psf (6.0 kPa), or cohesion factor $= 100 \times 125/(125 \times 50) = 2$, average friction $\tan \bar{\phi} = 0.8 \times \tan 30° + 0.2 \tan 35° = 0.602$, or $\bar{\phi} = 31°$, and $r_u = 0.33$. From Fig. 7.24, when $r_u = 0$, $F = 2.2$, when $r_u = 0.5$, $F = 1.1$. By interpolation, when $r_u = 0.33$, $F = 1.1 + (0.5 - 0.33) \times 1.1/0.5 = 1.47$.

Method 5: This method is tedious but gives more reliable results. Because the rock is located at some distance from the surface, case 3 applies. Figure 7.41 shows the location of two trial circles. Circle 1 is tangent to the rock, and the computation of average soil parameters is presented in Table

7.5. Circle 2 is about 10 ft (3 m) shallower, and the computation is presented in Table 7.6. For circle 1 with a center 45 ft (13.7 m) above the top of dam, from Fig. 7.30, $N_s = 7.4$ and $N_f = 2.0$. From Table 7.5 $\bar{c} = 119$ psf (5.7 kPa), tan $\bar{\phi} = 0.579$, and $r_u = 0.383$, so $F = 7.4 [119/(125 \times 50) + (1 - 0.383) \times 0.579/2.0] = 1.46$. For circle 2 with a center 85 ft (25.9 m) above the dam, $N_s = 8.8$, $N_f = 2.6$, $\bar{c} = 126$ psf (6.0 kPa), tan $\bar{\phi} = 0.587$, and $r_u = 0.329$, so $F = 8.8[126/(125 \times 50) + (1 - 0.329) \times 0.587/2.6] = 1.51$. Therefore, the most critical circle is tangent to the rock with a factor of safety of 1.46.

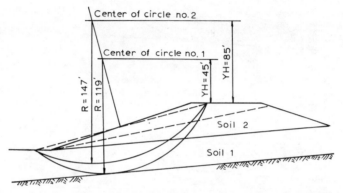

FIGURE 7.41   Determination of safety factor by method 5. (1 ft = 0.305 m)

It can be seen that methods 4 and 5 check closely while method 3 results in a factor of safety somewhat greater. This discrepancy is due to the fact that the most critical circle for $\bar{\phi} = 0$, as assumed in method 3, is quite different from the circle assumed in method 5. If more accurate results are desired, the use of method 5 is recommended.

Table 7.5 Average Soil Parameters for Circle No. 1

Case 3 S 3 H 50 ft HY 45 ft Y 0.9 R 119 ft r_u 0.383 γ 125 pcf

| ITEM (1) | L_c ft (2) | L_c/R (3) | L/R | | | COHESION $\bar{c}$ | A/R^2 | | WEIGHT | |
			CUMULATIVE (4)	INDIVIDUAL (5)	FRACTION (6)	psf (7)	CUMULATIVE (8)	INDIVIDUAL (9)	UNIT pcf (10)	TOTAL (11)
Phreatic	174	1.46		1.637				0.319	62.4	19.906
Soil 1	158	1.33		1.455	0.806	100		0.231	125.0	28.875
Soil 2	187	1.57	1.805	0.350	0.194	200	0.416	0.185	125.0	23.125
Total				1.805	1.000	119		0.416		52.000

FRICTION DISTRIBUTION BASED ON WEIGHT ABOVE FAILURE ARC

ITEM	FORMULA (12)	WEIGHT (13)	CORRECTION FACTOR (14)	CORRECTED WEIGHT (15)	FRACTION (16)	FRICTION ϕ DEGREE (17)	$\tan\bar{\phi}$ (18)
Soil 1	$28.875 + \frac{1}{2} \times 35 \times 142$ $\times 125/(119)^2$	50.810	1.1	46.191	0.986	30	0.577
Soil 2	$52.00 - 50.810$	1.190	1.8	0.661	0.014	35	0.700
Total		52.000		46.852	1.000		0.579

(1 ft=0.305 m. 1 psf=47.9 Pa. 1 pcf=157.1 N/m³)

Table 7.6 Average Soil Parameters for Circle No. 2

Case 3   S 3   H 50 ft   HY 85 ft   R 147 ft r_u 0.329 γ 125 pcf

ITEM (1)	L_c ft (2)	L_c/R (3)	L/R CUMULATIVE (4)	L/R INDIVIDUAL (5)	FRACTION (6)	COHESION $\bar{c}$ psf (7)	A/R^2 CUMULATIVE (8)	A/R^2 INDIVIDUAL (9)	WEIGHT UNIT pcf (10)	WEIGHT TOTAL (11)
Phreatic	165	1.12		1.189				0.130	62.4	8.112
Soil 1	143	0.97		1.013	0.736	100		0.082	125.0	10.250
Soil 2	187	1.27	1.376	0.363	0.264	200	0.197	0.115	125.0	14.375
Total				1.376	1.000	126		0.197		24.625

FRICTION DISTRIBUTION BASED ON WEIGHT ABOVE FAILURE ARC

	FORMULA (12)	WEIGHT (13)	CORRECTION FACTOR (14)	CORRECTED WEIGHT (15)	FRICTION FRACTION (16)	ϕ DEGREE (17)	$\tan\phi$ (18)
Soil 1	$10.25 + \tfrac{1}{2} \times 32 \times 127$ $\times\, 125/(147)^2$	22.004	1.05	20.956	0.918	30	0.577
Soil 2	$24.625 - 22.004$	2.621	1.4	1.872	0.082	35	0.700
Total		24.625		22.828	1.000		0.587

(1 ft=0.305 m. 1 psf=47.9 Pa. 1 pcf=157.1 N/m³)

Part III
Computerized Methods of
Stability Analysis

8
SWASE for Plane Failure

8.1 INTRODUCTORY REMARKS

The SWASE (Sliding Wedge Analysis of Sidehill Embankments) computer program can be used to determine the factor of safety of a slope when some planes of weakness exist within the slope. These failure planes may exist at the bottom of an embankment or at any other locations. However, the maximum number of failure planes is limited to three because the program can only handle three blocks. If there are more than three planes, they must be approximated by three planes in order to use this program. If no planes of weakness exist, the cylindrical failure surface will be more critical and the REAME computer program, as presented in Chap. 9, should be used instead. Otherwise, both programs should be used to determine which is more critical.

The program requires a storage space of 16K and has the following features:

1. If the failure plane starts from the toe or the surface of the slope and if the outslope is uniform, the program will calculate the weight of each block automatically from the input dimensions.
2. If the failure plane does not start from the toe or the surface of the slope or if the outslope is irregular, the weight of each block must be calculated by the user and then input into the computer. In such a case, the unit weight of soil must be assigned zero.
3. Each block has the same unit weight but the shear strength along the failure plane at the bottom of each block can be specified individually.
4. Either the static or the seismic factor of safety can be computed.
5. Seepage can be considered by specifying a pore pressure ratio.
6. More than one problem can be run at the same time. If different sets of failure planes are specified, the results can be printed in a tabular form, so that the set with the minimum factor of safety can be easily discerned.

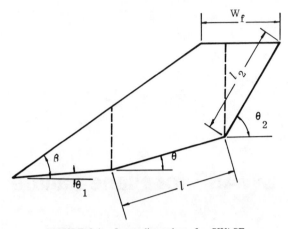

FIGURE 8.1. Input dimensions for SWASE.

A program listing in FORTRAN is presented in Appendix III and a listing in BASIC is in Appendix IV.

8.2 THEORETICAL DEVELOPMENT

Figure 8.1 shows the input dimensions that must be provided for a sidehill embankment consisting of three blocks. If there are only two failure planes, or two blocks, the length of natural slope at the middle block, l, must be specified as zero. The equations used to determine the factor of safety for three blocks are presented here. The equations for two blocks are much simpler and will not be presented.

Figure 8.2 is the free-body diagram showing the forces on each block. There are a total of six unknowns (P_1, P_2, $\bar{N}_1$, $\bar{N}_2$, $\bar{N}$, and factor of safety, F), which can be solved by six equilibrium equations, two for each block.

For the bottom block (block 1) summing all forces in the vertical directions

$$W_1 + P_1 \sin \phi_d - \bar{N}_1 \cos \theta_1 - r_u W_1 \cos^2 \theta_1 - T_1 \sin \theta_1 = 0 \quad (8.1)$$

in which $\phi_d = \tan^{-1} [(\tan \bar{\phi})/F]$ is the angle of internal friction of fill, and T_1 is the shear force on failure plane at bottom, which can be expressed as

$$T_1 = (\bar{c}_1 l_1 + \bar{N}_1 \tan \bar{\phi}_1)/F \quad (8.2)$$

Summing all forces in the horizontal direction

$$P_1 \cos \phi_d + \bar{N}_1 \sin \theta_1 + r_u W_1 \sin \theta_1 \cos \theta_1 + C_s W_1 - T_1 \cos \theta_1 = 0 \quad (8.3)$$

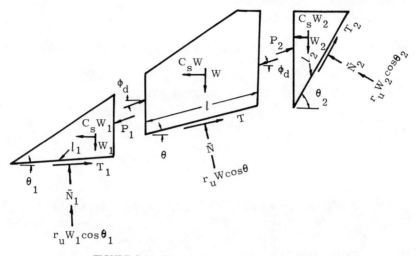

FIGURE 8.2. Free body diagram for three blocks.

From Eqs. 8.1, 8.2, and 8.3

$$\bar{N}_1 = \{W_1[\cos \phi_d - r_u \cos \theta_1 \cos (\phi_d - \theta_1) - C_S \sin \phi_d] + [\bar{c}_1 l_1 \sin (\phi_d - \theta_1)]/F\}/\{\cos (\phi_d - \theta_1) - [\tan \bar{\phi}_1 \sin (\phi_d - \theta_1)]/F\} \tag{8.4}$$

From Eq. 8.3

$$P_1 = (T_1 \cos \theta_1 - \bar{N}_1 \sin \theta_1 - r_u W_1 \sin \theta_1 \cos \theta_1 - C_S W_1)/\cos \phi_d \tag{8.5}$$

For the top block (block 2)

$$W_2 - P_2 \sin \phi_d - \bar{N}_2 \cos \theta_2 - r_u W_2 \cos^2\theta_2 - T_2 \sin \theta_2 = 0 \tag{8.6}$$

$$T_2 = (\bar{c}_2 l_2 + \bar{N}_2 \tan \bar{\phi}_2)/F \tag{8.7}$$

$$P_2 \cos \phi_d - \bar{N}_2 \sin \theta_2 - r_u W_2 \cos \theta_2 \sin \theta_2 - C_S W_2 + T_2 \cos \theta_2 = 0 \tag{8.8}$$

From Eqs. 8.6, 8.7, and 8.8

$$\bar{N}_2 = \{w_2[\cos \phi_d - r_u \cos \theta_2 \cos (\phi_d - \theta_2) - C_S \sin \phi_d] + \bar{c}_2 l_2 [\sin(\phi_d - \theta_2)]/F\}/\{\cos (\phi_d - \theta_2) - [\tan \bar{\phi}_2 \sin (\phi_d - \theta_2)]/F\} \tag{8.9}$$

From Eq. 8.8

$$P_2 = (\bar{N}_2 \sin \theta_2 - T_2 \cos \theta_2 + r_u W_2 \cos \theta_2 \sin \theta_2 + C_S W_2)/\cos \phi_d \tag{8.10}$$

For the middle block

$$W + P_2 \sin \phi_d - P_1 \sin \phi_d - \bar{N} \cos \theta - r_u W \cos^2 \theta - T \sin \theta = 0 \tag{8.11}$$

$$T = (\bar{c}l + \bar{N} \tan \bar{\phi})/F \tag{8.12}$$

$$-P_2 \cos \phi_d + P_1 \cos \phi_d - \bar{N} \sin \theta - r_u W \cos \theta \sin \theta + T \cos \theta - C_S W = 0 \tag{8.13}$$

From Eqs. 8.11, 8.12, and 8.13

$$\bar{N} = \{W(1 - r_u \cos^2 \theta) + (P_2 - P_1) \sin \phi_d - (\bar{c}l \sin \theta)/F\}/ \\ [\cos \theta + (\tan \bar{\phi} \sin \theta)/F] \tag{8.14}$$

Eq. 8.13 can be written as

$$\text{Function } (F) = (P_1 - P_2) \cos \phi_d - \bar{N} \sin \theta - r_u W \cos \theta \sin \theta + \\ T \cos \theta - C_S W = 0 \tag{8.15}$$

in which P_1, P_2, T, and $\bar{N}$ are a function of F and can be determined from Eqs. 8.5, 8.10, 8.12, and 8.14, respectively. A subroutine from the IBM 360 Scientific Subroutine Package was used to solve for F (International Business Machines Corporation, 1970).

8.3 DESCRIPTION OF PROGRAM

The SWASE program is composed of one main program, one function subprogram, and two subroutines. It was written both in FORTRAN IV and BASIC and requires a storage space of 16K.

The main program is used for input and output, the conversion of degrees into radians, and the computation of required lengths and weights.

Function subprogram FCT is used to determine T_1 (Eq. 8.2), $\bar{N}_1$ (Eq. 8.4), P_1 (Eq. 8.5), T_2 (Eq. 8.7), $\bar{N}_2$ (Eq. 8.9), P_2 (Eq. 8.10), T (Eq. 8.12), and $\bar{N}$ (Eq. 8.14). These variables are entered into Eq. 8.15 to form FCT(F).

Subroutine RTMI is used to solve a general nonlinear equation of the form FCT(F) = 0 by means of Mueller's iteration scheme of successive bisection and inverse parabolic interpolation (International Business Machines Corporation, 1970). To find the root F, it is necessary to assume a left limit or lower bound of F, F_l, and a right limit or upper bound of F, F_r, such that

$$\text{FCT}(F_l) \times \text{FCT}(F_r) < 0 \tag{8.16}$$

The lower bound is determined by subroutine FIXLI. The upper bound is increased by intervals of 0.5 until Eq. 8.16 is satisfied. If Eq. 8.16 is not satisfied even when $F_r = 10$, the factor of safety is greater than 10, and a message to this effect will be printed. This usually occurs when some of the input dimensions are erroneous, so the user should check the input data and make the necessary changes.

Subroutine FIXLI is used to determine the lower bound of F. Starting from $F = 0.1$, the subroutine calls function subprogram FCT and computes T_1, T_2, and T by Eqs. 8.2, 8.7, and 8.12. If any one of these three shear forces is negative, the subroutine increases F by 0.5 until all three become positive. The first value of F which results in positive T_1, T_2, and T is used as F_l.

Figure 8.3 is a simplified flow chart for SWASE.

8.4 DATA INPUT

The input and output parameters described here are applicable to the FORTRAN version of SWASE only. As the BASIC version is in the interactive mode, users who understand the FORTRAN version should have no difficulty

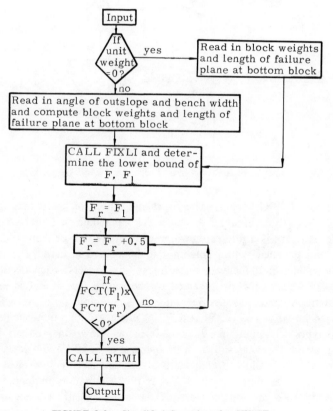

FIGURE 8.3.   Simplified flow chart for SWASE.

```
1ST  Card  TITLE (20A4)
2ND  Card  NSET, IPBW, IPUW (3I5)
3RD  Card  C(1), PHI(1), C(2), PHI(2), C(3), PHI(3), PHIMA (7F10.3)
4TH  Card  DSBM, DSTP, DSME, LNS2, LNS, RU, SEIC, GAMMA (8F10.3)
5TH  Card  WHEN GAMMA IS NOT ZERO
           ANOUT, BW (2F10.3)
5TH  Card  WHEN GAMMA IS ZERO
           W1, W2, W, Al1 (2F10.0, F10.3)
REPEAT 4TH TO 5TH CARDS NSET TIMES
```

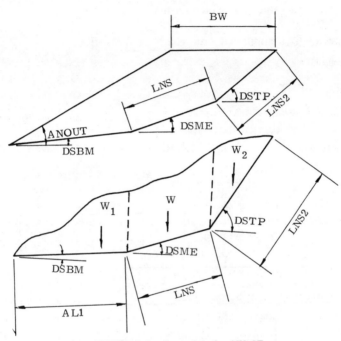

FIGURE 8.4.   Input data for SWASE.

in using the BASIC version. The input and output of basic version are presented in Sect. 8.6. Figure 8.4 shows the input data in FORTRAN.

The first card is a title card, which can be any title or comment related to the job and punched within column 1 to 80 of a data card.

The second card includes three integer parameters, each occupying five columns of a data card, as indicated by the format 3I5. NSET is the number of problems or data sets. In Section 8.5, six sample problems will be presented, so NSET = 6. IPBW and IPUW are used for output control. There are two types of output, one in long form, as shown in problem 1, and the other in tabular form, as shown in problems 2 to 6. IPBW indicates the initial problem when the block weights are given, and IPUW indicates the initial problem when the unit weight is given. If the output is in long form, both IPBW and IPUW are assigned zero. If several problems are run at the same time, the output can be arranged in tabular form. If the block weights are given, as indicated by an irregular slope, IPBW can be assigned as 1 and IPUW as 0, so that tabular form with block weights given will be used, beginning from the first problem. If the unit weight is given, which indicates a uniform slope, IPUW can be assigned as 1 and IPBW as 0, so that tabular form with unit weight given will be used, beginning from the first problem. For the sample problems presented in Sect. 8.5, IPBW = 2 and IPUW = 5, so that the first problem is printed in long form, problems 2 to 4 are printed in tabular form with block weights given, and problems 5 and 6 in tabular form with the unit weight given.

The third card indicates seven parameters, each occupying 10 columns of a data card, as indicated by the format 7F10.3. C(1) and PHI(1) are the cohesion and angle of internal friction of the material along the failure plane in the bottom block, while subscript 2 is for the top block and subscript 3 is for the middle block. If there are only two blocks, C(3) and PHI(3) can be left blank or assigned any values. PHIMA is the angle of internal friction between two blocks. The computed factor of safety becomes smaller as PHIMA is decreased, so the assumption of PHIMA = 0 is on the safe side and is recommended.

The fourth card includes eight parameters, each occupying 10 columns of a data card, as indicated by the format 8F10.3. DSBM is the degree of slope at bottom. DSTP is the degree of slope at top. DSME is the degree of slope at middle. LNS2 is the length of failure plane at top. LNS is the length of failure plane at middle. RU is the pore pressure ratio. SEIC is the seismic coefficient. GAMMA is the unit weight of soil.

The fifth card has two different sets of data, depending on whether the block weight or the unit weight is given. When the unit weight, GAMMA, is not zero, read in the angle of outslope, ANOUT, and the bench width, BW, each occupying 10 columns of a data card, as indicated by the format 2F10.3. When GAMMA is zero, read in the block weights, W_1, W_2, and W, and the length of failure plane at the bottom block, $AL1$, each occupying 10 columns of a data card, as indicated by the format 3F10.0, F10.3. Format F10.0 implies that the decimal point can be placed in any column as long as the data is within the 10 columns assigned.

The fourth and fifth cards can be repeated NSET times.

All input parameters will be printed as output together with the computed factor of safety.

8.5 SAMPLE PROBLEMS

Figures 8.5 to 8.10 are the six example problems involving a hollow fill, which is used extensively for the disposal of mine spoil. If the natural ground surface is not properly scalped, a layer of weak material may exist at the bottom of the fill and SWASE program can be used to determine the factor of safety of plane failure along the natural ground surface. The profile of the

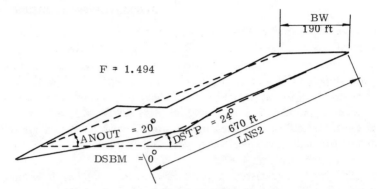

FIGURE 8.5. Problem no. 1 by SWASE. (1 ft = 0.305 m)

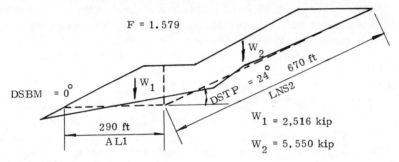

FIGURE 8.6. Problem no. 2 by SWASE. (1 ft = 0.305 m, 1 kip = 4.45 kN)

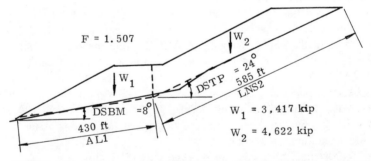

FIGURE 8.7. Problem no. 3 by SWASE. (1 ft = 0.305 m, 1 kip = 4.45 kN)

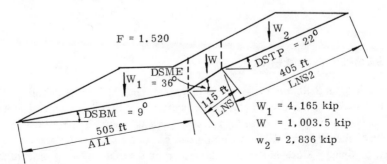

FIGURE 8.8. Problem no. 4 by SWASE. (1 ft = 0.305 m, 1 kip = 4.45 kN)

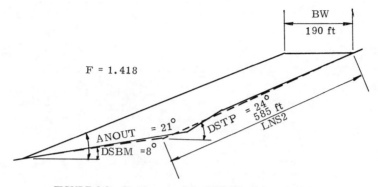

FIGURE 8.9.   Problem no. 5 by SWASE. (1 ft = 0.305 m)

ground surface at the bottom of the fill can be obtained from a topographical map along the center of the channel, which is usually considered as the most critical cross section. The cross section in problems 1 to 4 has a large bench at the middle of the slope, while that in problems 5 and 6 has a uniform slope. In the stability analysis, it is assumed that the material along the failure surface has an effective cohesion of 160 psf (7.7 kPa) and an effective friction angle of 24°. It is assumed that the fill has a total unit weight of 125 pcf (19.6 kN/m³), a pore pressure ratio of 0, and no friction between two sliding blocks.

In problem 1, Fig. 8.5, the slope is approximated by a straight line, so GAMMA is not assigned zero. The natural ground surface is approximated by a horizontal line and an inclined line. The factor of safety as determined by the SWASE program is 1.494. The same factor of safety can be obtained by the simplified method.

In problem 2, Fig. 8.6, the actual bench slope is used, so GAMMA is assigned zero, and the two block weights, W_1 and W_2, are specified. A hand calculation shows that W_1 is 2,516,000 lb (11,200 kN) and W_2 5,550,000 lb (24,700 kN). The factor of safety determined by SWASE is 1.579, which is

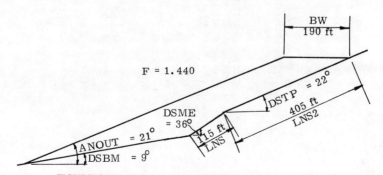

FIGURE 8.10.   Problem no. 6 by SWASE. (1 ft = 0.395 m)

SIX EXAMPLE PROBLEMS(OR DATA SETS) FOR HOLLOW FILLS

PROGRAM OPTIONS

NO. OF DATA SETS   6
STARTING FROM DATA SET NO. 2 USE TABULAR PRINTOUT WITH BLOCK WEIGHTS GIVEN
STARTING FROM DATA SET NO. 5 USE TABULAR PRINTOUT WITH UNIT WEIGHTS GIVEN

COHESION AT BOTTOM BLOCK   160.000   FRICTION ANGLE AT BOTTOM BLOCK   24.000
COHESION AT TOP BLOCK   160.000   FRICTION ANGLE AT TOP BLOCK   24.000
COHESION AT MIDDLE BLOCK   160.000   FRICTION ANGLE AT BOTTOM BLOCK   24.000
ANGLE OF INTERNAL FRICTION OF FILL   0.000

DATA SET   1

DEGREE OF SLOPE AT BOTTOM   0.000

DEGREE OF SLOPE AT TOP   24.000

DEGREE OF SLOPE AT MIDDLE   0.000

LENGTH OF NATURAL SLOPE AT TOP   670.000_

LENGTH OF NATURAL SLOPE AT MIDDLE   0.000

PORE PRESSURE RATIO   0.000

SEISMIC COEFFICIENT   0.000

UNIT WEIGHT   125.000

ANGLE OF OUTSLOPE   20.000

BENCH WIDTH   190.000
THE FACTOR OF SAFETY   1.494

SET	DSBM	DSTP	DSME	LNS2	LNS	RU	SEIC	W1	W2	W	AL1	F
2	0.000	24.000	0.000	670.000	0.000	0.000	0.000	2516000.000	5550000.000	0.000	290.000	1.579
3	8.000	24.000	0.000	585.000	0.000	0.000	0.000	3417000.000	4622000.000	0.000	430.000	1.507
4	9.000	22.000	36.000	405.000	115.000	0.000	0.000	4165000.000	2836000.000	1003500.000	505.000	1.520

SET	DSBM	DSTP	DSME	LNS2	LNS	RU	SEIC	GAMMA	ANOUT	BW	F
5	8.000	24.000	0.000	585.000	0.000	0.000	0.000	125.000	21.000	190.000	1.418
6	9.000	22.000	36.000	405.000	115.000	0.000	0.000	125.000	21.000	190.000	1.440

FIGURE 8.11. Output of example problems by SWASE.

slightly greater than the 1.494 obtained in problem 1. This indicates that the approximation of a benched slope by a straight line is on the safe side.

In problem 3, Fig. 8.7, the natural ground surface is approximated by two inclined lines. The factor of safety is 1.507 which is only slightly different from problem 2's 1.579 when the natural ground is approximated by a horizontal line and an inclined line.

In problem 4, Fig. 8.8, the natural ground surface is approximated by three straight lines. The factor of safety is 1.520, which is approximately the same as problem 3's 1.507 when two straight lines are used.

In problem 5, Fig. 8.9, the outslope is assumed to be uniform, so GAMMA is not zero. By approximating the natural ground surface by two inclined lines, the factor of safety is 1.418.

In problem 6, Figure 8.10, the natural slope is approximated by three straight lines, and a factor of safety of 1.440 is obtained.

In using the SWASE program, it is not necessary that the failure planes be located at the bottom of the fill. Any failure planes consisting of three straight lines or less can be assumed, and the factor of safety determined. The failure plane with the smallest factor of safety is the most critical.

Figure 8.11 shows the output of the six sample problems.

8.6 BASIC VERSION

The input data in BASIC version is the same as those in FORTRAN except that IPBW and IPUW are not used in the program. All output is in long form because the tabular form is too wide and cannot be accommodated within 72 columns. In order to compare several data sets, the set number and its factor of safety are printed at the end together with the minimum factor of safety among all sets. The input and output for problems 1 and 2 are shown in the following pages.

INPUT TITLE ?TWO EXAMPLE PROBLEMS BY SWASE

INPUT NO. OF DATA SETS TO BE RUN   ?2

COHESION ALONG FAILURE PLANE AT BOTTOM BLOCK ?160

FRIC. ANGLE ALONG FAILURE PLANE AT BOTTOM BLOCK ?24

COHESION ALONG FAILURE PLANE AT TOP BLOCK ?160

FRIC. ANGLE ALONG FAILURE PLANE AT TOP BLOCK ?24

COHESION ALONG FAILURE PLANE AT MIDDLE BLOCK ?160

FRIC. ANGLE ALONG FAILURE PLANE AT MIDDLE BLOCK ?24

ANGLE OF INTERNAL FRICTION OF FILL   ?0
DATA SET   1

DEGREE OF SLOPE AT BOTTOM   ?0

DEGREE OF SLOPE AT TOP   ?24

DEGREE OF SLOPE AT MIDDLE   ?0

LENGTH OF NATURAL SLOPE AT TOP   ?670

LENGTH OF NATURAL SLOPE AT MIDDLE   ?0

PORE PRESSURE RATIO   ?0

SEISMIC COEFFICIENT   ?0

UNIT WEIGHT OF FILL   ?125

ANGLE OF OUTSLOPE   ?20

BENCH WIDTH   ?190

THE FACTOR OF SAFETY   1.49367

```
DATA SET        2


DEGREE OF  SLOPE AT BOTTOM      ?0

DEGREE OF  SLOPE AT TOP         ?24

DEGREE OF  SLOPE AT MIDDLE      ?0

LENGTH OF NATURAL SLOPE AT TOP             ?670

LENGTH OF NATURAL SLOPE AT MIDDLE          ?0

PORE PRESSURE RATIO            ?0

SEISMIC COEFFICIENT            ?0

UNIT WEIGHT OF FILL            ?0

WEIGHT OF BOTTOM BLOCK         ?2516000

WEIGHT OF TOP BLOCK            ?5550000

WEIGHT OF MIDDLE BLOCK         ?0

LENGTH OF NATURAL SLOPE AT BOTTOM          ?290

THE FACTOR OF SAFETY           1.57887



              SET            FACTOR OF SAFETY

               1                1.49367

               2                1.57887



        MINIMUM FACTOR OF SAFETY =  1.49367  AT SET  1
```

9
REAME for Cylindrical Failure

9.1 INTRODUCTORY REMARKS

The REAME (Rotational Equilibrium Analysis of Multilayered Embankments) computer program can be used to determine the factor of safety of a slope based on a cylindrical failure surface (Huang, 1981b). A major advantage of REAME over other comparable computer programs is that it requires very little computer time to run. This is made possible by an efficient method of numbering the boundary lines for different soils and controlling the number of circles. This program is particularly useful to those who have very little experience in stability analysis. Without worrying about the computing cost, a large region can be searched, and the minimum factor of safety determined. Features of this program can be described briefly as follows:

1. Slopes of any configuration with a large number of different soil layers can be handled.
2. Seepage can be considered by specifying a piezometric surface or a pore pressure ratio. If necessary, several different seepage cases can be considered simultaneously to save the computer time.
3. Either the static or the seismic factors of safety can be computed.
4. Either the simplified Bishop method or the normal method can be used to determine the factor of safety. When the normal method is specified and the most critical circle is found, the factor of safety for this particular circle based on the simplified Bishop method is also printed.
5. More flexibility is allowed in radius control. One or more radius control zones can be set up, and the number of circles in each zone specified.
6. The factors of safety at a number of individual centers or at a group of centers, which form a grid, can be determined. By selecting one or more trial centers, a search routine can be activated to locate the minimum factor of safety. To obtain the minimum factor of safety in one

run, the automatic search can follow immediately after the grid, using the most critical center obtained from the grid as the initial trial center for the search.

7. To preclude the formation of shallow circles, a minimum depth may be specified. Any circle having the tallest slice smaller than the minimum depth will not be run.
8. A cross section of the slope can be plotted by the printer, including all soil layers, the piezometric surface, and the most critical circle.

9.2 THEORETICAL DEVELOPMENT

Either the normal method or the simplified Bishop method can be used to determine the factor of safety. When the pore pressure ratio is specified, Eq. 2.11, based on the normal method as described in Chap. 2, can be used. When a phreatic surface is specified, Eq. 2.11 becomes

$$F = \frac{\sum\limits_{i=1}^{n} [\bar{c}b_i \ \sec \ \theta_i \ + \ (\gamma h_i \ - \ \gamma_w h_{iw}) \ \cos \ \theta_i \ \tan \ \bar{\phi}]}{\sum\limits_{i=1}^{n} (W_i \ \sin \ \theta_i \ + \ C_s W_i a_i/R)} \qquad (9.1)$$

in which $W_i = \gamma h_i b_i$. The method in which Eq. 2.11 or 9.1 is applied is called the normal method.

In the simplified Bishop method, it is assumed that the forces on the sides of each slice are in a horizontal direction (Bishop, 1955). This assumption implies that there is no friction between two slices. The forces acting on the ith slice are shown in Fig. 9.1. Note that the mobilized shear stress along the failure surface is obtained by dividing the shear strength by a factor of safety. By summing the forces in the vertical direction to zero

$$\bar{N}_i \ \cos \ \theta_i \ + \ \gamma_w h_{iw} \ b_i \ + \ \left(\frac{\bar{c}b_i \ \sec \ \theta_i \ + \ \bar{N}_i \ \tan \ \bar{\phi}}{F} \right) \ \sin \ \theta_i \ - \ \gamma h_i b_i \ = \ 0$$

or

$$\bar{N}_i \ = \ \frac{b_i \ (\gamma h_i \ - \ \gamma_w h_{iw}) \ - \ (\bar{c}b_i \ \tan \ \theta_i)/F}{\cos \ \theta_i \ + \ (\sin \ \theta_i \ \tan \ \bar{\phi})/F} \qquad (9.2)$$

Combining Eqs. 2.10 and 9.2

$$F = \frac{\sum\limits_{i=1}^{n} \dfrac{\bar{c}b_i \ + \ b_i(\gamma h_i \ - \ \gamma_w h_{iw}) \ \tan \ \bar{\phi}}{\cos \ \theta_i \ + \ (\sin \ \theta_i \ \tan \ \bar{\phi})/F}}{\sum\limits_{i=1}^{n} (W_i \ \sin \ \theta_i \ + \ C_s W_i a_i/R)} \qquad (9.3)$$

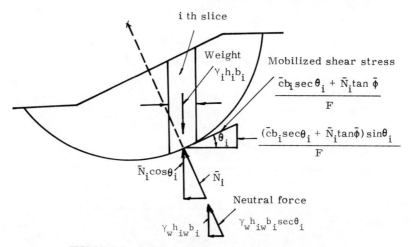

FIGURE 9.1.   Stability analysis by simplified Bishop method.

In terms of pore pressure ratio, Eq. 9.3 can be written as

$$F = \frac{\displaystyle\sum_{i=1}^{n} \frac{\bar{c}b_i + (1 - r_u)\ \gamma h_i b_i\ \tan\ \bar{\phi}}{\cos\ \theta_i + (\sin\ \theta_i\ \tan\ \bar{\phi})/F}}{\displaystyle\sum_{i=1}^{n} (W_i\ \sin\ \theta_i + C_s W_i a_i/R)} \tag{9.4}$$

The method in which Eq. 9.3 or 9.4 is applied is called the simplified Bishop method. Note that F appears on both sides of the equation, so a method of successive approximations must be used to solve Eq. 9.3 or 9.4.

A very effective procedure to solve these equations is by Newton's method of tangents, as shown in Fig. 9.2. Take Eq. 9.4 for example, which can be written as

$$f(F) = F \sum_{i=1}^{n} (W_i\ \sin\ \theta_i + C_s W_i a_i/R) - \sum_{i=1}^{n} \frac{\bar{c}b_i + (1 - r_u)\ \gamma h_i b_i\ \tan\ \bar{\phi}}{\cos\ \theta_i + (\sin\ \theta_i\ \tan\ \bar{\phi})/F} = 0 \tag{9.5}$$

The intersection F_{m+1} of the tangent to the curve $f(F)$ at $F = F_m$ with the F-axis is given by

$$F_{m+1} = F_m - \frac{F\ (F_m)}{f'(F_m)} \tag{9.6}$$

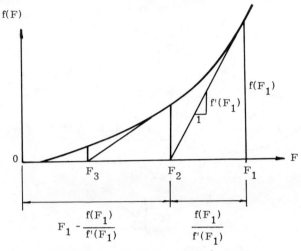

FIGURE 9.2. Newton's method of tangents.

in which $f'(F_m)$ is the first derivative of f with respect to F. Combining Eqs. 9.5 and 9.6

$$F_{m+1} = F_m \left\{ 1 - \frac{\sum\limits_{i=1}^{n}(W_i \sin \theta_i + C_s W_i a_i/R) - \sum\limits_{i=1}^{n} \dfrac{\bar{c}b_i + (1 - r_u) \gamma h_i b_i \tan \bar{\phi}}{F_m \cos \theta_i + \sin \theta_i \tan \bar{\phi}}}{\sum\limits_{i=1}^{n}(W_i \sin \theta_i + C_s W_i a_i/R) - \sum\limits_{i=1}^{n} \dfrac{[\bar{c}b_i + (1 - r_u) \gamma h_i b_i \tan \bar{\phi}] \sin \theta_i \tan \bar{\phi}}{(F_m \cos \theta_i + \sin \theta_i \tan \bar{\phi})^2}} \right\}$$

(9.7)

By using the factor of safety, F, obtained from the normal method, or Eq. 2.11, as the first trial value, Eq. 9.7 converges very rapidly usually within two or three iterations. An equation similar to Eq. 9.7 can be obtained when the phreatic surface is specified.

9.3 DESCRIPTION OF PROGRAM

The REAME program is composed of one main program and three subroutines. The FORTRAN version requires a storage space of 106K, of which 56K is for the object code and 50K for the array area. It can be applied to problems involving a maximum of 20 boundary lines (19 different soils), 50 points on each boundary line, 40 slices, 10 radius control zones, 9 boundary lines at the bottom of each radius control zone, and 5 cases of seepage. Each center can have as many as 90 trial radii. If these limits are exceeded, the corresponding dimensions should be changed accordingly. The BASIC version requires a storage of 51 blocks excluding the array area.

The storage can be reduced to 76K if the XYPLOT routine is deleted together with card MAIN0625 in the main program. The storage can be further reduced to 50K if the program is limited to 10 boundary lines, 10 points on each boundary line, 20 slices, 4 radius control zones, 1 case of seepage, and 20 radii at each center.

Complete listings of REAME in both FORTRAN and BASIC are presented in Appendices V and VI.

For smaller computers, a simplified version of REAME, called REAMES, was developed in both BASIC and FORTRAN (Huang, 1982). This simplified version requires a storage space of only 34K.

The main program is used for input and output, locating the center of circles, computing their radii, determining if the circle intersects the slope properly, adding more circles to make the last one as close to the specified minimum depth as possible, and searching for the minimum factor of safety.

Subroutine FSAFTY is used to determine the factor of safety of a circle when its center and radius are given by the main program. Subroutine SAVE is used to store the factors of safety of various circles for later printout, comparing them to determine which is the minimum, and adding more circles when needed. Subroutine XYPLOT is used to plot the slope for visual inspection. Figure 9.3 is a simplified flow chart for REAME.

9.4 GENERAL FEATURES

The program has many features which are similar to the ICES-LEASE computer program (Bailey and Christian, 1969). However, some of the features are quite unique in that they not only save the computer time but also ensure more accurate solutions.

Numbering of Soil Boundaries. To eliminate the necessity of renumbering soil boundaries and thus save a large amount of computer time, rules have been established for numbering the boundary lines and the points on each boundary line. A boundary line consists of one or more straight line segments, which separates two different soils. The boundary lines are numbered consecutively from bottom to top. The first one or more boundary lines are assigned to the boundary defined by the rock or stiff stratum at the bottom of the slope. The last boundary line is called the ground line and is assigned to the ground surface including the water surface, if any. When a ground line intersects the rock, it must terminate at the rock surface. No rock surface can be used as a ground line. Otherwise, the program will stop and print an error message. Any soil above a given boundary line has the same number as the boundary line. Consequently, the total number of soils is one less than the number of boundary lines.

Figure 9.4 shows the cross section of a slope. The slope is composed of four different soils, including soil 4 for water, and five boundary lines. The

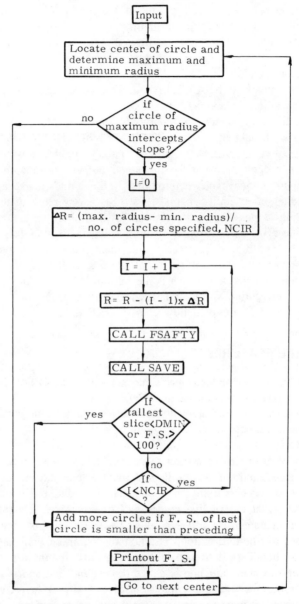

FIGURE 9.3. Simplified flow chart for REAME.

basic rule in numbering boundary lines is that a line with a lower number
must not lie above a line with a higher number. In other words, any vertical
line should intersect the boundary line with a lower number at a lower position. If this rule is violated, an error message will be printed and the computation stopped. Any body of water should be considered as a soil with a

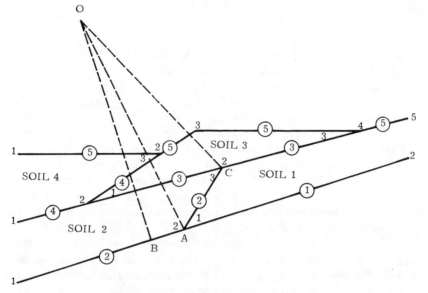

FIGURE 9.4.   Numbering of boundary lines.

cohesion and an angle of internal friction equal to zero and a unit weight equal to 62.4 lb/ft³, 1000 kg/m³, or 9.81 kN/m³. When the failure surface cuts through a body of water, the weight of water above the failure surface is ignored but the horizontal water pressure on the sliding mass is used in determining the overturning moment.

Each boundary line is defined by a number of points. These points are numbered consecutively from left to right, which is the positive direction of the x-coordinate. When two boundary lines intersect, the intersecting point has two numbers, one on each boundary line. For example, in Fig. 9.4, point 1 of line 1 is also point 2 of line 2, and point 3 of line 4 is also point 2 of line 5.

To be sure that all possible circles will be computed, it is desirable that the end point of the lowest and uppermost boundary lines, such as point 2 of line 1, point 1 of line 2, and points 1 and 5 of line 5 in Fig. 9.4, be extended as far out as possible. If the end point of ground line is not extended far enough, the maximum radius will be reduced, so that no circle will pass outside the ground line. If the circle lies above and outside the end point of the lowest boundary line, the soil outside the lowest boundary is not defined. The program will stop and print an error message. Although all other boundary lines can be of any length and terminate anywhere, it should be remembered that the length of a boundary line determines the demarcation between two soils. For a short and isolated boundary line, the soil above the boundary line within the region defined by the two vertical lines through the two end points is assigned the same number as the boundary line. The soil outside the two vertical lines has a different number, which is defined by the boundary line underneath. If point 1 of line 4 is not extended far enough, part of the water will be considered as soil 2 instead of soil 4.

Specification of Seepage. There are two methods for inputting seepage, one by specifying the coordinates of a piezometric surface and the other by specifying the pore pressure ratio. Unlike the ICES-LEASE program which allows each soil to have its own piezometric surface or pore pressure ratio, there is only one piezometric surface or one pore pressure ratio for the entire slope. This limitation should not affect the applicability of the program to most cases in engineering practice, unless some aquifer conditions exist.

The easiest way to consider seepage is to read in the coordinates of the phreatic surface. Theoretically, the phreatic surface is different from the piezometric surface. When the water is flowing, the phreatic surface should be higher than the piezometric surface. Therefore, the use of a phreatic surface is on the safe side.

The use of the pore pressure ratio is a convenient method for specifying seepage. Instead of inputing a number of coordinate values, a single parameter called the pore pressure ratio is all that is needed. This method is useful when equations or charts are used to determine the factor of safety or when the phreatic surface is not known.

Similar to the boundary lines, the piezometric line is also composed of a number of straight line segments. The points defining a piezometric line are also numbered consecutively from left to right. The piezometric line is completely independent of the boundary lines and can be located anywhere, either above or below the ground line. If a body of water exists, the water surface is a segment of the piezometric line. The two end points of a piezometric line should be extended as far out as that of the ground line. Otherwise, an error message will be printed and the computation stopped.

Control of Radius. Depending on the homogeneity of the slope, one or more radius control zones can be specified by the user. If the strength parameters for different soils in the slope do not change significantly, as is usually the case for effective stress analysis, the use of only one zone is sufficient. In this case, the maximum radius is controlled by the lowest boundary lines, or stiff stratum, and the minimum radius is governed by the uppermost boundary line, or ground line. Between the maximum and the minimum radii, any number of circles can be specified by the parameter $NCIR(1)$, where the subscript 1 indicates the first radius control zone which is the only zone in this case.

Since the last boundary line is the ground line and since every segment of the ground line is used for determining the minimum radius, it is not necessary for the user to specify the line number and the segment number in order to determine the minimum radius. For the case in Fig. 9.4 with only one radius control zone, the following information about the boundary line at the bottom of the radius control zone must be provided: total number of boundary lines $NOL(1) = 2$; for the first line, line number $LINO (1, 1) = 1$, beginning point number NBP(1, 1) = 1, and ending point number NEP(1, 1) = 2; for the second line, LINO(2, 1) = 2, NBP(2, 1) = 1, and NEP(2, 1) = 2. Note that when there are two subscripts, the first indicates the line sequence and the second indicates the zone number.

To illustrate how the maximum radius is determined, consider Fig. 9.4 with a center point at point O. Because boundary line 1 from point 1 to point 2 is specified, a line perpendicular to this segment is drawn from the center. As the intersection is outside the segment, the distance from the center to the nearest end point, OA, is selected as a tentative radius. Because the segment of line 2 from point 1 to point 2 is also specified, a line perpendicular to this segment is drawn from the center, and the distance OB is selected as a tentative radius. Finally by comparing OA and OB, the smaller radius, OB, is used as the maximum radius. If the ending point of line 2 is 3 instead of 2, a line perpendicular to the segment from 2 to 3 is also drawn from the center. As the intersection is outside the segment, the distance from the center to the nearest end point, OC, is also selected as a tentative radius. Since OC is smaller than OB, OC will be used as the maximum radius, so the circle is not tangent to the stiff stratum. This explains why the ending point for line 2 should be 2 instead of 3.

The same procedure is applied to determine the minimum radius. A line perpendicular to each segment of the ground line is drawn from the center, and a tentative radius is selected. The smallest of all the tentative radii is used as the minimum radius. No circle should have a radius equal to or smaller than the minimum radius.

The spacing of the circles between the maximum and the minimum radii is governed by the radius decrement, RDEC(1). If RDEC(1) is specified zero, the circles will be evenly spaced between the maximum and the minimum radii, or RDEC(1) = (maximum radius − minimum radius)/NCIR(1). If RDEC(1) is not zero, successive circles, starting from the maximum radius, with a radius decrement of RDEC(1) will be run until NCIR(1) circles are completed or until the radius becomes smaller than the minimum radius, whichever occurs first.

When RDEC(1) is not specified as zero, it is not necessary to begin from the first circle, or the circle with the maximum radius. A parameter INFC(1) can be used to identify the first circle to begin with. In most cases, INFC(1) should be set to 1, so the first circle to be run is the circle with the maximum radius. If INFC(1) is set to 2, the first circle to be run will be the second largest radius. This feature is particularly useful when the rock, or stiff stratum, is close to the surface of the slope. In such a case, the most critical circle is usually tangent to the rock, which is the first circle or the circle with the maximum radius. To be sure that this is true, a run can be made by specifying NCIR(1) to 2, INFC(1) to 1, and RDEC(1) to a small length, say 10 ft (3.05 m), so two circles will be run and the factor of safety for each circle printed. If, after inspecting the result, it is found that the circle with the maximum radius is not the most critical, a second run can be made by spec-

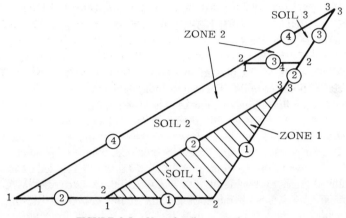

FIGURE 9.5. Use of radius control zones.

ifying NCIR(1) to 2, INFC(1) to 3, RDEC(1) to 10 ft (3.05 m), so two additional circles will be run and the result inspected. The process may be repeated until the user is pretty sure that the minimum factor of safety is obtained.

All the above discussion involves only one radius control zone. If it is known that the soils in a given zone are much weaker than the remaining soils, it may be more efficient to divide the cross section into two or more radius control zones, so that the number of circles in each zone can be specified separately. This situation is shown in Fig. 9.5, where soil 1 is much weaker than soils 2 and 3. In view of the fact that the most critical circle will not lie entirely within soils 2 and 3 but will definitely cut through soil 1, the cross section may be divided into two radius control zones with the lower zone designated as zone 1. The following are the input parameters:

Number of radius control zone, NRCZ = 2
Total number of lines at the bottom of zone 1, NOL(1) = 4 (although only one line, or line 1, is at the bottom of zone 1, portions of lines 2 and 3 are specified to avoid any circle cutting into the lowest boundary.)
Total number of lines at the bottom of zone 2, NOL(2) = 2
Radius decrement for zone 1, RDEC(1) = 0
Radius decrement for zone 2, RDEC(2) = 0
Number of circles in zone 1, NCIR(1) = 4
Line number for the first line at the bottom of zone 1, LINO(1,1) = 1
Beginning point for the first line at the bottom of zone 1, NBP(1,1) = 1
End point for the first line at the bottom of zone 1, NEP(1,1) = 3
Line number for the second line at the bottom of zone 1, LINO(2,1) = 2
Beginning point for the second line at the bottom of zone 1, NBP(2,1) = 1
End point for the second line at the bottom of zone 1, NEP(2,1) = 2

Line number for the third line at the bottom of zone 1, LINO(3,1) = 2
Beginning point for the third line at the bottom of zone 1, NBP(3,1) = 3
End point for the third line at the bottom of zone 1, NEP(3,1) = 4
Line number for the fourth line at the bottom of zone 1, LINO(4,1) = 3
Beginning point for the fourth line at the bottom of zone 1, NBP(4,1) = 2
End point for the fourth line at the bottom of zone 1, NEP(4,1) = 3
Number of circles in zone 2, NCIR(2) = 1
Line number for the first line at the bottom of zone 2, LINO(1,2) = 2
Beginning point for the first line at the bottom of zone 2, NBP(1,2) = 1
End point for the first line at the bottom of zone 2, NEP(1,2) = 4
Line number for the second line at the bottom of zone 2, LINO(2,2) = 3
Beginning point for the second line at the bottom of zone 2, NBP(2,2) = 2
End point for the second line at the bottom of zone 2, NEP(2,2) = 3

When more than one radius control zones is specified, the maximum radius for the second zone is used as the minimum radius for the first zone. If the maximum radius in a given zone is equal to or smaller than the minimum radius, no circle will be run in that zone. The minimum radius for the last zone is determined by the ground line.

In the above example, four circles are specified for zone 1, but only one for zone 2. If desired, NCIR(2) can be specified zero, so no circle will lie entirely in zone 2. However, it is preferable to specify one circle in zone 2 to be sure that the circle in zone 2 is not critical.

It should be pointed out that every segment of the boundary lines defined by the stiff stratum must be specified for radius control in one or more zones. If any segment is not specified, the circle may cut through the segment, and an error message will be printed. If a boundary line used for radius control is not continuous, more than one set of beginning and end points must be specified, as shown by line 2 for zone 1.

Number of Circles at Each Center. Although for a given center the user can specify any number of circles, NCIR(I), the number of circles actually run and printed out may be greater than NCIR(I). The reason that the number of circles may be greater than NCIR(I) is due to the parameter NK. This feature makes possible the use of a smaller number for NCIR(I), without sacrificing the accuracy of the results. No matter how many radius control zones are used, the circles will be numbered consecutively from the one with the maximum radius to the one with the smallest radius. Whenever a circle has a factor of safety smaller than the two adjacent circles, NK more circles will be added on each side of the given circle to locate the lowest factor of safety. Furthermore, if the last circle, or the circle with the smallest radius, has a factor of safety smaller than the previous circle, NK more circles with successively smaller radii will be added.

Figure 9.6 illustrates how the circles are added when the slope has only one radius control zone with NCIR(1) = 5 and NK = 3. Three circles are

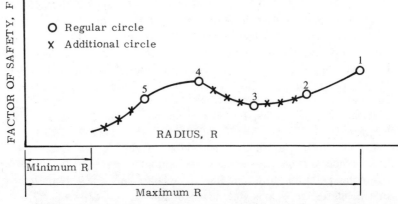

FIGURE 9.6. Location of additional circles for minimum factor of safety.

added on each side of circle 3 because it has a factor of safety smaller than both circles 2 and 4. Also three circles are added between circle 5 and the minimum radius, or ground line, because circle 5 has a factor of safety smaller than circle 4.

As in the ICES-LEASE program, the user can specify the minimum depth of tallest slice, DMIN. If a DMIN other than zero is specified, the factors of safety for all circles having the tallest slice greater than DMIN will be computed. If the tallest slice of the last one or more circles is smaller than DMIN, these circles will not be run. To ensure that one of the circles will have the tallest slice as close to DMIN as possible, even when a small number of circles is specified, a maximum of NK additional circles are added if the last circle has a factor of safety smaller than the previous circle. This is shown in Fig. 9.7 when five regular circles and three additional circles are specified. If the factors of safety for all five regular circles are computed, as shown in Fig. 9.7(a), three circles are added between circle 5 and the slope surface, as indicated by the minimum radius. Because the last of the three additional circles has a tallest slice smaller than DMIN, only the factor of safety of the first two circles is computed. If only four regular circles are run, as shown in Fig. 9.7(b), three circles are added between circle 4 and circle 5; the latter is not shown in the figure because its tallest slice is smaller than DMIN. Since two of the added circles have the tallest slice smaller than DMIN, only the factor of safety of one circle is computed. Due to geometric limitations, the number of circles actually run and printed out may be smaller than NCIR(I).

Method of Search. The method for specifying a grid and the procedure for search are the same as those in the ICES-LEASE program. However, if desired, search can follow immediately after creating the grid, using the center with the minimum factor of safety obtained from the grid as the initial

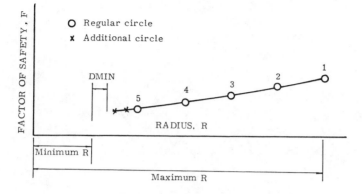

(a) ALL FIVE REGULAR CIRCLES ARE COMPUTED

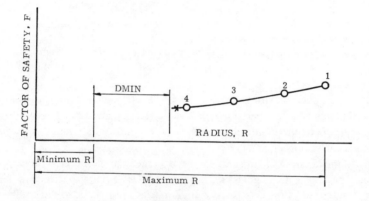

(b) NOT ALL REGULAR CIRCLES ARE COMPUTED

FIGURE 9.7.   Location of additional circles when minimum depth is specified.

center for search. Thus, the minimum factor of safety can be obtained in a single run.

The use of a grid and search does not require experience on the user. The grid can be extended as far out as possible, so that all critical regions can be covered. If any grid point is not properly located or no circle can be generated to intercept the slope, a message that no circle can intercept the slope or the circle is improper will be printed, and a large factor of safety will be assigned to that center. The program will go to the next center with very little loss in the computer time.

The location of the grid is defined by the x- and y-coordinates of three points, the number of divisions between points 1 and 2, NJ, and the number of divisions between points 2 and 3, NI. The three points from the two adjacent sides of a parallelogram. For the same three points, as shown in Fig. 9.8, the same grid is obtained in (a) and (b) but the grid in (c) is different.

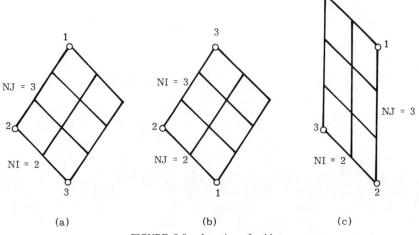

(a)   (b)   (c)

FIGURE 9.8.   Location of grid.

To use the search routine, it is necessary to specify the coordinates of a trial center, (XV, YV), and the increments XINC and YINC, in the horizontal and vertical directions respectively. If an improper center is selected for search, either due to wrong data or errors in coordinates, the program will stop automatically after searching for five successive centers of which no circle can be generated. Figure 9.9 shows how the search is conducted to locate the minimum factor of safety.

First, the factor of safety at the trial center, or point 1, is determined. Then proceeding to the right at a distance of XINC, point 2 is located. If the factor of safety at point 2 is smaller than that at point 1, continue to the right until the factor of safety at some point becomes greater than that at a previous point, i.e., until a smallest factor of safety is found in the horizontal direction. For the case shown in Fig. 9.9, the factor of safety at point 2 is greater than that at point 1, so point 3 at a distance of XINC on the left of point 1 is located. Because the factor of safety at point 3 is greater than that at point 1, point 1 has the smallest factor of safety in the horizontal direction. If the factor of safety at point 3 is smaller than at point 1, continue to the left until the smallest factor of safety in the horizontal direction is obtained. The search is then switched to the vertical direction. Proceeding upwards at a distance of YINC from the point with the smallest factor of safety, point 4 is located. Because the factor of safety at point 4 is smaller than that at point 1, continue to proceed upwards until a smallest factor of safety in the vertical direction is found. Then switch to the horizontal direction and repeat the process until the factor of safety at some point, such as point 4, is smaller than that at the four surrounding points, such as points 1, 5, 6 and 7. This is the minimum factor of safety based on the full increments of XINC and YINC.

Starting from point 4, the above process is repeated by using one-fourth

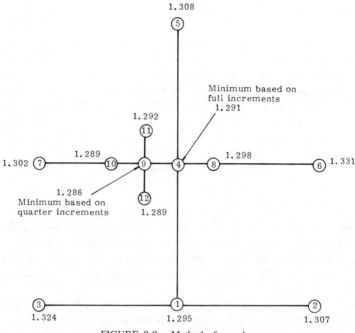

FIGURE 9.9. Method of search.

of XINC and YINC as the increments for search. The minimum factor of safety is found at point 9, which is smaller than the factor of safety at the four surrounding points, or points 4, 10, 11, and 12. The example shown in Fig. 9.9 is highly idealized. It usually takes many more steps to obtain a minimum factor of safety. However, the basic principle is the same. The search proceeds alternately in the horizontal and the vertical directions until a factor of safety smaller than the four surrounding points is obtained.

Subdivision of Slices. The initial number of slices is specified by the parameter NSLI. For a given circle, the intersection of the circle with the ground line is computed. The width of each slice is then equal to the horizontal distance between the two intersecting points divided by NSLI. If the center of the circle is lower than both the intersecting points, the circle should not be considered critical, and a large factor of safety is assigned.

At each breaking point of the ground line and at each intersection of the circle with any boundary line a subdivision of the original slice will be made. Depending on the complexity of the cross section, the actual number of slices may be much greater than NSLI.

Treatment of Tension Cracks. When the slope has a relatively high cohesion, tension cracks may often occur at the top of the failure surface. Although the REAME program does not consider tension cracks directly, they can be handled indirectly depending on whether the location of the failure circle is known or not. If the depth of the tension crack is assumed but the location of the failure circle is not known, a horizontal boundary line, as shown by line 2 in Fig. 9.10, is established across the top of the slope. Soil 2, which lies above line 2, is assigned a cohesion and a friction angle equal to zero. If the tension crack is filled with water, the unit weight of soil 2 is assigned 62.4 pcf (9.8 kN/m³); otherwise a unit weight of zero is assumed. If the location of the failure circle and tension crack is known, the end point of line 2 is located at the bottom of the tension crack, instead of at the slope surface. This end point can be used as a radius control, so that the circle will pass through the bottom of the crack. This case is illustrated by an example in Sect. 10.3.

Graphical Plot. A cross section of the slope can be plotted, including all boundary lines, the piezometric line, if any, and the location of the most critical circle. The plot routine will be activated only when NPLOT is assigned 1 and only after a search for the minimum factor of safety is performed.

Based on the coordinates of the cross section, the program will select proper scales for horizontal and vertical distances, so that the plot will fit onto one computer sheet. Usually the vertical dimension is plotted in a larger scale compared to the horizontal dimension, so the cross section is distorted. Numerals from 1 to 10 (10 is shown by a 0) are plotted to indicate the boundary lines. If there are more than 10 boundary lines, alphabetic letters will then be used. The letter P is used to indicate a piezometric line. If a portion of the piezometric line coincides with a boundary line, only the boundary line will be printed. The intersection of two boundary lines is indicated by an X. The failure surface is plotted by a series of asterisks,*.

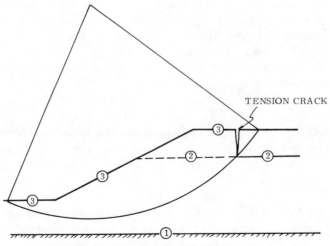

FIGURE 9.10. Treatment of tension crack.

9.5 DATA INPUT

The input parameters will be discussed according to the order in which they are read in. If the input parameter is an array, the variable defining the dimension of the array is also shown.

Any unit can be used for the parameters as long as it is consistent. In U.S. customary units, distance is in feet (ft), unit weight in pounds per cubic foot (lb/ft³), and cohesion in pounds per square foot (lb/ft²). In SI units, distance is in meters (m), unit weight in kilonewtons per cubic meter (kN/m³) and cohesion in kilonewtons per square meter (kN/m² or kPa). If the unit weight is in kilograms per cubic meter (kg/m³), the program will convert it automatically to kilonewtons per cubic meter (kN/m³) because any unit weight in terms of kilograms per cubic meter (kg/m³) is a large number and can be easily detected. To simulate surcharge in pounds per square foot (lb/ft²) or kilonewtons per square meter (kN/m²), be sure that the value of unit weight is not greater than 900. If the unit weight is greater than 900, it is presumed that the unit weight is in kilograms per cubic meter (kg/m³) and will be converted as such.

The input parameters are explained as follows:

TITLE—title. Any title or comment can be punched within columns 1 to 80 of a data card.

NCASE—number of cases. Any number of cases, each involving a different cross section or a different set or parameters, can be analyzed in the same run. If an error occurs in one case, the program will go to the next case.

NBL—number of boundary lines. The number of boundary lines is limited to 20.

NPBL(NBL)—number of points on each boundary line. The number of points on each boundary line is limited to 50. The maximum dimension is NPBL(20).

XBL(NPBL(NBL),NBL) and YBL(NPBL(NBL),NBL)—x- and y-coordinates of each point on a boundary line. The coordinates, which can be either positive or negative, must increase from left to right and from bottom to top. The maximum dimension is XBL (50,20) and YBL (50,20).

NRCZ—number of radius control zones. NRCZ may be assigned zero if the radius of a circle is specified. NRCZ should not be zero if grid (NSRCH = 0) or automatic search (NSRCH = 1) is used. The number of radius control zones is limited to 10.

NPLOT—condition of plot. NPLOT is assigned 1 if the cross section is to be plotted, and 0 if not. The plot routine will be activated only after an automatic search (NSRCH = 1) is performed.

NQ—number of seepage cases. Up to five seepage cases can be computed at the same time. If NSRCH = 1, NQ must be equal to 1, because only one seepage case can be considered at a time. If NQ is greater than 1 and an automatic search follows immediately after grid, all seepage cases will be searched, one by one. When there is no seepage, NQ should still be specified as 1. If zero is specified instead, the program will change it to 1.

NOL(NRCZ+1)—number of lines defining the lower boundary of each radius control zone. The number of lines for each radius control zone is limited to nine. The maximum dimension is NOL(11), which includes the ground line as the last line.

RDEC(NRCZ)—radius decrement. Assign zero if the circles are spaced uniformly over the entire zone; otherwise specify the actual length. When the radius decrement is zero, the spacing between circles will be determined by the computer based on the number of circles specified. The maximum dimension is RDEC(10).

NCIR(NRCZ)—number of circles in each radius control zone. If there is only one radius control zone, five to eight circles are usually sufficient. If there are more than one zone, three to five circles may be used in the weaker zone and one or more circles in the stronger zone. The maximum dimension is NCIR(10).

INFC(NRCZ)—identification number for first circle. When RDEC is assigned zero, INFC should always be 1, because the first circle, or the circle with the largest radius, should be computed. When RDEC is not zero, INFC may be assigned a number greater than 1, if the computation is a continuation of a previous run. The maximum dimension is INFC(10).

LINO(NOL(NRCZ+1), NRCZ+1)—boundary line number for each line sequence in a radius control zone. The maximum dimension is LINO(9,11).

NBP (NOL(NRCZ+1), NRCZ+1)—beginning point number for each line sequence in a radius control zone. The maximum dimension is NBP(9,11).

NEP (NOL(NRCZ+1), NRCZ+1)—end point number for each line sequence in a radius control zone. The maximum dimension is NEP (9,11).

C(NBL−1)—cohesion of each soil. The number of soils is one less than the number of boundary lines. Depending on the type of analysis, either total or effective cohesion can be used. The maximum dimension is C(19).

PHID(NBL−1)—angle of internal friction of each soil. Depending on the type of analysis, either total or effective angle of internal friction can be used. The angle is in degrees. The maximum dimension is PHID(19).

G(NBL−1)—unit weight of each soil. The unit weight is the total or mass unit weight. The maximum dimension is G(19).

METHOD—method used for determining the factor of safety. Assign zero for the normal method and 1 for the simplified Bishop method. The use of the simplified Bishop method is recommended.

NSPG—condition of seepage. Assign zero for no seepage, 1 for seepage defined by a piezometric line, and 2 for seepage defined by a pore pressure ratio.

NSRCH—condition of search. Assign zero for grid, 1 for automatic search, and 2 for individual centers.

NSLI—initial number of slices. In most cases, 10 slices are more than sufficient. Due to further subdivisions, the actual number of slices is greater than NSLI. The actual number of slices is limited to 40.

NK—number of additional circles. More circles are added if the factor of safety of a given circle is smaller than that of the two adjoining circles or if the factor of safety of the last circle is smaller than that of the previous circle. The use of 3 is usually sufficient. NK may be assigned zero if no additional circles are desired.

SEIC—seismic coefficient. The coefficient may range from 0 to 0.15 or more depending on geographical locations. When SEIC = 0, the static factor of safety is obtained.

DMIN—minimum depth of tallest slice. If the depth of the tallest slice is smaller than DMIN, the circle is not considered critical, and no factor of safety will be computed. If DMIN is assigned zero, all circles including those with a shallow depth will be computed.

GW—unit weight of water. The value of GW is used only when seepage is defined by a piezometric line, or NSPG = 1.

NPWT(NQ)—number of points on each piezometric line. This parameter must be specified when NSPG = 1. The number of points on a piezometric line is limited to 50. The maximum dimension is NPWT(5).

XWT(NPWT(NQ),NQ) and YWT(NPWT(NQ),NQ)—x- and y-coordinates of each point on a piezometric line. These parameters must be specified when NSPG = 1. The piezometric line must extend as far out as the ground line. The maximum dimension is XWT(50,5) and YWT(50,5).

RU(NQ)—pore pressure ratio. This parameter must be specified when NSPG = 2. The maximum dimension is RU(5).

X(3) and Y(3)—x- and y-coordinates of three points defining a grid. These parameters must be specified when NSRCH = 0.

XINC and YINC—x and y increments. Use only when NSRCH = 0 or 1. If NSRCH = 0 and both XINC and YINC are zero, the factors of safety will be determined by the grid only. If NSRCH = 0 and either XINC or YINC is not zero, automatic search using XINC and YINC as increments will follow immediately after grid. If NSRCH = 1, either XINC or YINC should not be zero. If one of the increments is zero, no search will be made in that direction.

NJ—number of divisions between point 1 and point 2. Use only when NSRCH = 0.

NI—number of divisions between point 2 and point 3. Use only when NSRCH = 0.

NP—total number of trial centers. This parameter is designed for NSRCH = 2 when the factors of safety at a number of individual centers are to be computed. It can also be used for NSRCH = 1 when several initial trial centers are run at the same time.

XV, YV—x- and y-coordinates of a trial center. Use only when NSRCH = 1 or 2.

R—radius of circle. Use only when NSRCH = 2. R can be assigned any given value. If R = 0 or NSRCH = 0 or 1, information on radius control zones must be provided so that R can be computed automatically.

The format used for data input is very simple. All integers are in I5 format, each occupying five columns of an 80 column card, while all real numbers are in F10.3 format, each occupying 10 columns. If the first alphabet of a parameter is I, J, K, L, M, or N, it is an integer, otherwise it is a real number. After a little practice, it is very easy to type in a number by memorizing the required spaces without using a coding sheet. For example, to type in a one-digit integer requires four spaces, a two-digit integer requires three spaces, a real number from 0.000 to 9.999 requires five spaces, etc. By counting the number of spaces, the data can be typed on a terminal just as easily as punched on a data card.

The input data deck is listed below. The number of cards indicated is correct only when the given data can be accommodated in an 80-column card. If the data require more than 80 columns, they should continue on the next card until the given data are exhausted.

1. 1 card (20A4) TITLE
2. 1 card (I5) NCASE
3. 1 card (16I5) NBL, (NPBL(I),I= 1, NBL)
   2 cards will be required if the number of boundary lines is greater than 15.
4. NBL cards (8F10.3) (XBL(I,J), YBL(I,J), I = 1, NPBLJ)
   Each boundary line, as indicated by subscript J, should begin with a new card. Each card can only accommodate four data points. NPBLJ is the total number of points on line J.
5. 1 card (3I5) NRCZ, NPLOT, NQ
6. If NRCZ = 0 go to (10)
7. 1 card (10I5) (NOL(I),I = 1, NRCZ)
8. 1 card (8F10.3) (RDEC(I), I = 1, NRCZ)
   2 cards will be required if the number of radius control zones is more than 8.
9. NRCZ cards (16I5) (NCIR(I),INFC(I),(LINO(J,I), NBP(J,I),NEP (J,I), J = 1, NOLI))
   Each radius control zone, as indicated by subscript I, should begin with a new card. NOLI is the total number of lines at the bottom of zone I.

10. NSOIL cards (3F10.3) C(I), PHID(I), G(I)
    Each soil, as indicated by subscript I, should be punched on a separate card.
11. 1 card (5I5) METHOD, NSPG, NSRCH, NSLI, NK
12. 1 card (3F10.3) SEIC, DMIN, GW
13. If NSPG = 0, go to (19)
14. If NSPG = 2, go to (18)
15. 1 card (5I5) (NPWT(I),I = 1,NQ)
16. NQ cards (8F10.3) (XWT(I,J), YWT(I,J), I = 1,NPWTJ)
    Each piezometric line, as indicated by subscript J, should begin with a new card. NPWTJ is the total number of points on line J.
17. Go to (19)
18. 1 card (5F10.3) (RU(I), I = 1, NQ)
19. If NSRCH = 1 or 2, go to (23)
20. 1 card (8F10.3) (X(I), Y(I), I = 1, 3), XINC, YINC
21. 1 card (2I5) NJ, NI
22. Go to (28)
23. 1 card (I5) NP
24. If NSRCH = 1, go to (27)
25. 1 card (3F10.3) XV, YV, R
26. Go to (28)
27. 1 card (4F10.3) XV, YV, XINC, YINC
28. If NSRCH = 1 or 2, go to (24) NP times
29. Go to (3) NCASE times

9.6 SAMPLE PROBLEMS

Two sample problems will be presented to illustrate the application of the computer program. The input data deck and the complete output are printed so that the user can employ any one of these problems in a trial run to check the correctness of the source program. Due to the versatility of the program, it is not feasible to present every possible case and option, so only the most general situation will be considered. It is hoped that, after the use of the program for some time, the user will become familiar with the program and apply the proper option to solve any practical problem most effectively.

Before applying REAME, it is advisable to plot a cross section of the slope on a graph paper, number the boundary lines, and enter the coordinates of all points on the boundary lines as well as on the phreatic line. A grid defined by three points is then drawn. For users with very little experience in stability analysis, the grid should extend as far out as possible, so that all critical regions will be covered. The lowest factor of safety at each grid point is determined, and the automatic search routine is activated, using the grid point with the smallest factor of safety as the initial center for search. To be sure that a minimum factor of safety is obtained, the factors of safety at all the grid points should be carefully inspected. If the possibility of a local minimum exists, an additional search should be made to locate the absolute minimum.

Due to space limitations, a relatively small grid is used in the two sample problems. All problems are only run once, with grid followed immediately by search, and no additional search is made. If the grid is properly located, a single run should result in a factor of safety very close to, if not equal to, the absolute minimum.

Example 1: Figure 9.11 shows the backfill on a strip-mined bench to restore the terrain to the original contour. The coordinates of the cross

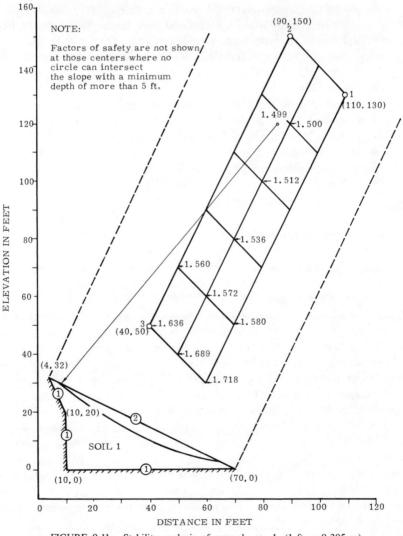

FIGURE 9.11. Stability analysis of example no. 1. (1 ft = 0.305 m)

section and the numbers assigned to the boundary lines are shown. The backfill has a cohesion of zero and an angle of internal friction of 35°. It is assumed that all surface water will be diverted from the fill and there is no seepage. To avoid the use of shallow circles, which cut a thin slice on the surface, a minimum depth of 5 ft (1.5 m) is specified. The complete output is printed and should be referred to during the following discussion.

Due to geometric limitations, it is not necessary to use a very large grid. Any trial center outside the region bounded by the two dashed lines shown in Fig. 9.11 cannot generate any circle which will intercept the slope. If a large grid extending outside this region is used, a message that the circle does not intercept the slope will be printed at those centers outside the region. The program still works and there is very little loss in the computer time. Even when a small grid near the center of the region is specified, many of these centers cannot have a circle with a tallest slice greater than 5 ft (1.5 m). Therefore, a message that the depth of the tallest slice is less than DMIN is printed at these centers and a large factor of safety is assigned.

At each center, the radius and the corresponding factor of safety are printed on two lines. The first line is for the regular circles and the second line for the additional circles. Although five regular circles, NCIR(1) = 5, and three additional circles, NK = 3, are specified, the number of circles printed is less because only those circles with a depth greater than 5 ft (1.5 m) are run.

The minimum factor of safety is 1.499. Note that at each center the shallowest circle with a depth slightly greater than 5 ft (1.5 m) is the most critical. The input data deck is shown below and the complete output is printed on pages 168–170.

```
1    5   10   15   20   25   30   35   40   45   50   55   60   65   70   75   80
******************************************************************************
*    *    *    *    *    *    *    *    *    *    *    *    *    *    *    *    *
******************************************************************************

EXAMPLE NO. 1
   1
   2    4    2
    4.000      32.000      10.000      20.000      10.000       0.000      70.000       0.000
    4.000      32.000      70.000       0.000
    1    1    1
   1
    0.000
    5    1    1    1    4
    0.000      35.000     125.000
    1    0    0   10    3
    0.000       5.000      62.400
  110.000     130.000      90.000     150.000      40.000      50.000       4.000       8.000
    2    5

******************************************************************************
*    *    *    *    *    *    *    *    *    *    *    *    *    *    *    *    *
******************************************************************************
1    5   10   15   20   25   30   35   40   45   50   55   60   65   70   75   80
```

CASE NUMBER   1

NUMBER OF BOUNDARY LINES=   2
NUMBER OF POINTS ON BOUNDARY LINES ARE: 4   2

ON BOUNDARY LINE NO. 1,POINT NO. AND COORDINATES ARE:
1   4.000   32.000   2   10.000   20.000   3   10.000   0.0   4   70.000   0.0

ON BOUNDARY LINE NO. 2,POINT NO. AND COORDINATES ARE:
1   4.000   32.000   2   70.000   0.0

LINE NO. AND SLOPE OF EACH SEGMENT ARE:
      1   -2.000 99999.000   0.0
      2   -0.485

NO. OF RADIUS CONTROL ZONES= 1   PLOT OR NO PLOT= 1   NO. OF SEEPAGE CASES= 1

TOTAL NO. OF LINES AT BOTTOM OF RADIUS CONTROL ZONES ARE: 1

FOR RAD. CONT. ZONE NO. 1   RADIUS DECREMENT=   0.0   NO. OF CIRCLES= 5   ID NO. FOR FIRST CIRCLE= 1
      LINE NO.= 1   BEGIN PT. NO.= 1   END PT. NO.= 4

SOIL NO.   COHESION   F. ANGLE   UNIT WT.
      1   0.0   35.000   125.000

SEISMIC COEFFICIENT=   0.0   MIN. DEPTH OF TALLEST SLICE=   5.000   UNIT WEIGHT OF WATER=   62.500

THE FACTORS OF SAFETY ARE DETERMINED BY THE SIMPLIFIED BISHOP METHOD

NSPG= 0 NSRCH= 0 NO. OF SLICES= 10 NO. OF ADD. RADII= 3

POINT1=(110.000,   130.000) POINT2=(90.000,   150.000) POINT3=(40.000,   50.000) NJ=   2 NI=   5
      AUTOMATIC SEARCH WILL FOLLOW AFTER GRID WITH XINC=   4.000 AND YINC=   8.000

AT POINT (110.000,   130.000) UNDER SEEPAGE   1 THE DEPTH OF TALLEST SLICE IS LESS THAN DMIN

AT POINT (100.000,   140.000) UNDER SEEPAGE   1 THE DEPTH OF TALLEST SLICE IS LESS THAN DMIN

AT POINT (90.000,   150.000) UNDER SEEPAGE   1 THE DEPTH OF TALLEST SLICE IS LESS THAN DMIN

AT POINT (100.000,   110.000) UNDER SEEPAGE   1 THE DEPTH OF TALLEST SLICE IS LESS THAN DMIN

AT POINT (90.000,   120.000) UNDER SEEPAGE   1,THE RADIUS AND THE CORRESPONDING FACTOR OF SAFETY ARE:
      121.655   1.503
      121.408   1.500
      LOWEST FACTOR OF SAFETY=   1.500 AND OCCURS AT RADIUS =   121.408

AT POINT (80.000,   130.000) UNDER SEEPAGE   1 THE DEPTH OF TALLEST SLICE IS LESS THAN DMIN

AT POINT (90.000,   90.000) UNDER SEEPAGE   1 THE DEPTH OF TALLEST SLICE IS LESS THAN DMIN

AT POINT (80.000,   100.000) UNDER SEEPAGE   1,THE RADIUS AND THE CORRESPONDING FACTOR OF SAFETY ARE:
      100.499   1.534   99.268   1.517
      98.960   1.512
      LOWEST FACTOR OF SAFETY=   1.512 AND OCCURS AT RADIUS =   98.960

AT POINT (70.000,   110.000) UNDER SEEPAGE   1 THE DEPTH OF TALLEST SLICE IS LESS THAN DMIN

AT POINT (80.000,   70.000) UNDER SEEPAGE   1 THE DEPTH OF TALLEST SLICE IS LESS THAN DMIN

AT POINT (70.000,   80.000) UNDER SEEPAGE   1,THE RADIUS AND THE CORRESPONDING FACTOR OF SAFETY ARE:
      80.000   1.595   78.397   1.566   76.794   1.536
      LOWEST FACTOR OF SAFETY=   1.536 AND OCCURS AT RADIUS =   76.794

AT POINT (60.000,   90.000) UNDER SEEPAGE   1 THE DEPTH OF TALLEST SLICE IS LESS THAN DMIN

AT POINT (70.000,   50.000) UNDER SEEPAGE   1,THE RADIUS AND THE CORRESPONDING FACTOR OF SAFETY ARE:
      50.000   1.595
      49.750   1.587   49.499   1.580
      LOWEST FACTOR OF SAFETY=   1.580 AND OCCURS AT RADIUS =   49.499

AT POINT (60.000,   60.000) UNDER SEEPAGE   1,THE RADIUS AND THE CORRESPONDING FACTOR OF SAFETY ARE:
      60.000   1.715   57.925   1.665   55.850   1.613
      55.332   1.599   54.813   1.586   54.294   1.572
      LOWEST FACTOR OF SAFETY=   1.572 AND OCCURS AT RADIUS =   54.294

AT POINT (50.000,   70.000) UNDER SEEPAGE   1,THE RADIUS AND THE CORRESPONDING FACTOR OF SAFETY ARE:
      59.666   1.579
      59.396   1.573   59.125   1.566   58.855   1.560
      LOWEST FACTOR OF SAFETY=   1.560 AND OCCURS AT RADIUS =   58.855

AT POINT (60.000,   30.000) UNDER SEEPAGE   1,THE RADIUS AND THE CORRESPONDING FACTOR OF SAFETY ARE:
      30.000   1.847   28.526   1.775
      28.158   1.756   27.789   1.737   27.421   1.718
      LOWEST FACTOR OF SAFETY=   1.718 AND OCCURS AT RADIUS =   27.421

AT POINT (50.000,   40.000) UNDER SEEPAGE   1,THE RADIUS AND THE CORRESPONDING FACTOR OF SAFETY ARE:
      40.000   1.990   37.453   1.897   34.907   1.797   32.360   1.689
      LOWEST FACTOR OF SAFETY=   1.689 AND OCCURS AT RADIUS = - - - 32.360

AT POINT (40.000,   50.000) UNDER SEEPAGE   1,THE RADIUS AND THE CORRESPONDING FACTOR OF SAFETY ARE:
      40.249   1.776   38.580   1.715   36.911   1.652
      36.493   1.636
      LOWEST FACTOR OF SAFETY=   1.636 AND OCCURS AT RADIUS =   36.493

AT POINT (90.000,   120.000),RADIUS 121.408
      THE MINIMUM FACTOR OF SAFETY IS   1.500

AT POINT (90.000,   120.000) UNDER SEEPAGE   1,THE RADIUS AND THE CORRESPONDING FACTOR OF SAFETY ARE:
      121.655   1.503
      121.408   1.500
      LOWEST FACTOR OF SAFETY=   1.500 AND OCCURS AT RADIUS =   121.408

AT POINT (94.000,   120.000) UNDER SEEPAGE   1 THE DEPTH OF TALLEST SLICE IS LESS THAN DMIN

AT POINT (86.000,   120.000) UNDER SEEPAGE   1,THE RADIUS AND THE CORRESPONDING FACTOR OF SAFETY ARE:
      120.283   1.509
      120.017   1.505   119.751   1.502   119.484   1.499
      LOWEST FACTOR OF SAFETY=   1.499 AND OCCURS AT RADIUS =   119.484

AT POINT (82.000,   120.000) UNDER SEEPAGE   1 THE DEPTH OF TALLEST SLICE IS LESS THAN DMIN

AT POINT (86.000,   128.000) UNDER SEEPAGE   1 THE DEPTH OF TALLEST SLICE IS LESS THAN DMIN

AT POINT (86.000,   112.000) UNDER SEEPAGE   1,THE RADIUS AND THE CORRESPONDING FACTOR OF SAFETY ARE:
      113.137   1.513
      112.868   1.510   112.599   1.507   112.330   1.503
      LOWEST FACTOR OF SAFETY=   1.503 AND OCCURS AT RADIUS =   112.330

AT POINT (87.000,   120.000) UNDER SEEPAGE   1,THE RADIUS AND THE CORRESPONDING FACTOR OF SAFETY ARE:
      120.967   1.511
      120.688   1.508   120.410   1.505   120.131   1.501
      LOWEST FACTOR OF SAFETY=   1.501 AND OCCURS AT RADIUS =   120.131

AT POINT (85.000,   120.000) UNDER SEEPAGE   1,THE RADIUS AND THE CORRESPONDING FACTOR OF SAFETY ARE:
      119.604   1.506
      119.349   1.503   119.095   1.500
      LOWEST FACTOR OF SAFETY=   1.500 AND OCCURS AT RADIUS =   119.095

AT POINT (86.000,   122.000) UNDER SEEPAGE   1,THE RADIUS AND THE CORRESPONDING FACTOR OF SAFETY ARE:
      121.754   1.504
      121.504   1.501
      LOWEST FACTOR OF SAFETY=   1.501 AND OCCURS AT RADIUS =   121.504

AT POINT (86.000,   118.000) UNDER SEEPAGE   1,THE RADIUS AND THE CORRESPONDING FACTOR OF SAFETY ARE:
      118.828   1.514   117.694   1.500
      LOWEST FACTOR OF SAFETY=   1.500 AND OCCURS AT RADIUS =   117.694

AT POINT (86.000,   120.000),RADIUS 119.484
      THE MINIMUM FACTOR OF SAFETY IS   1.499

```
CROSS SECTION IN DISTORTED SCALE
NUMERALS INDICATE BOUNDARY LINE NO. IF THERE ARE MORE THAN 10
BOUND. LINES,ALPHABETS WILL THEN BE USED. P INDICATES
PIZOMETRIC LINE. IF A PORTION OF PIEZOMETRIC LINE COINCIDES WITH
THE GROUND OR ANOTHER BOUNDARY LINE,ONLY THE GROUND OR BOUNDARY
LINE WILL BE SHOWN. X INDICATES INTERSECTION OF TWO BOUNDARY
LINES.  * INDICATES FAILURE SURFACE.
THE MINIMUM FACTOR OF SAFETY IS      1.499
3.600E 01 X++++++++++X++++++++++X++++++++++X++++++++++X++++++++++X
          +                                                     +
          +                                                     +
          +                                                     +
          +                                                     +
3.200E 01 X X2       +          +          +          +         X
          +    2                                                +
          +      2                                              +
          +   1    2                                            +
          +       *2                                            +
2.800E 01 X         2 +          +          +          +        X
          +         * 2                                         +
          +    1      * 2                                       +
          +             2                                       +
          +          *   2                                      +
2.400E 01 X          *   2       +          +          +        X
          +    1        2                                       +
          +          *   2                                      +
          +           *   2                                     +
          +                2                                    +
2.000E 01 X      1    +  *     2 +          +          +        X
          +            *      2                                 +
          +              *      2                               +
          +                *    2                               +
          +                 *    2                              +
1.600E 01 X          +        +  22       +          +          X
          +                 *      2                            +
          +                  *      2                           +
          +                   *      2                          +
          +                    *     2                          +
1.200E 01 X          +          + *      2+          +          X
          +                       *      2                      +
          +                        **     2                     +
          +                         *      2                     +
          +                          *      2                    +
8.000E 00 X          +          +    * +    2       +          X
          +                          **    2                   +
          +                           *     2                  +
          +                          **      2                 +
          +                            **     2                +
4.000E 00 X          +          +      ** 2+        X
          +                            **2                      +
          +                             *2                      +
          +                               2                     +
          +                                2                    +
0.0       X+++++1111X111111111X111111111X111111111X111X+++++X
          +                                                     +
          +                                                     +
          +                                                     +
          +                                                     +
-4.000E 00 X++++++++++X++++++++++X++++++++++X++++++++++X++++++++++X
          0.0        1.60E 01  3.20E 01  4.80E 01  6.40E 01  8.00E 01
```

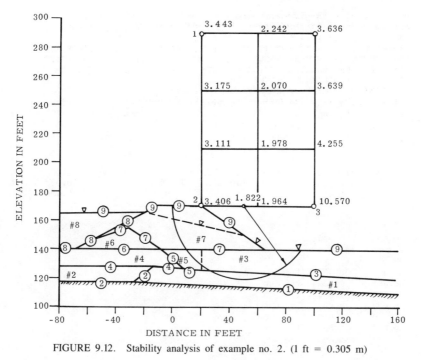

FIGURE 9.12. Stability analysis of example no. 2. (1 ft = 0.305 m)

Example 2: Figure 9.12 shows an earth dam with bedrock at a considerable distance from the surface. Because the soils in the foundation are weaker than those in the dam, two radius control zones are specified with five circles in zone 1 and no circles in zone 2. The coordinates of the boundary and piezometric lines and the shear strength of different soils are shown in the printout. The minimum factor of safety for the dam is 1.822.

A warning message that the maximum radius is limited by the end point of ground line appears at several centers. These centers are located on the two extreme sides of the grid, where the factor of safety is quite high, so the warning has no effect on the minimum factor of safety obtained. At center (20,250) the circle with the lowest factor of safety is not tangent to the rock. Therefore, the omission of the circle tangent to the rock has no effect on the lowest factor of safety at this center.

A warning message that either the overturning or the resisting moment is zero appears at three centers: (100,290), (100,250) and (100,210). When the radius is reduced to a certain value, the circle cuts the horizontal ground beyond the toe, so the overturning moment is zero and a large factor of safety is assumed. At center (100,170) when the radius is 47.036, the factor of safety is greater than 100, so no more circles are run.

The input data deck and the complete output are printed on pages 172–176.

```
1    5   10   15   20   25   30   35   40   45   50   55   60   65   70   75   80
*********************************************************************************
*    *    *    *    *    *    *    *    *    *    *    *    *    *    *    *    *
*********************************************************************************

EXAMPLE NO. 2
    1
    9    2    3    2    3    3    2    4    4    6
  -28.000  117.000  160.000  109.000
  -80.000  117.000  -28.000  117.000  -12.000  128.500
   20.000  126.000  160.000  118.000
  -80.000  128.000  -12.000  128.500    4.000  126.500
   -7.000  140.000    4.000  126.500   20.000  126.000
  -70.000  140.000   -7.000  140.000
  -50.000  150.000  -30.000  154.000   -7.000  140.000   65.000  140.000
  -80.000  140.000  -70.000  140.000  -50.000  150.000  -20.000  165.000
  -80.000  165.000  -20.000  165.000  -10.000  170.000   20.000  170.000
   65.000  140.000  160.000  140.000
    2    1    1
    2    3
    0.000    0.000
    5    1    2    1    2    1    1    2
    0    1    6    1    2    7    3    4    9    5    6
  100.000   28.000  125.000
  200.000   30.000  125.000
  150.000   29.000  125.000
  200.000   25.000  125.000
  250.000   18.000  130.000
 1500.000   10.000  130.000
 2000.000   15.000  130.000
    0.000    0.000   62.400
    1    1    0   10    2
    0.000    0.000   62.400
    5
  -80.000  165.000  -20.000  165.000   50.000  150.000   65.000  140.000
  160.000  140.000
   20.000  290.000   20.000  170.000  100.000  170.000   20.000   20.000
    3    2

*********************************************************************************
*    *    *    *    *    *    *    *    *    *    *    *    *    *    *    *    *
*********************************************************************************
1    5   10   15   20   25   30   35   40   45   50   55   60   65   70   75   80
```

CASE NUMBER   1

NUMBER OF BOUNDARY LINES=   9
NUMBER OF POINTS ON BOUNDARY LINES ARE: 2   3   2   3   3   2   4   4
      6

ON BOUNDARY LINE NO. 1,POINT NO. AND COORDINATES ARE:
1   -28.000   117.000   2   160.000   109.000

ON BOUNDARY LINE NO. 2,POINT NO. AND COORDINATES ARE:
1   -80.000   117.000   2   -28.000   117.000   3 -12.000   128.500

ON BOUNDARY LINE NO. 3,POINT NO. AND COORDINATES ARE:
1   20.000   126.000   2   160.000   118.000

ON BOUNDARY LINE NO. 4,POINT NO. AND COORDINATES ARE:
1   -80.000   128.000   2   -12.000   128.500   3   4.000   126.500

ON BOUNDARY LINE NO. 5,POINT NO. AND COORDINATES ARE:
1   -7.000   140.000   2   4.000   126.500   3   20.000   126.000

ON BOUNDARY LINE NO. 6,POINT NO. AND COORDINATES ARE:
1   -70.000   140.000   2   -7.000   140.000

ON BOUNDARY LINE NO. 7,POINT NO. AND COORDINATES ARE:
1   -50.000   150.000   2   -30.000   154.000   3   -7.000   140.000   4   65.000   140.000

ON BOUNDARY LINE NO. 8,POINT NO. AND COORDINATES ARE:
1   -80.000   140.000   2   -70.000   140.000   3 -50.000   150.000   4 -20.000   165.000

ON BOUNDARY LINE NO. 9,POINT NO. AND COORDINATES ARE:
1   -80.000   165.000   2   -20.000   165.000   3 -10.000   170.000   4   20.000   170.000   5   65.000   140.000
PUN FILE 6825 FROM OS370   COPY 001   NOHOLD
6   160.000   140.000

    LINE NO. AND SLOPE OF EACH SEGMENT ARE:
        1   -0.043
        2   0.0   0.719
        3   -0.057
        4   0.007   -0.125
        5   -1.227   -0.031
        6   0.0
        7   0.200   -0.609   0.0
        8   0.0   0.500   0.500
        9   0.0   0.500   0.0   -0.667   0.0

NO. OF RADIUS CONTROL ZONES= 2   PLOT OR NO PLOT= 1   NO. OF SEEPAGE CASES= 1

TOTAL NO. OF LINES AT BOTTOM OF RADIUS CONTROL ZONES ARE: 2   3

FOR RAD. CONT. ZONE NO. 1   RADIUS DECREMENT=   0.0   NO. OF CIRCLES= 5   ID NO. FOR FIRST CIRCLE= 1
    LINE NO.= 2   BEGIN PT. NO.= 1   END PT. NO.= 2
    LINE NO.= 1   BEGIN PT. NO.= 1   END PT. NO.= 2

FOR RAD. CONT. ZONE NO. 2   RADIUS DECREMENT=   0.0   NO. OF CIRCLES= 0   ID NO. FOR FIRST CIRCLE= 1
    LINE NO.= 6   BEGIN PT. NO.= 1   END PT. NO.= 2
    LINE NO.= 7   BEGIN PT. NO.= 3   END PT. NO.= 4
    LINE NO.= 9   BEGIN PT. NO.= 5   END PT. NO.= 6

SOIL NO.   COHESION   F. ANGLE   UNIT WT.
    1   100.000   28.000   125.000
    2   200.000   30.000   125.000
    3   150.000   29.000   125.000
    4   200.000   25.000   125.000
    5   250.000   18.000   130.000
    6   1500.000   10.000   130.000
    7   2000.000   15.000   130.000
    8   0.0   0.0   62.400

SEISMIC COEFFICIENT=   0.0   MIN. DEPTH OF TALLEST SLICE=   0.0   UNIT WEIGHT OF WATER=   62.400

THE FACTORS OF SAFETY ARE DETERMINED BY THE SIMPLIFIED BISHOP METHOD

```
NSPG= 1   NSRCH= 0   NO. OF SLICES= 10   NO. OF ADD. RADII= 2

NO. OF POINTS ON WATER TABLE FOR EACH CASE=   5

UNDER SEEPAGE CONDITION 1,POINT NO. AND COORDINATES OF WATER TABLE ARE:
 1  -80.000  165.000   2  -20.000  165.000   3   50.000  150.000   4   65.000  140.000   5  160.000  140.000
POINT1=(   20.000,   290.000) POINT2=(   20.000,   170.000) POINT3=(  100.000,   170.000) NJ=   3 NI=   2
   AUTOMATIC SEARCH WILL FOLLOW AFTER GRID WITH XINC=   20.000 AND YINC=   20.000

****WARNING AT NEXT CENTER**** AT RADIUS CONTROL ZONE NO. 1,MAXIMUM RADIUS IS LIMITED BY THE END POINT OF GROUND LINE

AT POINT (   20.000,  290.000) UNDER SEEPAGE     1,THE RADIUS AND THE CORRESPONDING FACTOR OF SAFETY ARE:
      160.078    3.443       158.062    3.587       156.047    3.898       154.031    4.537       152.016    5.613
      LOWEST FACTOR OF SAFETY=   3.443 AND OCCURS AT RADIUS =   160.078

****WARNING AT NEXT CENTER**** AT RADIUS CONTROL ZONE NO. 1,MAXIMUM RADIUS IS LIMITED BY THE END POINT OF GROUND LINE

AT POINT (   20.000,  250.000) UNDER SEEPAGE     1,THE RADIUS AND THE CORRESPONDING FACTOR OF SAFETY ARE:
      131.244    3.272       126.995    3.259       122.746    3.175       118.498    3.457       114.249    4.414
      125.579    3.292       124.163    3.263       121.330    3.228       119.914    3.315
      LOWEST FACTOR OF SAFETY=   3.175 AND OCCURS AT RADIUS =   122.746

AT POINT (   20.000,  210.000) UNDER SEEPAGE     1,THE RADIUS AND THE CORRESPONDING FACTOR OF SAFETY ARE:
       94.957    3.136        89.965    3.111        84.974    3.175        79.983    3.418        74.991    4.173
       93.293    3.123        91.629    3.123        88.301    3.111        86.638    3.134
      LOWEST FACTOR OF SAFETY=   3.111 AND OCCURS AT RADIUS =   88.301

AT POINT (   20.000,  170.000) UNDER SEEPAGE     1,THE RADIUS AND THE CORRESPONDING FACTOR OF SAFETY ARE:
       54.993    3.406        49.994    3.661        44.996    4.010        39.997    4.340        34.999    5.364
      LOWEST FACTOR OF SAFETY=   3.406 AND OCCURS AT RADIUS =   54.993

AT POINT (   60.000,  290.000) UNDER SEEPAGE     1,THE RADIUS AND THE CORRESPONDING FACTOR OF SAFETY ARE:
      176.585    2.531       171.268    2.360       165.951    2.267       160.634    2.572       155.317    3.000
      169.495    2.281       167.723    2.242       164.179    2.265       162.406    2.501
      LOWEST FACTOR OF SAFETY=   2.242 AND OCCURS AT RADIUS =   167.723

AT POINT (   60.000,  250.000) UNDER SEEPAGE     1,THE RADIUS AND THE CORRESPONDING FACTOR OF SAFETY ARE:
      136.621    2.274       131.297    2.076       125.973    2.222       120.648    2.358       115.324    2.703
      134.846    2.230       133.072    2.146       129.522    2.070       127.747    2.161
      LOWEST FACTOR OF SAFETY=   2.070 AND OCCURS AT RADIUS =   129.522

AT POINT (   60.000,  210.000) UNDER SEEPAGE     1,THE RADIUS AND THE CORRESPONDING FACTOR OF SAFETY ARE:
       96.657    2.059        91.326    1.978        85.994    2.067        80.663    2.230        75.331    2.529
       94.880    2.022        93.103    1.981        89.549    1.991        87.771    2.015
      LOWEST FACTOR OF SAFETY=   1.978 AND OCCURS AT RADIUS =   91.326

AT POINT (   60.000,  170.000) UNDER SEEPAGE     1,THE RADIUS AND THE CORRESPONDING FACTOR OF SAFETY ARE:
       56.693    2.002        51.355    1.964        46.016    2.070        40.677    2.260        35.339    2.827
       54.914    1.987        53.134    1.972        49.575    1.970        47.796    1.996
      LOWEST FACTOR OF SAFETY=   1.964 AND OCCURS AT RADIUS =   51.355

****WARNING AT NEXT CENTER**** AT RADIUS CONTROL ZONE NO. 1,MAXIMUM RADIUS IS LIMITED BY THE END POINT OF GROUND LINE

****WARNING AT NEXT CENTER**** WHEN RADIUS IS   152.311
EITHER THE OVERTURNING OR THE RESISTING MOMENT IS 0,SO A LARGE FACTOR OF SAFETY IS ASSIGNED

AT POINT (  100.000,  290.000) UNDER SEEPAGE     1,THE RADIUS AND THE CORRESPONDING FACTOR OF SAFETY ARE:
      161.555    3.636       159.244    4.073       156.933    4.809       154.622    6.164       152.311**********
      LOWEST FACTOR OF SAFETY=   3.636 AND OCCURS AT RADIUS =   161.555

****WARNING AT NEXT CENTER**** AT RADIUS CONTROL ZONE NO. 1,MAXIMUM RADIUS IS LIMITED BY THE END POINT OF GROUND LINE

****WARNING AT NEXT CENTER**** WHEN RADIUS IS   113.060
EITHER THE OVERTURNING OR THE RESISTING MOMENT IS 0,SO A LARGE FACTOR OF SAFETY IS ASSIGNED

AT POINT (  100.000,  250.000) UNDER SEEPAGE     1,THE RADIUS AND THE CORRESPONDING FACTOR OF SAFETY ARE:
      125.300    3.639       122.240    4.256       119.180    5.556       116.120    9.102       113.060**********
      LOWEST FACTOR OF SAFETY=   3.639 AND OCCURS AT RADIUS =   125.300

****WARNING AT NEXT CENTER**** AT RADIUS CONTROL ZONE NO. 1,MAXIMUM RADIUS IS LIMITED BY THE END POINT OF GROUND LINE

****WARNING AT NEXT CENTER**** WHEN RADIUS IS   74.439
EITHER THE OVERTURNING OR THE RESISTING MOMENT IS 0,SO A LARGE FACTOR OF SAFETY IS ASSIGNED

AT POINT (  100.000,  210.000) UNDER SEEPAGE     1,THE RADIUS AND THE CORRESPONDING FACTOR OF SAFETY ARE:
```

```
      92.195    4.255      87.756    5.663      83.317   9.283      78.878   44.835      74.439**********
   LOWEST FACTOR OF SAFETY=    4.255 AND OCCURS AT RADIUS =    92.195


AT POINT (   100.000,   170.000) UNDER SEEPAGE    1,THE RADIUS AND THE CORRESPONDING FACTOR OF SAFETY ARE:
      58.394   10.570      52.715   22.928      47.036 3808.736
   LOWEST FACTOR OF SAFETY=   10.570 AND OCCURS AT RADIUS =    58.394




FOR PIEZOMETRIC LINE NO.    1

AT POINT (    60.000,   170.000),RADIUS    51.355
   THE MINIMUM FACTOR OF SAFETY IS    1.964



AT POINT (    60.000,   170.000) UNDER SEEPAGE    1,THE RADIUS AND THE CORRESPONDING FACTOR OF SAFETY ARE:
      56.693    2.002      51.355    1.964      46.016   2.070      40.677    2.260      35.339    2.827
      54.914    1.987      53.134    1.972      49.575   1.970      47.796    1.996
   LOWEST FACTOR OF SAFETY=    1.964 AND OCCURS AT RADIUS =    51.355


AT POINT (    80.000,   170.000) UNDER SEEPAGE    1,THE RADIUS AND THE CORRESPONDING FACTOR OF SAFETY ARE:
      57.544    3.068      52.035    3.363      46.526   4.209      41.017    6.181      35.509   19.496
   LOWEST FACTOR OF SAFETY=    3.068 AND OCCURS AT RADIUS =    57.544


AT POINT (    40.000,   170.000) UNDER SEEPAGE    1,THE RADIUS AND THE CORRESPONDING FACTOR OF SAFETY ARE:
      55.843    2.101      50.674    1.956      45.506   1.946      40.337    2.100      35.169    2.645
      48.952    1.890      47.229    1.903      43.783   2.000      42.060    2.039
   LOWEST FACTOR OF SAFETY=    1.890 AND OCCURS AT RADIUS =    48.952


AT POINT (    20.000,   170.000) UNDER SEEPAGE    1,THE RADIUS AND THE CORRESPONDING FACTOR OF SAFETY ARE:
      54.993    3.406      49.994    3.661      44.996   4.010      39.997    4.340      34.999    5.364
   LOWEST FACTOR OF SAFETY=    3.406 AND OCCURS AT RADIUS =    54.993


AT POINT (    40.000,   190.000) UNDER SEEPAGE    1,THE RADIUS AND THE CORRESPONDING FACTOR OF SAFETY ARE:
      75.825    2.140      70.660    2.040      65.495   1.987      60.330    2.151      55.165    2.557
      68.938    2.007      67.217    1.973      63.773   2.034      62.052    2.082
   LOWEST FACTOR OF SAFETY=    1.973 AND OCCURS AT RADIUS =    67.217


AT POINT (    40.000,   150.000) UNDER SEEPAGE    1,THE RADIUS AND THE CORRESPONDING FACTOR OF SAFETY ARE:
      35.861    2.908      30.689    3.020      25.517   3.642      20.344    5.035      15.172    7.241
   LOWEST FACTOR OF SAFETY=    2.908 AND OCCURS AT RADIUS =    35.861

AT POINT (    45.000,   170.000) UNDER SEEPAGE    1,THE RADIUS AND THE CORRESPONDING FACTOR OF SAFETY ARE:
      56.056    1.987      50.845    1.828      45.633   1.883      40.422    2.002      35.211    2.336
      54.319    1.901      52.582    1.862      49.107   1.829      47.370    1.849
   LOWEST FACTOR OF SAFETY=    1.828 AND OCCURS AT RADIUS =    50.845


AT POINT (    50.000,   170.000) UNDER SEEPAGE    1,THE RADIUS AND THE CORRESPONDING FACTOR OF SAFETY ARE:
      56.268    1.912      51.015    1.822      45.761   1.876      40.507    1.967      35.254    2.258
      54.517    1.868      52.766    1.830      49.263   1.826      47.512    1.839
   LOWEST FACTOR OF SAFETY=    1.822 AND OCCURS AT RADIUS =    51.015


AT POINT (    55.000,   170.000) UNDER SEEPAGE    1,THE RADIUS AND THE CORRESPONDING FACTOR OF SAFETY ARE:
      56.481    1.915      51.185    1.869      45.888   1.938      40.592    2.031      35.296    2.480
      54.715    1.892      52.950    1.876      49.419   1.875      47.654    1.893
   LOWEST FACTOR OF SAFETY=    1.869 AND OCCURS AT RADIUS =    51.185


AT POINT (    50.000,   175.000) UNDER SEEPAGE    1,THE RADIUS AND THE CORRESPONDING FACTOR OF SAFETY ARE:
      61.264    1.954      56.011    1.847      50.758   1.906      45.505    2.012      40.253    2.288
      59.513    1.882      57.762    1.854      54.260   1.854      52.509    1.870
   LOWEST FACTOR OF SAFETY=    1.847 AND OCCURS AT RADIUS =    56.011


AT POINT (    50.000,   165.000) UNDER SEEPAGE    1,THE RADIUS AND THE CORRESPONDING FACTOR OF SAFETY ARE:
      51.273    2.040      46.018    1.972      40.764   2.036      35.509    2.169      30.255    2.640
      49.521    1.997      47.770    1.980      44.267   1.970      42.515    1.986
   LOWEST FACTOR OF SAFETY=    1.970 AND OCCURS AT RADIUS =    44.267



FOR PIEZOMETRIC LINE NO.    1

AT POINT (    50.000,   170.000),RADIUS    51.015
   THE MINIMUM FACTOR OF SAFETY IS    1.822
```

```
CROSS SECTION IN DISTORTED SCALE
NUMERALS INDICATE BOUNDARY LINE NO. IF THERE ARE MORE THAN 10
BOUND. LINES,ALPHABETS WILL THEN BE USED. P INDICATES
PIZOMETRIC LINE. IF A PORTION OF PIEZOMETRIC LINE COINCIDES WITH
THE GROUND OR ANOTHER BOUNDARY LINE,ONLY THE GROUND OR BOUNDARY
LINE WILL BE SHOWN. X INDICATES INTERSECTION OF TWO BOUNDARY
LINES.  * INDICATES FAILURE SURFACE.
THE MINIMUM FACTOR OF SAFETY IS      1.822
1.840E 02 X+++++++++X+++++++++X+++++++++X+++++++++X+++++++++X
          +                  +                            +
          +                  +                            +
          +                  +                            +
          +                  +                            +
1.760E 02 X        +         X         +         +         X
          +                  +                            +
          +                  +                            +
          +                  +                            +
          +                9999                           +
1.680E 02 X        +         X         +         +         X
          +              9 +                              +
          +        9999999X  +  9                         +
          +               P +                             +
          +               P+                              +
1.600E 02 X        +    8   X  9       +         +         X
          +                +P                             +
          +                + P                            +
          +           8    +  P 9                         +
          +            7   +   P                          +
1.520E 02 X        +    7   X   P       +         +        X
          +            X    +     9                       +
          +             7  +                              +
          +               +*                              +
          +          8    +         9                     +
1.440E 02 X        +    7 X         +         +            X
          +          8   +                                +
          +                +                              +
          +        8X6666666X77777777X999999999999        +
          +                +                              +
1.360E 02 X        +        X *      +         +           X
          +                +                              +
          +                +                              +
          +                +                              +
          +                +                              +
1.280E 02 X      44444444X4X   *     +         +           X
          +        X5X333    *                            +
          +        +      3333                            +
          +        +   *   *  333                         +
          +     2  +    *  *    3333                      +
1.200E 02 X        +    X    * * +       333+             X
          +        +    *                   3             +
          +    222222X1111                               +
          +        +11111                                +
          +        +    11111                             +
1.120E 02 X        +    X      +11111    +                X
          +        +                1111                  +
          +        +                   1                  +
          +        +                                      +
          +        +                                      +
1.040E 02 X+++++++++X+++++++++X+++++++++X+++++++++X+++++++++X
       -1.60E 02 -8.00E 01  0.0        8.00E 01  1.60E 02  2.40E 02
```

9.7 BASIC VERSION

To save computer storage space and avoid repeating input data for both static and seismic cases, the BASIC version is different from the FORTRAN version in the following ways.

1. Either the normal method or the simplified Bishop method can be used in FORTRAN but only the simplified Bishop method is employed in BASIC.
2. Several seepage conditions can be considered at the same time in FORTRAN but only one condition at a time in BASIC.
3. Several cases of different slopes can be analyzed in one computer run in FORTRAN but only one case at a time in BASIC. However, BASIC can analyze both static and seismic factors of safety at the same time.
4. Grid, automatic search, and one or more centers can be specified in FORTRAN but only grid and automatic search in BASIC.
5. Every circle with its radius and factor of safety is printed in FORTRAN but only portions of the circles are printed in BASIC.

The BASIC program requires a storage of 51 blocks, excluding the array area. The storage can be reduced if the plot subroutine is removed by simply deleting statements 2735 to 2790 and 4725 to 6255.

The BASIC program can be used in either the interactive or the batch mode. No matter which mode is used, a file name consisting of not more than six characters must be assigned. When the computer asks "READ FROM FILE?," always input zero when no data has been previously supplied to the file, so the interactive mode will be activated. In the interactive mode, a series of questions are asked by the computer so the user can enter the necessary data. These data will be placed in the specified file. If the user makes a mistake, he or she can still continue inputting the data into the file until the computer asks "CONTINUE?" The user can then enter 0 to stop the program, correct a mistake in the file, and run the program again. When the computer asks "READ FROM FILE?" in the second run, always enter 1 so the batch mode will be used. The use of a file can save a great deal of time in repeating the input data. Batch mode can be used whenever data are stored in a file.

When the computer asks "NO. OF STATIC AND SEISMIC CASES?," enter 1 if only the static or only one seismic case is desired. Otherwise, input the number of cases including both static and seismic. Before starting each case, the program will ask for seismic coefficient. Input 0 if the static factor of safety is desired.

The input and output for example 1, using the interactive mode and including both the static and the seismic cases, and for example 2, using the batch mode for the static case, are listed in the following pages.

```
TITLE - ?EXAMPLE NO. 1 BY REAME

FILE NAME ?CASE1
READ FROM FILE?(ENTER 1 WHEN READ FROM FILE & 0 WHEN NOT) ?0

NO. OF STATIC AND SEISMIC CASES- ?2

CASE NO. 1    SEISMIC COEFFICIENT= ?0

NUMBER OF BOUNDARY LINES - ?2

NO. OF POINTS ON BOUNDARY LINE 1 = ?4

BOUNDARY LINE - 1
  1 X-COORDINATE= ?4
    Y-COORDINATE= ?32
  2 X-COORDINATE= ?10
    Y-COORDINATE= ?20
  3 X-COORDINATE= ?10
    Y-COORDINATE= ?0
  4 X-COORDINATE= ?70
    Y-COORDINATE= ?0

NO. OF POINTS ON BOUNDARY LINE 2 = ?2

BOUNDARY LINE - 2
  1 X-COORDINATE= ?4
    Y-COORDINATE= ?32
  2 X-COORDINATE= ?70
    Y-COORDINATE= ?0


LINE NO. AND SLOPE OF EACH SEGMENT ARE:
  1           -2  99999  0
  2           -0.484848

MIN. DEPTH OF TALLEST SLICE= ?5

NO. OF RADIUS CONTROL ZONES- ?1

RADIUS DECREMENT FOR ZONE 1 = ?0

NO. OF CIRCLE FOR ZONE 1 = ?5

ID NO. FOR FIRST CIRCLE FOR ZONE 1 = ?1

NO. OF BOTTOM LINES FOR ZONE 1 = ?1

INPUT LINE NO.,BEGIN PT. NO.,AND END PT. NO. FOR ZONE 1
EACH LINE ON ONE LINE & EACH ENTRY SEPARATED BY COMMA
  ?1,1,4

INPUT COHESION, FRIC. ANGLE, UNIT WT. OF SOIL
EACH SOIL ON ONE LINE & EACH ENTRY SEPARATED BY COMMA
  ?0,35,125

ANY SEEPAGE? (ENTER 0 WITHOUT SEEPAGE, 1 WITH PHREATIC
SURFACE, AND 2 WITH PORE PRESSURE RATIO) ?0

ANY SEARCH?(ENTER 0 WITH GRID AND 1 WITH SEARCH) ?0

NO. OF SLICES= ?10

NO. OF ADD. RADII= ?3
```

INPUT COORD. OF GRID POINTS 1,2,AND 3

POINT 1 X-COORDINATE = ?110
        Y-COORDINATE = ?130

POINT 2 X-COORDINATE = ?90
        Y-COORDINATE = ?150

POINT 3 X-COORDINATE = ?40
        Y-COORDINATE = ?50

X INCREMENT= ?4

Y INCREMENT= ?8

NO. OF DIVISIONS BETWEEN POINTS 1 AND 2= ?2

NO. OF DIVISIONS BETWEEN POINTS 2 AND 3= ?5
CONTINUE?(ENTER 1 FOR CONTINUING AND 0 FOR STOP ?1

AUTOMATIC SEARCH WILL FOLLOW AFTER GRID

AT POINT (110 130)THE RADIUS AND FACTOR OF SAFETY ARE:
 136.015   1000000
LOWEST FACTOR OF SAFETY = 1000000 AND OCCURS AT RADIUS = 136.015

AT POINT (100 140)THE RADIUS AND FACTOR OF SAFETY ARE:
 143.178   1000000
LOWEST FACTOR OF SAFETY = 1000000 AND OCCURS AT RADIUS = 143.178

AT POINT (90 150)THE RADIUS AND FACTOR OF SAFETY ARE:
 146.014   1000000
LOWEST FACTOR OF SAFETY = 1000000 AND OCCURS AT RADIUS = 146.014

AT POINT (100 110)THE RADIUS AND FACTOR OF SAFETY ARE:
 114.018   1000000
LOWEST FACTOR OF SAFETY = 1000000 AND OCCURS AT RADIUS = 114.018

AT POINT (90 120)THE RADIUS AND FACTOR OF SAFETY ARE:
 121.655   1.50328
 120.665   1000000
LOWEST FACTOR OF SAFETY = 1.50038 AND OCCURS AT RADIUS = 121.408

AT POINT (80 130)THE RADIUS AND FACTOR OF SAFETY ARE:
 124.016   1000000
LOWEST FACTOR OF SAFETY = 1000000 AND OCCURS AT RADIUS = 124.016

```
AT POINT ( 90   90 )THE RADIUS AND FACTOR OF SAFETY ARE:
 92.1954      1000000
LOWEST FACTOR OF SAFETY = 1000000 AND OCCURS AT RADIUS = 92.1954




AT POINT ( 80  100 )THE RADIUS AND FACTOR OF SAFETY ARE:
 100.499      1.53412
 99.2678      1.51655
 98.0369      1000000
LOWEST FACTOR OF SAFETY = 1.51213 AND OCCURS AT RADIUS = 98.9601




AT POINT ( 70  110 )THE RADIUS AND FACTOR OF SAFETY ARE:
 102.176      1000000
LOWEST FACTOR OF SAFETY = 1000000 AND OCCURS AT RADIUS = 102.176




AT POINT ( 80   70 )THE RADIUS AND FACTOR OF SAFETY ARE:
 70.7107      1000000
LOWEST FACTOR OF SAFETY = 1000000 AND OCCURS AT RADIUS = 70.7107




AT POINT ( 70   80 )THE RADIUS AND FACTOR OF SAFETY ARE:
 80           1.59461
 78.397       1.5657
 76.7941      1.53622
 75.1911      1000000
LOWEST FACTOR OF SAFETY = 1.53622 AND OCCURS AT RADIUS = 76.7941




AT POINT ( 60   90 )THE RADIUS AND FACTOR OF SAFETY ARE:
 80.6226      1000000
LOWEST FACTOR OF SAFETY = 1000000 AND OCCURS AT RADIUS = 80.6226




AT POINT ( 70   50 )THE RADIUS AND FACTOR OF SAFETY ARE:
 50           1.59461
 48.9981      1000000
LOWEST FACTOR OF SAFETY = 1.58022 AND OCCURS AT RADIUS = 49.4991




AT POINT ( 60   60 )THE RADIUS AND FACTOR OF SAFETY ARE:
 60           1.71512
 57.9252      1.66477
 55.8504      1.61261
 53.7757      1000000
LOWEST FACTOR OF SAFETY = 1.57223 AND OCCURS AT RADIUS = 54.2944
```

```
AT POINT ( 50   70 )THE RADIUS AND FACTOR OF SAFETY ARE:
  59.6657         1.57943
  58.5849         1000000
LOWEST FACTOR OF SAFETY = 1.5599 AND OCCURS AT RADIUS = 58.8551



AT POINT ( 60   30 )THE RADIUS AND FACTOR OF SAFETY ARE:
  30              1.84658
  28.5263         1.77467
  27.0527         1000000
LOWEST FACTOR OF SAFETY = 1.71817 AND OCCURS AT RADIUS = 27.4211



AT POINT ( 50   40 )THE RADIUS AND FACTOR OF SAFETY ARE:
  40              1.99034
  37.4534         1.89721
  34.9068         1.79686
  32.3603         1.68854
  29.8137         1000000
LOWEST FACTOR OF SAFETY = 1.68854 AND OCCURS AT RADIUS = 32.3603



AT POINT ( 40   50 )THE RADIUS AND FACTOR OF SAFETY ARE:
  40.2492         1.776
  38.5799         1.71543
  36.9105         1.65215
  35.2412         1000000
LOWEST FACTOR OF SAFETY = 1.63588 AND OCCURS AT RADIUS = 36.4932


AT POINT ( 90   120 )RADIUS 121.408


THE MINIMUM FACTOR OF SAFETY IS 1.50038



AT POINT ( 90   120 )THE RADIUS AND FACTOR OF SAFETY ARE:
  121.655         1.50328
  120.665         1000000
LOWEST FACTOR OF SAFETY = 1.50038 AND OCCURS AT RADIUS = 121.408



AT POINT ( 94   120 )THE RADIUS AND FACTOR OF SAFETY ARE:
  122.376         1000000
LOWEST FACTOR OF SAFETY = 1000000 AND OCCURS AT RADIUS = 122.376



AT POINT ( 86   120 )THE RADIUS AND FACTOR OF SAFETY ARE:
  120.283         1.50858
  119.218         1000000
LOWEST FACTOR OF SAFETY = 1.49909 AND OCCURS AT RADIUS = 119.484
```

```
AT POINT ( 82  120 )THE RADIUS AND FACTOR OF SAFETY ARE:
  117.593       1000000
LOWEST FACTOR OF SAFETY = 1000000 AND OCCURS AT RADIUS = 117.593



AT POINT ( 86  128 )THE RADIUS AND FACTOR OF SAFETY ARE:
  126.254       1000000
LOWEST FACTOR OF SAFETY = 1000000 AND OCCURS AT RADIUS = 126.254



AT POINT ( 86  112 )THE RADIUS AND FACTOR OF SAFETY ARE:
  113.137       1.51345
  112.062       1000000
LOWEST FACTOR OF SAFETY = 1.50326 AND OCCURS AT RADIUS = 112.33



AT POINT ( 87  120 )THE RADIUS AND FACTOR OF SAFETY ARE:
  120.967       1.51126
  119.852       1000000
LOWEST FACTOR OF SAFETY = 1.50139 AND OCCURS AT RADIUS = 120.131



AT POINT ( 85  120 )THE RADIUS AND FACTOR OF SAFETY ARE:
  119.604       1.50593
  118.587       1000000
LOWEST FACTOR OF SAFETY = 1.49986 AND OCCURS AT RADIUS = 119.095



AT POINT ( 86  122 )THE RADIUS AND FACTOR OF SAFETY ARE:
  121.754       1.50377
  120.755       1000000
LOWEST FACTOR OF SAFETY = 1.50084 AND OCCURS AT RADIUS = 121.504



AT POINT ( 86  118 )THE RADIUS AND FACTOR OF SAFETY ARE:
  118.828       1.51372
  117.694       1.50006
  116.56        1000000
LOWEST FACTOR OF SAFETY = 1.50006 AND OCCURS AT RADIUS = 117.694


AT POINT ( 86  120 )RADIUS 119.484


THE MINIMUM FACTOR OF SAFETY IS 1.49909


ANY PLOT?(ENTER 0 FOR NO PLOT AND 1 FOR PLOT) ?1

YOU MAY LIKE TO ADVANCE PAPER TO THE TOP OF NEXT PAGE
SO THE ENTIRE PLOT WILL FIT IN ONE SINGLE PAGE.
FOR THE PROGRAM TO PROCEED, HIT THE RETURN KEY.
AFTER PLOT, YOU MAY LIKE TO ADVANCE PAPER TO NEXT PAGE
AND HIT THE RETURN KEY AGAIN
  ?
```

```
EXAMPLE NO. 1 BY REAME
FOR SEISMIC COEFFICIENT OF 0
AT POINT ( 86  120 )RADIUS 119.484
THE MINIMUM FACTOR OF SAFETY IS 1.49909


   36    X+++++++++X+++++++++X+++++++++X+++++++++X+++++++++X
         +                                                 +
         +                                                 +
         +                                                 +
         +                                                 +
   32    X   X2     +         +         +         +         X
         +     2                                           +
         +    *2                                           +
         +      2                                          +
         +    * 2                                          +
   28    X   1     2+        +         +         +          X
         +      *  2                                        +
         +       *  2                                       +
         +          2                                       +
         +       *  2                                       +
   24    X   1    *   2      +         +         +          X
         +              2                                   +
         +       *    2                                     +
         +        *    2                                    +
         +             2                                    +
   20    X     1   +  *    2+        +         +            X
         +            *    2                                +
         +            *     2                               +
         +             *     2                              +
         +              *    2                              +
   16    X         +       + 2       +         +            X
         +               *    2                             +
         +              *      2                            +
         +              *     2                             +
         +               *   2                              +
   12    X         +       + *   2+        +                X
         +                  *    2                          +
         +                 **    2                          +
         +                  *     2                         +
         +                 *     2                          +
    8    X         +         + * +   2     +                X
         +                    **    2                       +
         +                   *    2                         +
         +                  **    2                         +
         +                  **  2                           +
    4    X         +         +       ** 2+        X
         +                              **2                 +
         +                              *2                  +
         +                               2                  +
         +                               2                  +
    0    X+++++1111X111111111X111111111X111111111X111X++++X
         +                                                 +
         +                                                 +
         +                                                 +
         +                                                 +
   -4    X+++++++++X+++++++++X+++++++++X+++++++++X+++++++++X
         0        16        32        48        64        80
```

```
CASE NO. 2    SEISMIC COEFFICIENT= ?0.1


AUTOMATIC SEARCH WILL FOLLOW AFTER GRID


AT POINT ( 110   130 )THE RADIUS AND FACTOR OF SAFETY ARE:
   136.015       1000000
LOWEST FACTOR OF SAFETY = 1000000 AND OCCURS AT RADIUS = 136.015



AT POINT ( 100   140 )THE RADIUS AND FACTOR OF SAFETY ARE:
   143.178       1000000
LOWEST FACTOR OF SAFETY = 1000000 AND OCCURS AT RADIUS = 143.178



AT POINT ( 90   150 )THE RADIUS AND FACTOR OF SAFETY ARE:
   146.014       1000000
LOWEST FACTOR OF SAFETY = 1000000 AND OCCURS AT RADIUS = 146.014



AT POINT ( 100   110 )THE RADIUS AND FACTOR OF SAFETY ARE:
   114.018       1000000
LOWEST FACTOR OF SAFETY = 1000000 AND OCCURS AT RADIUS = 114.018



AT POINT ( 90   120 )THE RADIUS AND FACTOR OF SAFETY ARE:
   121.655       1.18965
   120.665       1000000
LOWEST FACTOR OF SAFETY = 1.18718 AND OCCURS AT RADIUS = 121.408



AT POINT ( 80   130 )THE RADIUS AND FACTOR OF SAFETY ARE:
   124.016       1000000
LOWEST FACTOR OF SAFETY = 1000000 AND OCCURS AT RADIUS = 124.016


AT POINT ( 90   90 )THE RADIUS AND FACTOR OF SAFETY ARE:
   92.1954       1000000
LOWEST FACTOR OF SAFETY = 1000000 AND OCCURS AT RADIUS = 92.1954



AT POINT ( 80   100 )THE RADIUS AND FACTOR OF SAFETY ARE:
   100.499       1.21593
   99.2678       1.20096
   98.0369       1000000
LOWEST FACTOR OF SAFETY = 1.19719 AND OCCURS AT RADIUS = 98.9601
```

```
AT POINT ( 70   110 )THE RADIUS AND FACTOR OF SAFETY ARE:
  102.176         1000000
LOWEST FACTOR OF SAFETY = 1000000 AND OCCURS AT RADIUS = 102.176




AT POINT ( 80   70 )THE RADIUS AND FACTOR OF SAFETY ARE:
  70.7107         1000000
LOWEST FACTOR OF SAFETY = 1000000 AND OCCURS AT RADIUS = 70.7107




AT POINT ( 70   80 )THE RADIUS AND FACTOR OF SAFETY ARE:
  80              1.26736
  78.397          1.24279
  76.7941         1.21771
  75.1911         1000000
LOWEST FACTOR OF SAFETY = 1.21771 AND OCCURS AT RADIUS = 76.7941




AT POINT ( 60   90 )THE RADIUS AND FACTOR OF SAFETY ARE:
  80.6226         1000000
LOWEST FACTOR OF SAFETY = 1000000 AND OCCURS AT RADIUS = 80.6226




AT POINT ( 70   50 )THE RADIUS AND FACTOR OF SAFETY ARE:
  50              1.26736
  48.9981         1000000
LOWEST FACTOR OF SAFETY = 1.25514 AND OCCURS AT RADIUS = 49.4991




AT POINT ( 60   60 )THE RADIUS AND FACTOR OF SAFETY ARE:
  60              1.3695
  57.9252         1.32687
  55.8504         1.28264
  53.7757         1000000
LOWEST FACTOR OF SAFETY = 1.24835 AND OCCURS AT RADIUS = 54.2944




AT POINT ( 50   70 )THE RADIUS AND FACTOR OF SAFETY ARE:
  59.6657         1.25446
  58.5849         1000000
LOWEST FACTOR OF SAFETY = 1.23786 AND OCCURS AT RADIUS = 58.8551




AT POINT ( 60   30 )THE RADIUS AND FACTOR OF SAFETY ARE:
  30              1.48056
  28.5263         1.41985
  27.0527         1000000
LOWEST FACTOR OF SAFETY = 1.37208 AND OCCURS AT RADIUS = 27.4211
```

```
AT POINT ( 50   40 )THE RADIUS AND FACTOR OF SAFETY ARE:
  40              1.60175
  37.4534         1.52326
  34.9068         1.43859
  32.3603         1.347
  29.8137         1000000
LOWEST FACTOR OF SAFETY = 1.347 AND OCCURS AT RADIUS = 32.3603
```

```
AT POINT ( 40   50 )THE RADIUS AND FACTOR OF SAFETY ARE:
  40.2492         1.42097
  38.5799         1.36976
  36.9105         1.31617
  35.2412         1000000
LOWEST FACTOR OF SAFETY = 1.30239 AND OCCURS AT RADIUS = 36.4932
```

```
AT POINT ( 90   120 )RADIUS 121.408
```

```
THE MINIMUM FACTOR OF SAFETY IS 1.18718
```

```
AT POINT ( 90   120 )THE RADIUS AND FACTOR OF SAFETY ARE:
  121.655         1.18965
  120.665         1000000
LOWEST FACTOR OF SAFETY = 1.18718 AND OCCURS AT RADIUS = 121.408
```

```
AT POINT ( 94   120 )THE RADIUS AND FACTOR OF SAFETY ARE:
  122.376         1000000
LOWEST FACTOR OF SAFETY = 1000000 AND OCCURS AT RADIUS = 122.376
```

```
AT POINT ( 86   120 )THE RADIUS AND FACTOR OF SAFETY ARE:
  120.283         1.19417
  119.218         1000000
LOWEST FACTOR OF SAFETY = 1.18608 AND OCCURS AT RADIUS = 119.484
```

```
AT POINT ( 82   120 )THE RADIUS AND FACTOR OF SAFETY ARE:
  117.593         1000000
LOWEST FACTOR OF SAFETY = 1000000 AND OCCURS AT RADIUS = 117.593
```

```
AT POINT ( 86   128 )THE RADIUS AND FACTOR OF SAFETY ARE:
  126.254         1000000
LOWEST FACTOR OF SAFETY = 1000000 AND OCCURS AT RADIUS = 126.254
```

```
AT POINT ( 86  112 )THE RADIUS AND FACTOR OF SAFETY ARE:
   113.137        1.19832
   112.062        1000000
LOWEST FACTOR OF SAFETY = 1.18964 AND OCCURS AT RADIUS = 112.33




AT POINT ( 87  120 )THE RADIUS AND FACTOR OF SAFETY ARE:
   120.967        1.19646
   119.852        1000000
LOWEST FACTOR OF SAFETY = 1.18804 AND OCCURS AT RADIUS = 120.131




AT POINT ( 85  120 )THE RADIUS AND FACTOR OF SAFETY ARE:
   119.604      _ 1.19191
   118.587        1000000
LOWEST FACTOR OF SAFETY = 1.18674 AND OCCURS AT RADIUS = 119.095




AT POINT ( 86  122 )THE RADIUS AND FACTOR OF SAFETY ARE:
   121.754        1.19007
   120.755        1000000
LOWEST FACTOR OF SAFETY = 1.18758 AND OCCURS AT RADIUS = 121.504




AT POINT ( 86  118 )THE RADIUS AND FACTOR OF SAFETY ARE:
   118.828        1.19855
   117.694        1.18691
   116.56         1000000
LOWEST FACTOR OF SAFETY = 1.18691 AND OCCURS AT RADIUS = 117.694


AT POINT ( 86  120 )RADIUS 119.484


THE MINIMUM FACTOR OF SAFETY IS 1.18608


ANY PLOT?(ENTER 0 FOR NO PLOT AND 1 FOR PLOT) ?0


TITLE - ?EXAMPLE NO. 2 BY REAME

FILE NAME ?CASE2
READ FROM FILE?(ENTER 1 WHEN READ FROM FILE & 0 WHEN NOT) ?1

NO. OF STATIC AND SEISMIC CASES- ?1

CASE NO. 1    SEISMIC COEFFICIENT= ?0

NO. OF BOUNDARY LINES= 9

NO. OF POINTS ON BOUNDARY LINE 1 = 2
```

```
BOUNDARY LINE - 1
 1 X COORD.=-28              Y COORD.= 117
 2 X COORD.= 160             Y COORD.= 109

NO. OF POINTS ON BOUNDARY LINE 2 = 3

BOUNDARY LINE - 2
 1 X COORD.=-80              Y COORD.= 117
 2 X COORD.=-28             Y COORD.= 117
 3 X COORD.=-12             Y COORD.= 128.5

NO. OF POINTS ON BOUNDARY LINE 3 = 2

BOUNDARY LINE - 3
 1 X COORD.= 20             Y COORD.= 126
 2 X COORD.= 160            Y COORD.= 118

NO. OF POINTS ON BOUNDARY LINE 4 = 3

BOUNDARY LINE - 4
 1 X COORD.=-80             Y COORD.= 128
 2 X COORD.=-12             Y COORD.= 128.5
 3 X COORD.= 4              Y COORD.= 126.5

NO. OF POINTS ON BOUNDARY LINE 5 = 3

BOUNDARY LINE - 5
 1 X COORD.=-7              Y COORD.= 140
 2 X COORD.= 4              Y COORD.= 126.5
 3 X COORD.= 20             Y COORD.= 126

NO. OF POINTS ON BOUNDARY LINE 6 = 2

BOUNDARY LINE - 6
 1 X COORD.=-70             Y COORD.= 140

 2 X COORD.=-7              Y COORD.= 140

NO. OF POINTS ON BOUNDARY LINE 7 = 4

BOUNDARY LINE - 7
 1 X COORD.=-50             Y COORD.= 150
 2 X COORD.=-30             Y COORD.= 154
 3 X COORD.=-7              Y COORD.= 140
 4 X COORD.= 65             Y COORD.= 140

NO. OF POINTS ON BOUNDARY LINE 8 = 4

BOUNDARY LINE - 8
 1 X COORD.=-80             Y COORD.= 140
 2 X COORD.=-70             Y COORD.= 140
 3 X COORD.=-50             Y COORD.= 150
 4 X COORD.=-20             Y COORD.= 165

NO. OF POINTS ON BOUNDARY LINE 9 = 6

BOUNDARY LINE - 9
 1 X COORD.=-80             Y COORD.= 165
 2 X COORD.=-20             Y COORD.= 165
 3 X COORD.=-10             Y COORD.= 170
 4 X COORD.= 20             Y COORD.= 170
 5 X COORD.= 65             Y COORD.= 140
 6 X COORD.= 160            Y COORD.= 140
```

```
LINE NO. AND SLOPE OF EACH SEGMENT ARE:
  1                -4.25532E-2
  2                 0  0.71875
  3                -5.71429E-2
  4                 7.35294E-3 -0.125
  5                -1.22727 -0.03125
  6                 0
  7                 0.2 -0.608696  0
  8                 0  0.5  0.5
  9                 0  0.5  0 -0.666667  0
```

MIN. DEPTH OF TALLEST SLICE= 0
NO. OF RADIUS CONTROL ZONES= 2

RADIUS DECREMENT FOR ZONE 1 = 0
NO. OF CIRCLES FOR ZONE 1 = 5
ID NO. FOR FIRST CIRCLE FOR ZONE 1 = 1
NO. OF BOTTOM LINES FOR ZONE 1 = 2

FOR ZONE 1   LINE SEQUENCE 1
LINE NO.= 2   BEG. NO.= 1   END NO.= 2
FOR ZONE 1   LINE SEQUENCE 2
LINE NO.= 1   BEG. NO.= 1   END NO.= 2

RADIUS DECREMENT FOR ZONE 2 = 0
NO. OF CIRCLES FOR ZONE 2 = 0
ID NO. FOR FIRST CIRCLE FOR ZONE 2 = 1
NO. OF BOTTOM LINES FOR ZONE 2 = 3

FOR ZONE 2   LINE SEQUENCE 1
LINE NO.= 6   BEG. NO.= 1   END NO.= 2
FOR ZONE 2   LINE SEQUENCE 2
LINE NO.= 7   BEG. NO.= 3   END NO.= 4
FOR ZONE 2   LINE SEQUENCE 3
LINE NO.= 9   BEG. NO.= 5   END NO.= 6

SOIL NO.	COHESION	FRIC. ANGLE	UNIT WEIGHT
1	100	28	125
2	200	30	125
3	150	29	125
4	200	25	125
5	250	18	130
6	1500	10	130
7	2000	15	130
8	0	0	62.4

USE PHREATIC SURFACE
UNIT WEIGHT OF WATER= 62.4
USE GRID
NO. OF SLICES= 10   NO. OF ADD. RADII= 2

NO. OF POINTS ON WATER TABLE= 5
 1 X COORD.=-80   Y COORD.= 165
 2 X COORD.=-20   Y COORD.= 165
 3 X COORD.= 50   Y COORD.= 150
 4 X COORD.= 65   Y COORD.= 140
 5 X COORD.= 160   Y COORD.= 140

INPUT COORD. OF GRID POINTS 1,2,AND 3

POINT 1 X COORD.= 20 Y COORD.= 290
POINT 2 X COORD.= 20 Y COORD.= 170
POINT 3 X COORD.= 100 Y COORD.= 170

```
X INCREMENT= 20            Y INCREMENT= 20
NO. OF DIVISIONS BETWEEN POINTS 1 AND 2= 3
NO. OF DIVISIONS BETWEEN POINTS 2 AND 3= 2

AUTOMATIC SEARCH WILL FOLLOW AFTER GRID

****  WARNING AT NEXT CENTER  ****
MAXIMUM RADIUS IS LIMITED BY END POINT OF GROUND LINE



AT POINT ( 20   290 )THE RADIUS AND FACTOR OF SAFETY ARE:
   160.078        3.44306
   158.062        3.58652
   156.047        3.89805
   154.031        4.53717
   152.016        5.61341
LOWEST FACTOR OF SAFETY = 3.44306 AND OCCURS AT RADIUS = 160.078


****  WARNING AT NEXT CENTER  ****
MAXIMUM RADIUS IS LIMITED BY END POINT OF GROUND LINE



AT POINT ( 20   250 )THE RADIUS AND FACTOR OF SAFETY ARE:
   131.244        3.27185
   126.995        3.25883
   122.746        3.17457
   118.498        3.45714
   114.249        4.41394
LOWEST FACTOR OF SAFETY = 3.17457 AND OCCURS AT RADIUS = 122.746



AT POINT ( 20   210 )THE RADIUS AND FACTOR OF SAFETY ARE:
    94.9566       3.13644
    89.9653       3.11109
    84.974        3.17457
    79.9826       3.41777
    74.9913       4.17271
LOWEST FACTOR OF SAFETY = 3.11068 AND OCCURS AT RADIUS = 88.3015



AT POINT ( 20   170 )THE RADIUS AND FACTOR OF SAFETY ARE:
    54.9928       3.40591
    49.9942       3.661
    44.9957       4.01036
    39.9971       4.34045
    34.9986       5.36366
LOWEST FACTOR OF SAFETY = 3.40591 AND OCCURS AT RADIUS = 54.9928



AT POINT ( 60   290 )THE RADIUS AND FACTOR OF SAFETY ARE:
   176.585        2.53128
   171.268        2.35968
   165.951        2.26671
   160.634        2.57168
   155.317        3.00026
LOWEST FACTOR OF SAFETY = 2.24199 AND OCCURS AT RADIUS = 167.723
```

```
AT POINT ( 60   250 )THE RADIUS AND FACTOR OF SAFETY ARE:
   136.621        2.27377
   131.297        2.07579
   125.973        2.22164
   120.648        2.35839
   115.324        2.70294
LOWEST FACTOR OF SAFETY = 2.07006 AND OCCURS AT RADIUS = 129.522



AT POINT ( 60   210 )THE RADIUS AND FACTOR OF SAFETY ARE:
   96.6572        2.05874
   91.3258        1.97813
   85.9943        2.06741
   80.6629        2.22988
   75.3314        2.52918
LOWEST FACTOR OF SAFETY = 1.97813 AND OCCURS AT RADIUS = 91.3258



AT POINT ( 60   170 )THE RADIUS AND FACTOR OF SAFETY ARE:
   56.6934        2.00237
   51.3547        1.96402
   46.016         2.06954
   40.6774        2.26018
   35.3387        2.82682
LOWEST FACTOR OF SAFETY = 1.96402 AND OCCURS AT RADIUS = 51.3547


****  WARNING AT NEXT CENTER   ****
MAXIMUM RADIUS IS LIMITED BY END POINT OF GROUND LINE



AT POINT ( 100   290 )THE RADIUS AND FACTOR OF SAFETY ARE:
   161.555        3.63633
   159.244        4.07313
   156.933        4.80853
   154.622        6.16424
   152.311        1000000
LOWEST FACTOR OF SAFETY = 3.63633 AND OCCURS AT RADIUS = 161.555

****  WARNING AT NEXT CENTER   ****
MAXIMUM RADIUS IS LIMITED BY END POINT OF GROUND LINE



AT POINT ( 100   250 )THE RADIUS AND FACTOR OF SAFETY ARE:
   125.3          3.63881
   122.24         4.25594
   119.18         5.55569
   116.12         9.10152
   113.06         1000000
LOWEST FACTOR OF SAFETY = 3.63881 AND OCCURS AT RADIUS = 125.3


****  WARNING AT NEXT CENTER   ****
MAXIMUM RADIUS IS LIMITED BY END POINT OF GROUND LINE
```

```
AT POINT ( 100   210 )THE RADIUS AND FACTOR OF SAFETY ARE:
92.1954        4.25523
87.7564        5.6629
83.3173        9.2831
78.8782        44.8327
74.4391        1000000
LOWEST FACTOR OF SAFETY = 4.25523 AND OCCURS AT RADIUS = 92.1954




AT POINT ( 100   170 )THE RADIUS AND FACTOR OF SAFETY ARE:
58.394         10.5704
52.7152        22.9279
47.0364        3802.76
LOWEST FACTOR OF SAFETY = 10.5704 AND OCCURS AT RADIUS = 58.394


AT POINT ( 60   170 )RADIUS 51.3547


THE MINIMUM FACTOR OF SAFETY IS 1.96402


AT POINT ( 60   170 )THE RADIUS AND FACTOR OF SAFETY ARE:
56.6934        2.00237
51.3547        1.96402
46.016         2.06954
40.6774        2.26018
35.3387        2.82682
LOWEST FACTOR OF SAFETY = 1.96402 AND OCCURS AT RADIUS = 51.3547




AT POINT ( 80   170 )THE RADIUS AND FACTOR OF SAFETY ARE:
57.5437        3.06843
52.0349        3.36259
46.5262        4.20907
41.0175        6.18048
35.5087        19.495
LOWEST FACTOR OF SAFETY = 3.06843 AND OCCURS AT RADIUS = 57.5437




AT POINT ( 40   170 )THE RADIUS AND FACTOR OF SAFETY ARE:
55.8431        2.10144
50.6745        1.95592
45.5058        1.9463
40.3372        2.09956
35.1686        2.64477
LOWEST FACTOR OF SAFETY = 1.88964 AND OCCURS AT RADIUS = 48.9516




AT POINT ( 20   170 )THE RADIUS AND FACTOR OF SAFETY ARE:
54.9928        3.40591
49.9942        3.661
44.9957        4.01036
39.9971        4.34045
34.9986        5.36366
LOWEST FACTOR OF SAFETY = 3.40591 AND OCCURS AT RADIUS = 54.9928
```

```
AT POINT ( 40   190 )THE RADIUS AND FACTOR OF SAFETY ARE:
   75.825        2.14034
   70.66         2.03963
   65.495        1.98703
   60.33         2.15061
   55.165        2.55691
LOWEST FACTOR OF SAFETY = 1.9734 AND OCCURS AT RADIUS = 67.2167
```

```
AT POINT ( 40   150 )THE RADIUS AND FACTOR OF SAFETY ARE:
   35.8612       2.90805
   30.6889       3.02045
   25.5167       3.64239
   20.3445       5.03478
   15.1722       7.24139
LOWEST FACTOR OF SAFETY = 2.90805 AND OCCURS AT RADIUS = 35.8612
```

```
AT POINT ( 45   170 )THE RADIUS AND FACTOR OF SAFETY ARE:
   56.0557       1.98651
   50.8445       1.8283
   45.6334       1.88271
   40.4223       2.00221
   35.2111       2.3364
LOWEST FACTOR OF SAFETY = 1.8283 AND OCCURS AT RADIUS = 50.8445
```

```
AT POINT ( 50   170 )THE RADIUS AND FACTOR OF SAFETY ARE:
   56.2682       1.91171
   51.0146       1.82199
   45.7609       1.87587
   40.5073       1.96687
   35.2536       2.25819
LOWEST FACTOR OF SAFETY = 1.82199 AND OCCURS AT RADIUS = 51.0146
```

```
AT POINT ( 55   170 )THE RADIUS AND FACTOR OF SAFETY ARE:
   56.4808       1.91511
   51.1846       1.86873
   45.8885       1.93838
   40.5923       2.03107
   35.2962       2.47961
LOWEST FACTOR OF SAFETY = 1.86873 AND OCCURS AT RADIUS = 51.1846
```

```
AT POINT ( 50   175 )THE RADIUS AND FACTOR OF SAFETY ARE:
   61.2637       1.95391
   56.011        1.8471
   50.7582       1.90595
   45.5055       2.01218
   40.2527       2.28809
LOWEST FACTOR OF SAFETY = 1.8471 AND OCCURS AT RADIUS = 56.011
```

```
AT POINT ( 50   165 )THE RADIUS AND FACTOR OF SAFETY ARE:
51.2727        2.04047
46.0182        1.9715
40.7636        2.03561
35.5091        2.16943
30.2546        2.64037
LOWEST FACTOR OF SAFETY = 1.97044 AND OCCURS AT RADIUS = 44.2667


AT POINT ( 50   170 )RADIUS 51.0146


THE MINIMUM FACTOR OF SAFETY IS 1.82199


ANY PLOT?(ENTER 0 FOR NO PLOT AND 1 FOR PLOT) ?0
```

10
Practical Examples

10.1 APPLICATIONS TO SURFACE MINING

Both the simplified and the computerized methods presented in this book were initially developed for use in surface mining. The methods have been applied to three types of fills: bench fills, hollow fills, and refuse embankments.

The Surface Mining Control and Reclamation Act of 1977 requires the return of disturbed land to its original contours. Therefore, the bench created by surface mining must be backfilled and restored to the original slope. These benches have a horizontal solid surface and a nearly vertical highwall. As the backfill material is quite homogeneous, the most critical failure surface is cylindrical, and the REAME program applies. For such a simple case, the simplified methods can also be used to determine the minimum factor of safety.

The hollow fills include head-of-hollow fills and valley fills. A head-of-hollow fill is located at the head of a hollow, where the drainage area above the fill is minimum and water can be easily diverted away from the fill. The drainage area above a valley fill may be quite large, and diversion ditches of sufficient capacity must be installed to prevent water from entering into the fill. As long as the water is diverted from the fill, the method of stability analysis for both fills is the same. The surface mining regulations require the scalping of ground surface and the removal of top soils or other weak materials before placing the hollow fill. In eastern Kentucky where the soil overburden is thin and rock appears a few feet or even several inches below the ground surface, most of the hollow fills are placed directly on a rock surface. Consequently, the most critical failure surface is cylindrical, and the REAME program applies. Of course, this case can easily be analyzed by the simplified methods. If a thin layer of weak materials exist at the bottom of the fill, the SWASE program should be used to determine the factor of safety for plane failure and the result should be compared with that obtained from REAME.

A refuse embankment can be an embankment with no water impoundment or a refuse dam. The method of stability analysis for the former is the same as for a hollow fill, while that for the latter is similar to an earth dam. The refuse dam is simply a dam used to impound the fine refuse in the form of slurry discharged from a coal preparation plant. After the fine refuse settles, the clarified water may be reclaimed and reused in the coal preparation plant. The coarse refuse is usually used to construct the dam, although borrow soil materials may also be used. As no weak layer is allowed to exist at the bottom of a refuse dam, the most critical failure surface is cylindrical and the REAME program applies.

In this chapter, a number of examples, all related to surface mining, are presented to illustrate the use of the computer programs. For simple cases with no seepage, the factor of safety obtained by the simplified methods will also be indicated.

During the past few years, the author has served as a consultant to various mining companies in Kentucky, Virginia, and West Virginia on the stability analysis of slopes. Most of the examples presented here are real cases, even though the specific location of the project and the company involved are not mentioned for confidentiality. Due to the mountainous terrain and the good drainage, none of these fills are constructed on weak and saturated foundations. Therefore, only the effective stress analysis for long term stability is needed. The total stress analysis is presented only when the effective stress analysis fails to provide the required factor of safety.

10.2 USE OF SWASE

Only two examples will be given to illustrate the applications of the SWASE program. The surface mining regulations require the scalping of the ground surface, so there is usually no weak layer at the bottom of the fill. In most cases, it is not necessary to run the SWASE program and the use of the REAME program is sufficient.

Hollow Fill with Poor Workmanship. A hollow fill shown in Fig. 10.1 had been constructed before the Surface Mining Control and Reclamation Act went into effect. In 1980, a failure occurred in the fill, as indicated by several tension cracks and the accompanying settlements. No diversion ditch was constructed surrounding the fill, and there was a spring at the bottom of the fill. A spoil sample taken from the failure area and subjected to a direct shear test gave an effective cohesion of 160 psf (7.7 kPa) and an effective friction angle of 26.8°. The degree of natural slope is quite flat and the angle of outslope of the fill is only 16°. By using the REAME program based on the cylindrical failure surface and the shear strength from the direct shear test, a factor of safety of 1.224 is obtained, even under a pore pressure ratio of 0.5. A pore pressure ratio of 0.5 indicates that the entire fill is under water, which is quite improbable. As the factor of safety is greater than 1, the failure is not due to

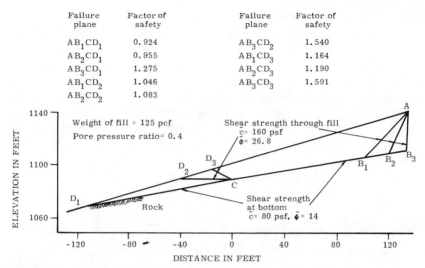

Failure plane	Factor of safety	Failure plane	Factor of safety
AB_1CD_1	0.924	AB_3CD_2	1.540
AB_2CD_1	0.955	AB_1CD_3	1.164
AB_3CD_1	1.275	AB_2CD_3	1.190
AB_1CD_2	1.046	AB_3CD_3	1.591
AB_2CD_2	1.083		

FIGURE 10.1.   Stability analysis by SWASE for fill with poor workmanship. (1 ft = 0.305 m, 1 psf = 47.9 Pa, 1 pcf = 157.1 N/m³)

improper design but rather due to poor workmanship. It was reported that, at the failure area, the original ground surface was not properly scalped. By assuming that the fill material has a cohesion of 160 psf (7.7 kPa) and a friction angle of 26.8° and that the material at the bottom of the fill has a cohesion of 80 psf (3.8 kPa) and a friction angle of 14°, or a shear strength about 50 percent of the fill material, the minimum factor of safety based on the plane failure was determined by the SWASE program. Starting from the tension crack at point A, three different failure planes are assumed, as shown by AB_1, AB_2, and AB_3. Also at point C, three failure planes shown by CD_1, CD_2, and CD_3 are assumed, thus resulting in a total of nine combinations. The shear strength along B_1C and CD_1 is assumed to be only 50 percent of that along other failure planes. Among these nine combinations, it was found that the failure surface AB_1CD_1 was most critical because of the lower shear strength along CD_1 as compared to CD_2 and CD_3. Should the same shear strength be used, the most critical failure surface would be AB_1CD_2. By assuming a pore pressure ratio of 0.4, the minimum factor of safety along AB_1CD_1 is 0.924. This may indicate that failure is not due to cylindrical failure but due to the weak material at the bottom of the fill. If the ground surface is not properly scalped or leaves and branches are left at the bottom, there should be lots of voids which will reduce greatly the shear strength. When the factor of safety is reduced to 1, failure will just begin to take place.

Because these voids will be reduced as settlements occur, it was suggested that the diversion ditch be constructed around the fill and that the surface be graded and the cracks sealed so that the pore pressure can be reduced and the factor of safety increased.

Hollow Fill on Weak Foundation. Figure 10.2 shows a hollow fill on a weak foundation. The fill material has an effective cohesion of 200 psf (9.6 kPa) and an effective friction angle of 37°, while the foundation material has an effective cohesion of 100 psf (4.8 kPa) and an effective friction angle of only 25°. The unit weight of both materials is 125 pcf (19.6 kN/m³), and there is no seepage. Because the foundation material is thin and is much weaker than the fill material, the most critical failure surface will be a series of planes through the foundation material, and the SWASE program can be applied. By assuming the failure surface as $A_1B_1C_1$, the SWASE program, based on two sliding blocks, yields a safety factor of 1.532. If the failure surface is assumed along $A_2B_2C_2D$, the factor of safety obtained by SWASE based on the weights of the three sliding blocks is 1.554. The factor of safety obtained by REAME based on the cylindrical failure is 1.629, which is not as critical as the plane failure.

10.3 USE OF REAME

Seven examples are given to illustrate the applications of the REAME program under various situations.

Three Types of Bench Fill. A stability analysis was made on three types of bench fill for possible use by a mining company, as shown in Fig. 10.3. The solid bench is assumed to be horizontal with a width of 120 ft (36.6 m), and the highwall is assumed to be vertical with a height of 67 ft (20.4 m). According to the regulations, the highwall must be backfilled to restore the terrain to the original contour. However, to support the post-mining land use plan, a 38-ft (11.6-m) wide road must be constructed along the bench. Depending on the location of the road, three different types of fill are possible. In Type-I fill, the road is located on the top of fill, leaving the top 25 ft (7.6 m) of highwall uncovered; the fill has an outslope of approximately 2h:1v, or 27°. In Type-II fill, the road is located on the solid bench at the toe of the fill, and the fill is placed on an outslope of approximately 1.25h:1v, or 39°, to the top of the highwall. In Type-III fill, the road is located at the midheight; the fill is placed on an outslope of 1.25h:1v, or 39°, below the road and 1.2h:1v, or 40°, above the road. The purpose here was to determine the factor of safety of these three types of fill.

The stability analysis was based on the following assumptions: (1) The spoil used for the fill has an effective cohesion of 200 psf (9.6 kPa), an effective friction angle of 30°, and a total unit weight of 125 pcf (196.kN/m³). (2) An effective surface drainage system was provided so there is no seepage within the fill. The REAME program based on the Fellenius or normal method was used. The factors of safety thus determined were checked by the simplified method.

Figure 10.4 shows the factors of safety of Type-I fill at nine grid points. Starting from the grid point with the lowest factor of safety of 1.669, a search was performed and a safety factor of 1.668 was found.

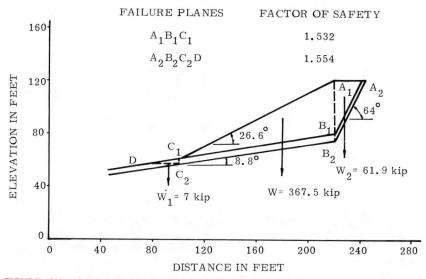

FIGURE 10.2.   Stability analysis by SWASE for fill on weak foundation. (1 ft = 0.305 m, 1 kip = 4.45 kPa)

The simplified method can be used easily to determine the factor of safety. With $\alpha = 0$, $\beta = 27°$, and $H = 42$ ft (12.8 m), the factor of safety determined from Figs. 7.12 and 7.13 is 1.71, which checks closely with the 1.67 given by REAME.

Figure 10.5 shows the factors of safety for Type-II fill. Starting from the grid point with a minimum factor of safety of 1.145, the search results in a minimum factor of safety of 1.069. The minimum factor of safety can also be determined by the simplified method using Figs. 7.12 and 7.13. The factor of safety is found to be 1.05, which also checks with the 1.07 given by REAME.

Strictly speaking, Figs. 7.12 and 7.13 cannot be applied to this case because the top of highwall limits the formation of the most critical circle. However, the difference in the factor of safety is quite small and is on the safe side.

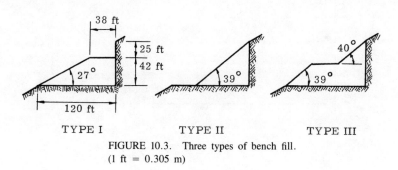

FIGURE 10.3.   Three types of bench fill. (1 ft = 0.305 m)

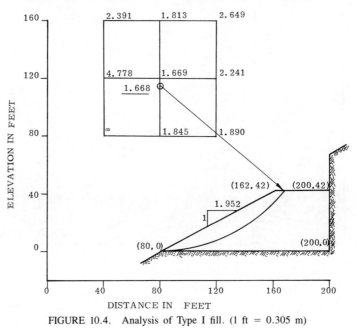

FIGURE 10.4.   Analysis of Type I fill. (1 ft = 0.305 m)

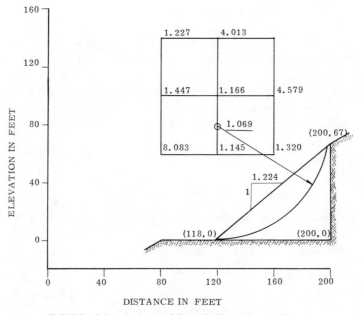

FIGURE 10.5.   Analysis of Type II fill. (1 ft = 0.305 m)

Figure 10.6 shows the factors of safety for Type-III fill at 20 grid points. Starting from the grid point with a lowest factor of safety of 1.406, a search was performed resulting in a minimum safety factor of 1.328. This is the minimum for the lower slope. Another search from the grid point with a factor of safety of 1.558 results in a minimum safety factor of 1.313. This is the absolute minimum of the embankment and also the minimum for the upper slope. It can be seen that when the two slopes are located at some distance apart, the failure surface through the entire embankment is not as critical as that through each individual slope alone.

If the minimum factor of safety is to be determined by the simplified method, two situations, one involving the upper slope alone and the other involving the two slopes combined, must be considered separately. For the upper slope with $\alpha = 0$, $\beta = 40°$, and $H = 34$ ft (10.4 m), a factor of safety of 1.27 is obtained, which checks with the 1.31 obtained from REAME. For the combined slopes, the original benched surface is approximated by a uniform slope with $\alpha = 0$, $\beta = 29°$, and $H = 67$ ft (20.4 m), a factor of safety of 1.40 is obtained, which is not as critical as the 1.27 for the upper slope.

Based on the above analysis, it can be seen that only Type-I fill achieves a safety factor greater than the 1.5 required and is therefore recommended.

Effect of Seepage on Bench Fill. A study was made on the stability of a bench fill, 40 ft (12.2 m) wide and 22 ft (6.7 m) high. The reason for conducting such a study was due to a complaint by a landlady, who lives near to the mining site, that the bench fill was unsafe because a pore pressure of zero was assumed in the stability analysis. A formal hearing was held in which there was sufficient testimony to indicate that field tests were possible and feasible on this site to get a site specific pore pressure value. The purpose of this study is to find the relationship between the height of phreatic surface and the factor of safety. The REAME computer program is particularly suited for this purpose because it can determine the factor of safety for each seepage case, all at the same time. The result of the study shows that the assumption of zero pore pressure is reasonable and that if there is a phreatic surface at the bottom of the fill it can be easily detected by field inspection.

Figure 10.7 shows the stability analysis of the bench fill. The factors of safety were determined by REAME using the simplified Bishop method. Five cases of seepage were assumed. Based on an effective cohesion of 200 psf (9.6 kPa), an effective friction angle of 30°, and a total unit weight of 125 pcf (19.6 kN/m³), the minimum factors of safety for phreatic surfaces of 0, 2.5, 5, 7.5, and 10 ft (0, 0.8, 1.5, 2.3 and 3.0 m) are 2.075, 2.056, 1.942, 1.707, and 1.590, respectively.

It is concluded that the assumption of zero pore pressure for bench fills is reasonable because the factor of safety is about the same whether there is no seepage or the line of seepage is 2.5 or 5 ft (0.8 or 1.5 m) above the bottom of fill and that a factor of safety greater than 1.5 can be achieved for this particular fill, even if 10 ft (3 m) of water accumulates at the bottom.

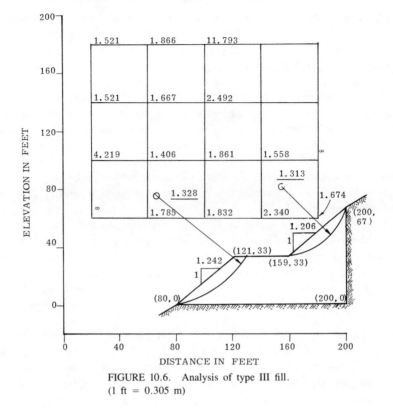

FIGURE 10.6.   Analysis of type III fill.
(1 ft = 0.305 m)

Total Stress Analysis for Bench Fill.   A stability analysis was made on a bench fill consisting of two benches. The upper bench is 55-ft (16.8-m) wide to accommodate a deep mine opening, and the lower bench is 30-ft (9.1-m) wide to allow for a coal stockpile, as shown in Fig. 10.8. After completing the deep mine operation, the benches are backfilled and restored to an original slope of 34°. It is now necessary to determine the factor of safety of the bench fill.

With an outslope of 34°, it is impossible to make an effective stress analysis while maintaining a factor of safety greater than 1.5. Therefore, a combined analysis, in which the mine spoil in the lower bench and in the top 10 ft (3.0 m) of the upper bench is compacted to at least 92 percent of the maximum proctor density, was proposed. The stability analysis is shown in Fig. 10.9. The numbers without parentheses are the factors of safety when the compaction is not controlled. It can be seen that with compaction control the minimum factor of safety is 1.518, while without compaction control the factor of safety is only 1.209.

In the analysis, the effective strength parameters with a cohesion of 80 psf (3.8 kPa) and a friction angle of 31° are used for the uncompacted zone, but the total strength parameters with a cohesion of 560 psf (26.9 kPa) and a

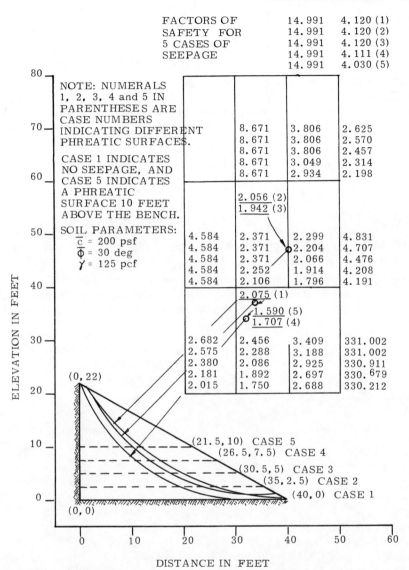

FIGURE 10.7. Effect of seepage on factor of safety. (1 ft = 0.305 m, 1 spf = 47.9 Pa, 1 pcf = 157.1 N/m³)

friction angle of 5.7° are used for the compacted zone. The Mohr's envelopes for these two analyses, one obtained from the direct shear test and the other from the unconfined and triaxial tests, are shown again in Fig. 10.10. It can be seen that when the normal stress is less than 0.5 tsf (47.9 kPa), or about 8 ft (2.4 m) of overburden, the shear strength based on the total stress is greater than that based on the effective stress, while the reverse is true when the overburden is greater than 8 ft (2.4 m). This is the reason why the total

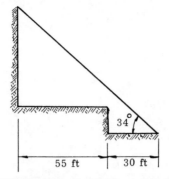

FIGURE 10.8.   Bench fill on steep slope.
(1 ft = 0.305 m)

strength can be used for the top 10 ft (3.0 m) of fill, and the effective strength used underneath. It should be pointed out that the compaction has a large effect on the total strength but little effect on the effective strength.

The effective strength parameters have been most frequently used in the stability analysis of bench fills. The parameters are based on the assumption

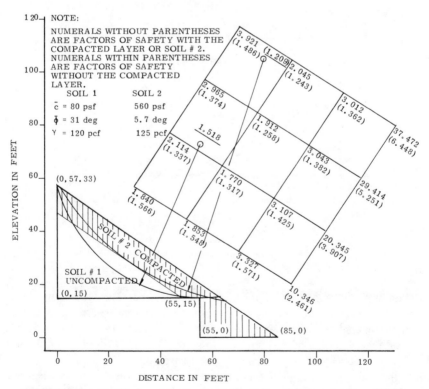

FIGURE 10.9.   Combined analysis for bench fill. (1 ft = 0.305 m, 1 psf = 47.9 Pa, 1 pcf = 157.1 N/m³)

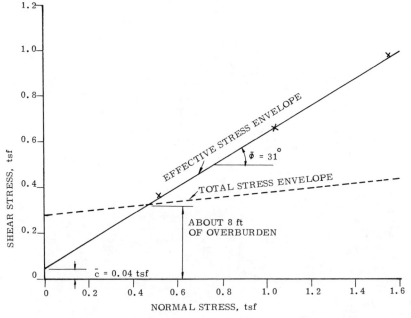

FIGURE 10.10. Shear strength of mine spoil. (1 ft = 0.305, 1 tsf = 95.8 kPa)

that the soil is completely saturated and subject to a given pore pressure. If the soil is well compacted and there is no chance that a pore pressure, other than the negative pore pressure from compaction, will be developed, a total stress analysis may be used. For this reason, a combined analysis, in which the total strength parameters are used for the compacted zone and the effective strength parameters for the uncompacted zone, is suggested.

Housing Project on Hollow Fill. A mining company planned to build multiple and single housing on a hollow fill. To ensure that the slope had an adequate factor of safety under a design pressure of 0.5 tsf (47.9 kPa), the REAME computer program was used.

Figure 10.11 shows the cross section of the hollow fill together with the results of stability analysis. A surcharge load of 1000 psf, or 0.5 tsf (47.9 kPa), was applied on the surface of the fill, where the proposed buildings are located, as indicated by the hatched lines. This surcharge was simulated by changing the ground line 2 ft (0.6 m) above the original surface.

The 2 ft (0.6 m) of material above the original surface was assumed to have a unit weight of 500 pcf (78.6 kN/m³), an effective cohesion of 0, and an effective friction angle of 0. The shear strength of the mine spoil as determined from the direct shear test was 140 psf (6.7 kPa) for cohesion and 30° for internal friction. The total unit weight is assumed to be 130 pcf (20.4 kN/m³). The results of the analysis show that the fill has a static factor of safety of 1.908 and a seismic factor of safety of 1.506. The minimum factor of safety occurs when the circle is shallow and is not affected by the housing project.

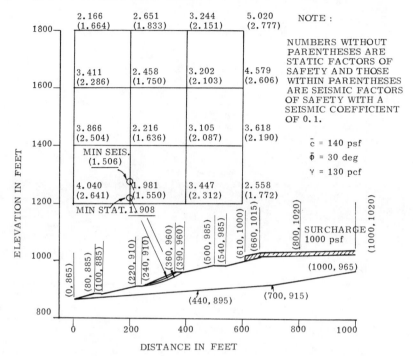

FIGURE 10.11. Stability analysis of housing project on hollow fill. (1 ft = 0.305 m, 1 psf = 47.9 Pa, 1 pcf = 157.1 N/m³)

Placement of Coarse and Dewatered Fine Refuse. A stability analysis was made on a proposed refuse embankment. The coal refuse to be disposed of includes the coarse refuse as well as the dewatered fine refuse. It was originally proposed that the coarse and fine refuse be mixed as a combined refuse to facilitate hauling. However, in view of the fact that the combined refuse has a low shear strength and cannot be made stable unless a very flat outslope or a very small height is used, it is suggested that the coarse refuse be placed on the surface of the outslope while the fine refuse be placed away from the surface. It is therefore necessary to determine the minimum width of coarse refuse on the surface so that a seismic factor of safety of 1.2 can be obtained. The seismic case is more critical because the fine refuse may be subjected to liquefaction, so an effective friction angle of only 20° is assumed.

The construction is divided into two stages. Figure 10.12 shows the stability analysis at the end of Stage II. There is a 20-ft (6.1-m) bench at the middle of the slope. The refuse area below the bench is constructed during the first stage, while that above the bench is constructed during the second

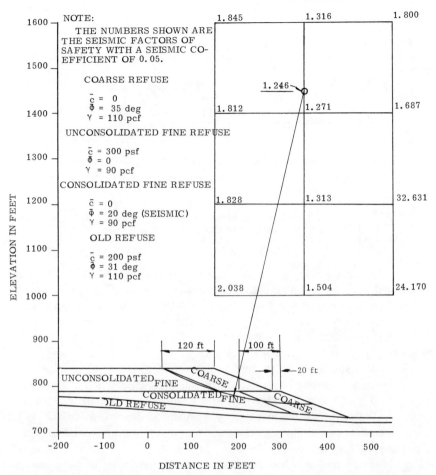

FIGURE 10.12. Stability analysis at the end of stage II. (1 ft = 0.305 m, 1 psf = 47.9 Pa, 1 pcf = 157.1 N/m³)

stage. There are four different materials in the embankment: coarse refuse, unconsolidated fine refuse, consolidated fine refuse, and old refuse located at the bottom. The shear strengths and unit weights for stability analysis are shown in the figure.

To facilitate the consolidation of the fine refuse placed during Stage I and provide drainage, it is suggested that a drainage blanket be placed at the bottom of the fill. The result of the analysis by REAME using the simplified Bishop method shows that when the coarse refuse has a width of 120 ft (36.5 m) for Stage II and 100 ft (30.5 m) for Stage I, a seismic factor of safety of 1.246 is obtained.

Rapid Drawdown. Figure 10.13 shows the cross section of a refuse fill along a creek. The fill was constructed several years ago before the proclama-

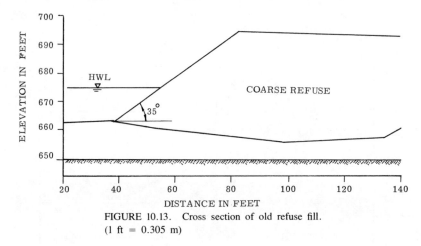

FIGURE 10.13.   Cross section of old refuse fill.
(1 ft = 0.305 m)

tion of the Surface Mining Control and Reclamation Act of 1977. The out-slope of the fill is 35°, which is much steeper than the 27° required by the current law. The height of the fill is only 32 ft (9.8 m), and the highest water level in the creek is 12 ft (3.6 m) above the toe. It was requested by the regulatory agency that a stability analysis of the fill be made to determine whether any remedial action was needed.

The shear strength of the refuse was determined from a direct shear test. An effective cohesion of 180 psf (8.6 kPa), an effective friction angle of 34°, and a total unit weight of 130 pcf (20.4 kN/m³) were used in the stability analysis. The results of the analysis by REAME are shown in Fig. 10.14. In view of the fact that the most critical situation occurs during rapid drawdown, five cases of seepage conditions were considered. Case 1 with a factor of safety of 1.619 applies when the water level in the creek is quite low, as indicated by a phreatic surface at an elevation of 658 ft (200 m); while case 5 with a factor of safety of 1.292 applies to full rapid drawdown. Between cases 1 and 5 are the partial drawdown situations, as indicated by different phreatic surfaces. As can be seen from the figure, the minimum factor of safety are 1.619 for case 1 with no rapid drawdown, 1.427 for case 2, 1.392 for case 3, 1.368 for case 4, and 1.292 for case 5 with full rapid drawdown.

In considering rapid drawdown, the phreatic surface within the slope is shown by the dashed line, while that outside the slope is along the slope and the ground surface. Whether the case of rapid drawdown should be taken into consideration depends on the permeability of the fill and the duration of the flood. If the fill is very permeable, the phreatic surface will recede as fast as the flood and no rapid drawdown need be considered. On the other hand, if the fill is very impermeable and the duration of the flood is very short, a phreatic surface cannot be developed within the slope, and the situation of rapid drawdown may never occur. The stability analysis has shown that the

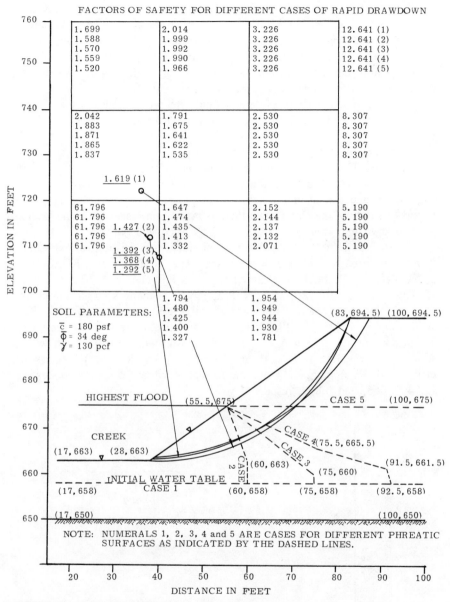

FIGURE 10.14. Stability analysis of fill subject to rapid drawdown. (1 f t = 0.305 m, 1 psf = 47.9 Pa, 1 pcf = 157.1 N/m³)

factor of safety is 1.619 without rapid drawdown but is reduced to 1.292 with full rapid drawdown. The factor of safety will lie between these two extremes, depending on the degree of rapid drawdown.

The permeability test on a specimen compacted to the same density as the refuse in the field shows that the refuse has a permeability of 0.08 ft/day (2.8 × 10⁻⁵ cm/sec), which is classified as a low permeability material. A simple calculation based on the actual hydraulic gradient indicates that it takes about 30 days for the water to travel 8 ft (2.4 m) so that the phreatic surface of case 2 can be fully developed within the most critical circle. As the duration of flood only lasts for 5 or 6 days, the case 2 phreatic surface cannot be fully developed. The actual factor of safety should lie between cases 1 and 2 with an average of 1.523, which is greater than the 1.5 required.

Tension Cracks. Figure 10.15 shows a slope 40 ft (12.2 m) high with a tension crack 20-ft (6.1 m) deep. The coordinates at the center of the failure circle are (84, 88). The soil has a cohesion of 800 psf (38.3 kPa), a friction angle of 0, and a total unit weight of 125 pcf (19.6 kN/m³). Determine the factors of safety of the circle when there is no tension crack, when the tension crack is dry, and when the tension crack is filled with water.

Beginning at the bottom of the tension crack, line 2 is drawn. Both lines 1 and 2 are used for radius control by specifying NOL (1) = 2, RDEC(1) = 0, NCIR(1) = 1, INFC(1) = 1, LINO(1,1) = 1, NBP(1,1) = 1, NEP(1,1) = 2, LINO(2,1) = 2, NBP (2,1) = 1, NEP(2,1) = 2. If there is no tension crack, soil 2 is assigned the same parameters as soil 1, i.e., c = 800 psf (38.3 kPa), ϕ = 0 and γ = 125 pcf (19.6 kN/m³). If the tension crack is dry, soil 2 is assigned c = 0, ϕ = 0, and γ = 0. If the tension crack is filled with water, soil 2 is assigned c = 0, ϕ = 0, and γ = 62.4 pcf (9.8 kN/m³). The results of the REAME program show that the factors of safety for the three cases are 1.024, 0.966, and 0.899, respectively.

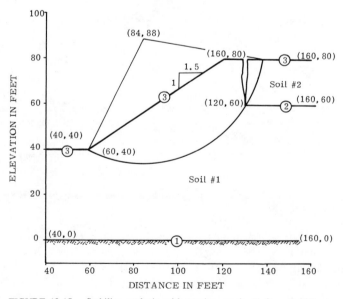

FIGURE 10.15. Stability analysis with tension crack. (1 ft = 0.305 m)

Part IV
Other Methods of Stability Analysis

11
Methods for Homogeneous Slopes

11.1 FRICTION CIRCLE METHOD

The normal and the simplified Bishop methods presented so far are based on the method of slices. However, the friction circle method, originally proposed by Taylor (1937), considers the stability of the entire sliding mass as a whole. The disadvantage of the friction circle method is that it can only be applied to a homogeneous slope with a given angle of internal friction. Although this method is of limited utility, an understanding of it will give insight into the problem of stability analysis.

Figure 11.1 shows the forces in a stability analysis by the friction circle method. A circular failure arc of radius R and a concentric circle of radius $R \sin \phi_d$ are shown, where ϕ_d is the developed friction angle. Any line tangent to the inner circle must intersect the main failure circle at an obliquity ϕ_d. This inner circle is called friction circle. The forces considered in this analysis include the driving force, D, which may consist of weight, seismic force, and neutral force; the resultant force due to cohesion, C_d; and the resultant of normal and frictional forces along the failure arc, P. The magnitude and the line of application of D are known. The magnitude of C_d is cL_c/F, where L_c is the chord length and F is the unknown factor of safety. The line of application of C_d is parallel to chord AB at a distance of RL/L_c from the center of circle, where L is the arc length. To satisfy moment equilibrium, the three forces C_d, D, and P must meet at the same point. The problem is how to determine the direction of P. Once the direction of P is known, a parallelogram can be drawn and the magnitude of C_d, as well as P, determined. The direction of P cannot be determined from statics, unless a distribution of the normal stress along the failure arc is assumed.

One possible, although somewhat trivial, assumption is that all of the normal stress is concentrated at a single point along the failure arc. In such a case P is tangent to the friction circle and a lower bound of F is obtained. Another assumption is that the normal stress is concentrated entirely at the

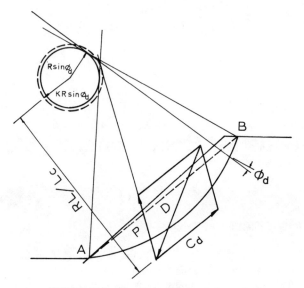

FIGURE 11.1. Forces in friction circle method.

two end points of the failure arc. In this case, the resultant of these two end forces is tangent to a circle slightly larger than the friction circle with a radius of $KR \sin \phi_d$, where K is a coefficient greater than unity, and an upper bound of F is obtained. Taylor (1937) computed the factor of safety by assuming the normal stress distributed uniformly or as a half sine curve. He found that the coefficient K depends on the central angle as shown in Fig. 11.2. Intuitively, the use of a half sine curve with a maximum normal stress at center and zero stress at both ends should provide a quite realistic value for the safety factor.

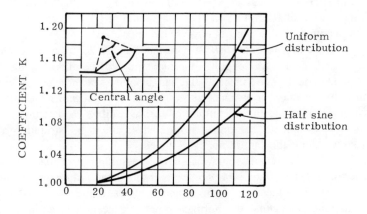

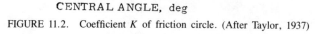

CENTRAL ANGLE, deg

FIGURE 11.2. Coefficient K of friction circle. (After Taylor, 1937)

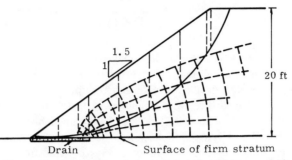

FIGURE 11.3. Slope analyzed by friction circle method. (After Whitman and Moore, 1963; 1 ft = 0.305 m)

The direction of P shown in Fig. 11.1 is based on the assumption that the forces are concentrated at the two end points.

Whitman and Moore (1963) applied different normal stress assumptions to determine the factor of safety of the slope shown in Fig. 11.3 by the friction circle method. By assuming that the soil has an effective cohesion of 90 psf (4.3 kPa), an effective friction angle of 32° and a total unit weight of 125 pcf (19.6 kN/m³), they found that the upper and lower bounds for the factor of safety are 1.60 and 1.27, respectively. Assuming that the normal effective stresses are distributed as a half sine curve, the factor of safety is 1.34.

Either an algebraic or a graphical method can be used to determine the safety factor at a given circle. First, a safety factor with respect to the friction angle, F_ϕ, is assumed and the developed friction angle determined by

$$\phi_d = \tan^{-1}(\tan \phi/F_\phi) \qquad (11.1)$$

Based on the central angle, the friction circle with a radius of $K \sin \phi_d$ can be constructed and the magnitude of C_d determined. The safety factor with respect to cohesion is

$$F_c = \frac{cL_c}{C_d} \qquad (11.2)$$

A trial-and-error procedure can be used until $F_\phi = F_d$.

11.2 LOGARITHMIC SPIRAL METHOD

In using the friction circle method or the method of slices, the distribution of forces along the failure arc or on both sides of a slice must be arbitrarily assumed. This difficulty can be overcome if a logarithmic spiral is used as a failure surface. No mater what the magnitude of normal forces on the failure surface may be, the property of the logarithmic spiral is such that the resultant of the normal and the frictional forces will always pass through the origin of the spiral. Consequently, when a moment is taken about the origin, the combined effect of normal and frictional forces is nil, and only the weight and the cohesion moments need be considered. This logarithmic spiral method was used by Taylor (1937) for the stability analysis of slopes.

In Taylor's method, the angle of internal friction of the soil was assumed fully mobilized and the developed cohesion, or the cohesion actually developed along the failure surface, was computed. A factor of safety was obtained by dividing the effective cohesion, or the maximum cohesion that could be mobilized, with the developed cohesion. The major disadvantages of this method are twofold. First, the factor of safety determined by Taylor is that with respect to cohesion, instead of that with respect to strength, i.e., both cohesion and angle of internal friction. Second, when the angle of internal friction of the soil exceeds the angle of slope, the developed cohesion becomes negative and a usable factor of safety cannot be obtained. To overcome these difficulties, Huang and Avery (1976) modified Taylor's method and determined the factor of safety with respect to shear strength. Using a factor of safety for internal friction as well as cohesion, the equation of a logarithmic spiral in polar coordinates can be expressed as

$$r = r_0 e^{\theta \tan \phi/F} \tag{11.3}$$

in which r is the radius from origin to logarithmic spiral; r_0 is the initial radius; θ is the angle between initial radius and radius r, in radians; and F is the factor of safety, which is assumed equal to 1 by Taylor but equal to or greater than 1 by Huang and Avery. If F is less than 1, Eq. 11.3 shows that the developed friction angle wil be greater than the effective friction angle, which is impossible. Therefore, if the analysis yields a factor of safety smaller than 1, the logarithmic spiral method cannot be used to determine the factor of safety with respect to strength, and Taylor's method should be used instead. This does not impose a serious limitation because a factor of safety less than 1 is unsafe and should not be used in design.

Figure 11.4 shows a logarithmic spiral passing through the toe of a simple slope with an angle, β, and a height, H. The origin, O, of the logarithmic spiral is located by two arbitrary angles t and θ. Following the procedure by Taylor (1937) and Huang and Avery (1976), the moment, M_w, about the origin due to the weight of sliding mass can be expressed as

$$M_w = \text{Function } (F, \ \theta, \ t, \ H, \ \beta, \ \gamma, \ \phi) \tag{11.4}$$

in which γ is the unit weight of soil. The moment, M_c, due to the cohesion along the logarithmic spiral failure surface is

$$M_c = \text{Function } (F, \ \theta, \ t, \ H, \ \phi, \ c) \tag{11.5}$$

If one lets $M_w = M_c$, an equation in terms of F is obtained. An iterative procedure can be used to solve F.

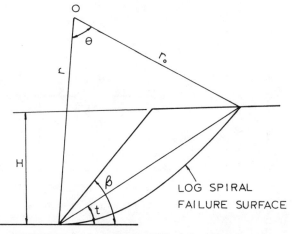

FIGURE 11.4. Logarithmic spiral passing through toe.

The logarithmic spiral passing through the toe may not have the lowest factor of safety. It is therefore necessary to examine a failure surface passing below the toe, as shown in Fig. 11.5. It can be easily proved that the most dangerous situation occurs when the origin of the logarithmic spiral lies vertically above the midheight of the slope. This is the most critical spiral for the given t and θ. If the slope is shifted to the left, as indicated by the dashed lines in Fig. 11.5, the overturning moment due to the weight of soil mass can be determined from Eq. 11.4. Because the actual slope is on the right of the dashed lines, the increase in overturning moment due to the removal of the additional weight, M'_w, can be determined in terms of F. Letting $M_w + M'_w = M_c$, an equation in the following form is obtained for solving the factor of safety, F.

$$F = \text{Function } (F, \ \theta, \ t, \ H, \ \beta, \ \gamma, \ \phi, \ c) \tag{11.6}$$

Huang and Avery (1976) developed a computer program for solving Eq. 11.6. The program consists of three DO loops; one for angle t, one for angle θ, and another for F. For a given t and θ, a value of F was assumed and a new F was computed from Eq. 11.6. By using the new F as the assumed F, another new F was obtained. The process was repeated until the difference between the new F and the assumed F became negligible. In the program, angle t starts from large to small at a specified interval, with the first angle slightly smaller than the angle of the slope. After a starting angle, θ, and a specified interval are assigned, the factors of safety for the starting θ and the next decreasing θ will be computed. If the latter is smaller than the former, the factor of safety for each successive decreasing θ will be determined until a lowest value is obtained. If greater, the movement will be in the opposite direction. Using the θ with the lowest factor of safety as a new starting angle

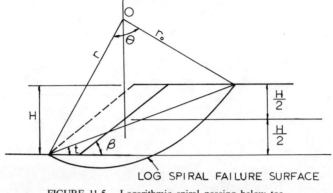

LOG SPIRAL FAILURE SURFACE

FIGURE 11.5. Logarithmic spiral passing below toe.

and an interval one-fourth of the original, the process is repeated until a new lowest factor of safety is obtained. If the lowest factor of safety for a given t is smaller than that for the previous t, computaton will proceed to the next t until a minimum factor of safety is obtained.

12
Methods for Nonhomogeneous Slopes

12.1 Earth Pressure Method

The application of earth pressure theory can be illustrated by the simple example shown in Fig. 12.1. By assuming the active force, P_A, and the passive force, P_p, as horizontal, it can be easily proved by Rankine's or Couloumb's theory that the failure plane inclines at an angle of $45° + \phi/2$ for the active wedge and $45° - \phi/2$ for the passive wedge. From basic soil mechanics

$$P_A = \tfrac{1}{2} \gamma H_A^2 \tan^2 \left(45° - \frac{\phi}{2}\right) - 2cH_A \tan \left(45° - \frac{\phi}{2}\right) \qquad (12.1)$$

$$P_P = \tfrac{1}{2} \gamma H_P^2 \tan^2 \left(45° + \frac{\phi}{2}\right) + 2cH_P \tan \left(45° + \frac{\phi}{2}\right) \qquad (12.2)$$

in which H_A and H_p are the height of active and passive wedges, respectively. The factor of safety can be determined by

$$F = \frac{cl + W \tan \phi}{P_A - P_p} \qquad (12.3)$$

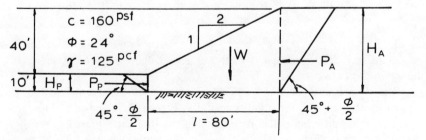

FIGURE 12.1. Active and passive earth pressures — simple case. (1 ft = 0.305 m, 1 psf = 47.9 Pa, 1 pcf = 157.1 N/m³)

in which l is the length of failure surface at middle block, and W is the weight of middle block.

For the case shown in Fig. 12.1 with $c = 160$ psf (7.7 kPa), $\phi = 24°$, $\gamma = 125$ pcf (19.6 kN/m³), $H_A = 50$ ft (15.2 m), $H_B = 10$ ft (3.0 m), and $l = 80$ ft (24.4 m)

$$W = \tfrac{1}{2} \times 80 \times (10 + 50) \times 125 = 300,000 \text{ lb } (1335 \text{ kN})$$

$$P_A = \tfrac{1}{2} \times 125 \times (50)^2 \times 0.422 - 2 \times 160 \times 50 \times \sqrt{0.422} = 55,540 \text{ lb } (247 \text{ kN})$$

$$P_P = \tfrac{1}{2} \times 125 \times (10)^2 \times 2.371 + 2 \times 160 \times 10 \times \sqrt{2.371} = 19,750 \text{ lb } (88 \text{ kN})$$

From Eq. 12.3

$$F = \frac{160 \times 80 + 300,000 \tan 24°}{55,540 - 19,750} = \frac{146,369}{35,790} = 4.090$$

The factor of safety obtained from the SWASE computer program, based on horizontal earth pressures, is 1.880, which is much smaller than the 4.090 obtained above. As shown in Fig. 8.2, the SWASE program also uses the concept of active and passive earth pressures. When $\phi_d = 0$, P_1 is equivalent to P_p, and P_2 is equivalent to P_A. However, in the SWASE program the shear resistance along the failure plane is reduced by a factor of safety, whereas no such a reduction is assumed in the above method.

The above concept was also employed by Mendez (1971) for solving a more complex case, as shown in Fig. 12.2. A computer program was developed in which the inclination of failure planes, θ_A and θ_p, and the inclination of earth pressures, α_A and α_p, can be varied. The wedge ABC is used to determine the active force, P_{A1}, in soil 1. The wedge CDF together with the weight of $BCDE$ and the active force P_{A1} is used to determine the active force, P_{A2}, in soil 2. The same procedure is applied to the passive wedge, as shown by P_{p1}, P_{p2}, and α_p in the figure. The factor of safety is determined by

$$F = \frac{cl + [W \cos \theta + P_A \sin (\alpha_A - \theta) + P_P \sin (\theta - \alpha_p)] \tan \phi}{W \sin \theta + P_A \cos (\alpha_A - \theta) - P_P \cos (\theta - \alpha_p)} \quad (12.4)$$

in which $P_A = P_{A1} + P_{A2}$, $P_p = P_{p1} + P_{p2}$, and θ is the angle of inclination of failure surface at middle block. It was found that the factor of safety is a minimum when $\alpha_A = \alpha_p = 0$.

The Department of Navy (1971) also suggests the use of active and passive earth pressures for stability analysis. Figure 12.3a shows a slope composed of three different soils. By assuming that the earth pressures are horizontal, the inclinations of failure planes in the active wedges are $45° + \phi/2$

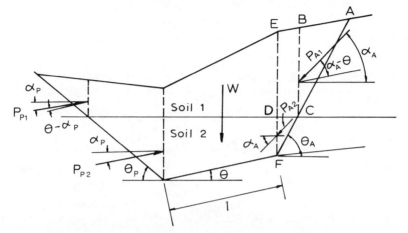

FIGURE 12.2. Active and passive earth pressures — complex case.

and those in the passive wedges are $45° - \phi/2$. For each active wedge, the driving force, D_A, based on no shear resistance along the failure plane, and the active force, P_A, are determined, as shown in Fig. 12.3b. The shear resistance, R_A, is defined as $D_A - P_A$. For each passive wedge, as shown in Fig. 12.3c, $R_p = D_p - P_p$. The factor of safety is determined by

$$F = \frac{\Sigma R_A + \Sigma R_P + c_3 l + W \tan \phi_3}{\Sigma D_A - \Sigma D_P} \qquad (12.5)$$

in which Σ is the summation over all wedges. For the case shown in Fig. 12.3a, there are three active wedges and two passive wedges.

By applying Eq. 12.5 to the case shown in Fig. 12.1, which consists of one soil with one active wedge and one passive wedge

$$W_A = \tfrac{1}{2} \times 50 \times 32.5 \times 125 = 101,560 \text{ lb } (465 \text{ kN})$$

$$D_A = 101,560 \tan (45° + 12°) = 156,390 \text{ lb } (696 \text{ kN})$$

$$W_P = \tfrac{1}{2} \times 10 \times 15.4 \times 125 = 9630 \text{ lb } (42.9 \text{ kN})$$

$$D_P = 9630 \tan (45° - 12°) = 6250 \text{ lb } (27.8 \text{ kN})$$

As determined previously, $P_A = 55,540$ lb (247 kN) and $P_p = 19,750$ lb (88 kN), so $R_A = 156,390 - 55,540 = 100,850$ lb (449 kN) and $R_p = 19,750 - 6250 = 13,550$ lb (60.1 kN). From Eq. 12.5

$$F = \frac{100,850 + 13,550 + 160 \times 80 + 300,000 \tan 24°}{156,390 - 6,250}$$

$$= \frac{260,769}{150,140} = 1.737$$

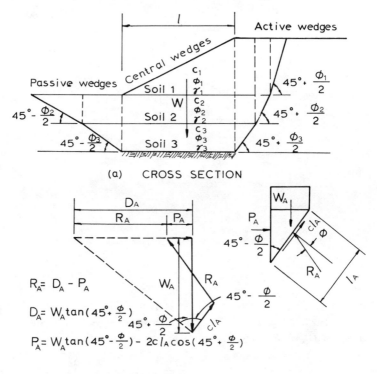

(a)   CROSS SECTION

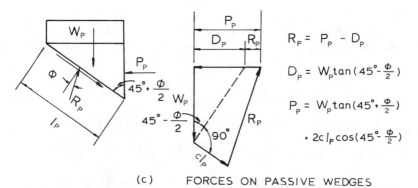

$R_A = D_A - P_A$

$D_A = W_A \tan(45° + \frac{\phi}{2})$

$P_A = W_A \tan(45° - \frac{\phi}{2}) - 2c\, l_A \cos(45° + \frac{\phi}{2})$

(b)   FORCES ON ACTIVE WEDGES

$R_P = P_P - D_P$

$D_P = W_P \tan(45° - \frac{\phi}{2})$

$P_P = W_P \tan(45° + \frac{\phi}{2})$

$+ 2c\, l_P \cos(45° - \frac{\phi}{2})$

(c)   FORCES ON PASSIVE WEDGES

FIGURE 12.3. Navy's method for plane failure. (After Department of Navy, 1971)

It can be seen that the Navy's method checks more closely with the SWASE computer program compared to the first method. The Navy's method is more reasonable because the driving and resisting forces on every block are used for determining the factor of safety, whereas the first method considers only the forces on the middle block. Drnevich (1972) developed a computer program based on the Navy's method with the exception that the active and passive forces were calculated for more general conditions. He considered the middle block as an active or passive wedge depending on the slope of the failure surface beneath it.

12.2 Janbu's Method

The method suggested by Janbu (1954, 1973) satisfies both force and moment equilibrium and is applicable to failure surfaces of any shape. Figure 12.4 shows the forces on a slice. If the location of the line of thrust is assumed, then h_t and δ can be determined, and the problem becomes statically determinate. If there are a total of n slices, the number of unknowns is $3n$, as tabulated below:

UNKNOWN	NUMBER
F (related to T)	1
N	n
ΔE	n
S	n - 1
Total	$3n$

The number of equations is also $3n$, so they can be solved statically. Based on equilibrium of vertical forces

$$N\cos\,\theta\,=\,W\,+\,\Delta S\,-\,T\,\sin\,\theta$$

or

$$N\,=\,(W\,+\,\Delta S)\,\sec\,\theta\,-\,T\,\tan\,\theta \qquad (12.6)$$

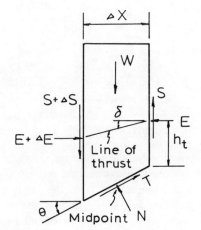

FIGURE 12.4. Forces on slice by Janbu's method.

Based on equilibrium of horizontal forces

$$\Delta E = N \sin \theta - T \cos \theta \qquad (12.7)$$

Substituting Eq. 12.6 into Eq. 12.7

$$\Delta E = (W + \Delta S) \tan \theta - T \sec \theta \qquad (12.8)$$

Based on moment equilibrium at the midpoint and considering Δx as infinitesimally small

$$S \Delta x = -E \Delta x \tan \delta + h_t \Delta E$$

or
$$S = -E \tan \delta + h_t \frac{\Delta E}{\Delta x} \qquad (12.9)$$

The requirement for overall horizontal equilibrium is

$$\Sigma \ \Delta E = 0 \qquad (12.10)$$

Substituting Eq. 12.8 into Eq. 12.10

$$\Sigma \ (W + \Delta S) \tan \theta - \Sigma \ T \sec \theta = 0 \qquad (12.11)$$

Since

$$T = \frac{c \Delta x \sec \theta + N \tan \phi}{F} \qquad (12.12)$$

Equation 12.11 becomes

$$F = \frac{\Sigma \ (c \Delta x \sec \theta + N \tan \phi) \sec \theta}{\Sigma \ (W + \Delta S) \tan \theta} \qquad (12.13)$$

Substituting Eq. 12.6 into Eq. 12.13

$$F = \frac{\Sigma \ \{c \Delta x \sec \theta + [(W + \Delta S) \sec \theta - T \tan \theta] \tan \phi\} \sec \theta}{\Sigma \ (W + \Delta S) \tan \theta} \qquad (12.14)$$

The solution procedure is as follows:

1. Assume $\Delta S = 0$, and solve the factor of safety, F, from Eq. 12.14 by a method of successive approximations. A factor of safety is first assumed, and T is computed from Eqs. 12.6 and 12.12 for each slice. Substituting T into Eq. 12.14, a new factor of safety is determined. The process is repeated until the assumed and the new factors of safety are nearly equal.

2. Substitute the final T into Eq. 12.8 to determine ΔE. Knowing ΔE, the value of E at each side of a slice can be determined by summation. Assume a reasonable line of thrust for E and find δ and h_t at each side of slice. Determine S from Eq. 12.9 and find ΔS by difference.

3. Using the ΔS thus determined, repeat step 1 and find a new factor of safety for the given ΔS.

4. Repeat steps 2 and 3 until the factor of safety converges to a specified tolerance.

Janbu's method was incorporated in a computer program developed by Purdue University for the Indiana State Highway Commission. The original program, STABL, was written by Siegel (1975) and was later revised by Boutrap (1977) and renamed STABL2. The program can generate circular failure surfaces, surfaces of sliding block character, and more general irregular surfaces of random shape. It can handle heterogeneous soil systems, anisotropic soil strength properties, excess pore water pressure due to shear, static groundwater and surface water, pseudo-static earthquake loading, and surcharge boundary loading. The original program was based on Janbu's method but the simplified Bishop method was later added for circular failure surfaces.

12.3 Morgenstern and Price's Method

The method developed by Morgenstern and Price (1965) considers not only the normal and tangential equilibrium but also the moment equilibrium for each slice. Figure 12.5 shows an arbitrary slice within a failure mass. Equilibrium is established by setting moments about the base to zero and having a

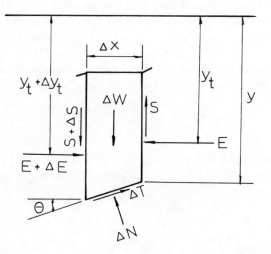

FIGURE 12.5. Forces on slice by Morgenstern and Price's method.

zero sum of the forces which are normal and tangential to the base of the slice. These relationships are

$$S = y \frac{dE}{dx} - \frac{d}{dx}(Ey_t) \tag{12.15}$$

$$\Delta N = (\Delta W - \Delta S) \cos \theta + \Delta E \sin \theta \tag{12.16}$$

$$\Delta T = (\Delta W + \Delta S) \sin \theta - \Delta E \cos \theta \tag{12.17}$$

The Mohr-Coulomb failure criterion is

$$\Delta T = \frac{c \Delta x \, \sec \theta + \Delta N (\tan \phi)}{F} \tag{12.18}$$

Combining Eqs. 12.16, 12.17, and 12.18 and let $x \to 0$

$$\frac{dE}{dx} \left(1 + \frac{\tan \phi}{F} \frac{dy}{dx} \right) + \frac{dS}{dx} \left(\frac{\tan \phi}{F} - \frac{dy}{dx} \right) = - \frac{c}{F} \left[1 + \left(\frac{dy}{dx} \right)^2 \right]$$

$$+ \frac{dW}{dx} \left(\frac{\tan \phi}{F} + \frac{dy}{dx} \right) \tag{12.19}$$

Equations 12.15 and 12.19 provide two differential equations for solving the unknown functions, E, S, and y_t. In order to complete the system of equations, it is assumed that

$$S = \lambda f(x) E \tag{12.20}$$

in which f(x) is a function of x and λ is a constant. The problem is now fully specified, and λ and F can be determined by solving Eqs. 12.15 and 12.19 that satisfy the appropriate boundary conditions. The function $f(x)$ is assumed to be linear by specifying a numerical value for f at each vertical side. An iterative procedure can be used to determine a unique value of λ, which satisfies the governing equations. The use of Eq. 12.20 makes the problem statically determinate because the number of unknowns is reduced to $3n$ as tabulated below:

UNKNOWN	NUMBER
F	1
λ	1
N	n
E	$n - 1$
y_t	$n - 1$
Total	$3n$

Morgenstern and Price's method was used in a computer program called MALE (Schiffman, 1972). The program can be used for analyzing the stability of earth and rock slopes when the failure surface is composed of a series of straight line segments. The slope may be zoned or layered with different materials in an arbitary arrangement. The program performs an effective stress stability analysis. It is possible to define an internal piezometric level as well as upstream and downstream external water surfaces. Pore water pressure calculations are based upon the piezometric level, the external water surfaces, or pore pressure ratios, r_u.

12.4 Spencer's Method

Spencer's method (1967, 1973, 1981) is very similar to Morgenstern and Price's method in that two equations are used to solve the factor of safety, F, and the angle of inclination of the interslice forces, δ. It is assumed that δ is constant for every slice. The method was used and extended by Wright (1969, 1974). Figure 12.6 shows the forces on an arbitrary slice. Summing forces along and normal to the base of the slice

$$T = W \sin \theta + (Z_2 - Z_1) \cos (\delta - \theta) \tag{12.21}$$

$$N = W \cos \theta + (Z_2 - Z_1) \sin (\delta - \theta) \tag{12.22}$$

From Mohr-Coulomb theory

$$T = \frac{c\Delta x \sec \theta + N \tan \phi}{F} \tag{12.23}$$

Substituting Eq. 12.23 into Eq. 12.21

$$\frac{c\Delta x \sec \theta + N \tan \phi}{F} = W \sin \theta + (Z_2 - Z_1) \cos (\delta - \theta) \tag{12.24}$$

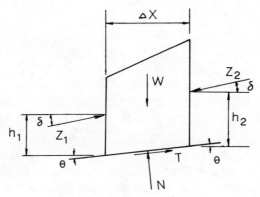

FIGURE 12.6. Forces on slice by Spencer's method.

Eliminating N from Eqs. 12.22 and 12.24 and solving for Z_2

$$Z_2 = Z_1 + \frac{c\Delta x \sec \theta - F \, W \sin \theta + W \cos \theta \tan \phi}{\cos (\delta - \theta) \, [F - \tan (\delta - \theta) \tan \phi]} \qquad (12.25)$$

Summing moments of forces about the midpoint at the base of slice

$$Z_1 \cos \delta \left(h_1 - \frac{\Delta x}{2} \tan \theta \right) + Z_1 \sin \delta \left(\frac{\Delta x}{2} \right) + Z_2 \sin \delta \left(\frac{\Delta x}{2} \right)$$

$$= Z_2 \cos \delta \left(h_2 + \frac{\Delta x}{2} \tan \theta \right) \qquad (12.26)$$

Solving Eq. 12.26 for h_2

$$h_2 = \left(\frac{Z_1}{Z_2} \right) h_1 + \frac{\Delta x}{2} (\tan \delta - \tan \theta) \left(1 + \frac{Z_1}{Z_2} \right) \qquad (12.27)$$

The boundary conditions are defined by Z_1 and h_1 for the first slice and Z_2 and h_2 for the last slice. In many cases, these values are zero. By using an assumed value for the solution parameters, F and δ, and the known boundary conditions, Z_1 and h_1, it becomes possible to use Eqs. 12.25 and 12.27 in a recursive manner, slice by slice, and evaluate Z_2 and h_2 for the last slice. The calculated values of Z_2 and h_2 at the boundary are compared with the given values. An adjustment is made to the assumed values of F and δ, and the procedure is repeated. The iterations are terminated when the calculated values of Z_2 and h_2 are within an acceptable tolerance to the known values of Z_2 and h_2 at the boundary.

This method is statically determinate because the number of unknowns is $3n$ as tabulated below:

UNKNOWN	NUMBER
F	1
δ	1
N	n
Z	$n - 1$
h_t	$n - 1$
Total	$3n$

A general computer program, SSTAB1, was developed by Wright (1974). The program can be used to compute the factor of safety for both circular and noncircular failure surfaces. For circular failure surfaces, an automatic search is available for locating a critical circle corresponding to a minimum factor of safety. The undrained shear strength and the pore pressure may be varied within the slope, and interpolations are made to determine their values along the failure surface. The program was modified by the Bureau of Reclamation and renamed SSTAB2 (Chugh, 1981). Major changes include the use of a new equation solver to facilitate covergence and the calculation of yield acceleration of a sliding mass.

12.5 Finite-Element Method

The Bureau of Mines (Wang et al., 1972) has developed a computer program for pit slope stability analysis by finite-element stress analysis and the limiting equilibrium method. Finite-element stress analysis satisfies static equilibrium and can account for the stress changes due to varied elastic properties, nonhomogeneity, and geometric shapes. In addition to the stress analysis, the program includes finite-element model mesh generation, plotting of the model mesh and stress contours, calculation of a factor of safety along a circular or a plane failure surface, and location of a critical circular failure surface.

The stress field in the slope is determined by two-dimensional plane strain analysis using triangular finite elements. Figure 12.7 shows the finite-element grid used for the analysis of a slope. Rigid boundaries are assumed at a considerable distance from the slope, so their existence has no effect on the stresses in the slope. In the finite-element method, the element stiffness matrix, which relates nodal forces to nodal displacements, is derived based on the minimization of total potential energy. These element stiffness matrices are then superimposed to form the overall stiffness matrix of the system. Given the forces or displacements at each node, the system of simultaneous equations based on the overall stiffness matrix can be solved for the displacement at each node. After the displacements are found, the stresses in each element can be determined. Details of the finite element method can be found in most textbooks on this subject (Zienkiewicz, 1971; Desai and Abel, 1972).

Figure 12.8 shows the stresses at a point on a failure surface. The stresses σ_x, σ_y, and τ_{xy} are calculated by the finite-element method. If the failure

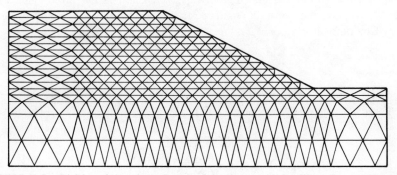

FIGURE 12.7.   Division of slope into triangular finite elements. (After Wang, Sun and Ropchan, 1972)

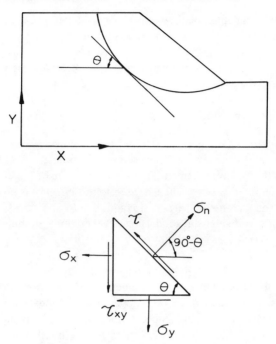

FIGURE 12.8. Stress at a point on failure surface.

plane makes an angle θ with the horizontal, the normal and shear stresses on the failure plane can be calculated by

$$\sigma_n = \tfrac{1}{2}\,(\sigma_x + \sigma_y) - \tfrac{1}{2}\,(\sigma_x - \sigma_y)\cos 2\theta + \tau_{xy}\sin 2\theta \qquad (12.28)$$

$$\tau = -\tau_{xy}\cos 2\theta - \tfrac{1}{2}\,(\sigma_x - \sigma_y)\sin 2\theta \qquad (12.29)$$

Because all the stresses along the failure surface are known from the finite-element analysis, the normal and shear stresses at every point along the failure surface can be calculated from Eqs. 12.28 and 12.29. From calculated normal stresses, the shear strength, s at all points may be obtained from Mohr-Coulomb theory, or

$$s = c + \sigma_n \tan \phi \qquad (12.30)$$

The total shear strength and the total shear force can be found by summing the shear strength and stress at all points along the failure surface. The factor of safety is

$$F = \frac{\Sigma(c + \sigma_n \tan \phi)\,\Delta l}{\Sigma\,\tau \Delta l} \qquad (12.31)$$

in which Δl is the incremental length.

In the computer program, the failure surface is divided into a number of segments of equal length. Each segment consists of two end points. The horizontal, vertical, and shear stresses at the midpoint of each segment are found by interpolating the stresses of the surrounding elements. The normal and shear stresses for each segment are then calculated from the stresses at its midpoint and the angle θ of the segment.

12.6 Probabilistic Method

Due to the large variations in the shear strength of soils, the probabilistic method has gained popularity during the past few years. The major disadvantage of the method is that a large number of tests is required to ascertain the variability of the shear strength. It is only after the variability is determined or assumed that the probability of failure can be determined.

In order to use the probabilistic method, the following statistical terms must be defined:

1. Sample mean: If x_i is the value of the ith sample and there are n sample units, then the average $\bar{X}$ is

$$\bar{X} = \frac{\sum\limits_{i=1}^{n} x_i}{n} \tag{12.32}$$

This value is called the sample mean and is the best estimate of the true or population mean, μ.

2. Variance: The variance of x, $V[x]$, is determined by

$$V[x] = \frac{\sum\limits_{i=1}^{n}(x_i - \bar{X})^2}{n-1} \tag{12.33}$$

The variance of $af_1(x) + bf_2(x)$, where a and b are constants and $f_1(x)$ and $f_2(x)$ are functions of x, is

$$V[af_1(x) + bf_2(x)] = a^2 V[f_1(x)] + b^2 V[f_2(x)] \tag{12.34}$$

3. Standard deviation: The square root of the variance is called the standard deviation, σ, or

$$\sigma = \sqrt{V[x]} \tag{12.35}$$

The standard deviation is the most common measure of sample dispersion. The larger the standard deviation, the greater the dispersion of the data about the mean.

4. Coefficient of variation: The coefficient of variation, C_v, is generally used in percentile form and is expressed as

$$C_v[x] = \left(\frac{\sigma}{\bar{X}}\right) \times 100 \qquad (12.36)$$

Table 12.1 shows the mean, standard deviation, and coefficient of variation of shear strength parameters from different sources as reported by Harr (1977). It can be seen that the variation of the friction angle in sand and gravel is much smaller than the variation of unconfined compressive strength in clay.

The distribution function most frequently used as a probability model is called the normal or Gaussian distribution. Although this symmetrical and bell-shaped distribution is very important, it should be realized that it is not the only type of distribution that can be used in the probabilistic method. The mathematical equation of normal distribution expressing the frequency of occurrence of the random variable x is

$$f(x) = \frac{1}{\sigma\sqrt{2\pi}} \exp\left[-\frac{1}{2}\left(\frac{x - \mu}{\sigma}\right)^2\right] \qquad (12.37)$$

Figure 12.9 shows a plot of normal distribution with $\sigma = 1$ and $\mu = 0$ and 4, respectively. Note that both curves are similar except that the x-coordinate is displaced by a constant distance. If the x-coordinate at peak is not equal to zero, it can be made zero by a simple displacement.

The cumulative distribution function for a normally distributed random

Table 12.1 Variability of Shear Strength Parameters

| MATERIAL | PARAMETER | | NUMBER OF SAMPLES | MEAN | STANDARD DEVIATION | COEFFICIENT OF VARIATION % |
	FRICTION ANGLE deg	UNCONFINED COMPRESSIVE STRENGTH, tsf				
Gravel	X		38	36.22	2.16	6.0
Gravelly Sand	X		81	37.33	1.97	5.3
Sand	X		73	38.80	2.80	7.0
Sand	X		136	36.40	4.05	11.0
Sand	X		30	40.52	4.56	11.0
Clay		X	279	2.08	1.02	49.1
Clay		X	295	1.68	0.69	40.9
Clay		X	187	1.49	0.59	39.6
Clay		X	53	1.30	0.62	47.7
Clay		X	231	0.97	0.26	29.0

(After Harr, 1977; 1 tsf = 95.8 kPa)

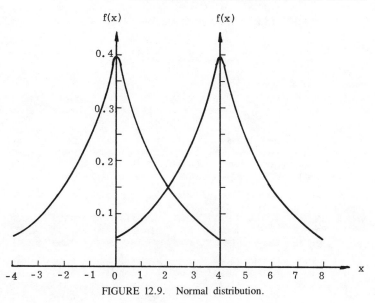

FIGURE 12.9. Normal distribution.

variable is tabulated in Table 12.2. Let u be the normal deviate, or

$$u = \frac{x - \mu}{\sigma} \qquad (12.38)$$

The area under the standardized normal distribution curve $f(u)$ between 0 and z, as shown in Fig. 12.10, can be written as

$$\psi(z) = \frac{1}{\sqrt{2\pi}} \int_0^z \exp\left(-\frac{u^2}{2}\right) du \qquad (12.39)$$

Using Eq. 12.39 and recognizing that the area under half of the standardized normal curve is $\frac{1}{2}$, the probability associated with the value of the random variable being less than any specified value can be determined.

In the deterministic methods of stability analysis, the factor of safety is defined as

$$F = \frac{s}{\tau} \qquad (12.40)$$

The shear strength, s, may be the average strength or the strength somewhat smaller than the average, and the shear stress, τ, may be the average stress or the stress somewhat greater than the average. In the probabilistic method, the stability is represented by a safety margin, defined as ρ

$$\rho = s - \tau \qquad (12.41)$$

Table 12.2   Area $\psi(z)$ under Normal Curve

z	.00	.01	.02	.03	.04	.05	.06	.07	.08	.09
0	0	.003969	.007978	.011966	.015953	.019939	.023922	.027903	.031881	.035856
.1	.039828	.043795	.047758	.051717	.055670	.059618	.063559	.067495	.071424	.075345
.2	.079260	.083166	.087064	.090954	.094835	.098706	.102568	.106420	.110251	.114092
.3	.117911	.121720	.125516	.129300	.133072	.136831	.140576	.144309	.148027	.151732
.4	.155422	.159097	.162757	.166402	.170031	.173645	.177242	.180822	.184386	.187933
.5	.191462	.194974	.198466	.201944	.205401	.208840	.212260	.215661	.219043	.222405
.6	.225747	.229069	.232371	.235653	.234914	.242154	.245373	.248571	.251748	.254903
.7	.258036	.261148	.264238	.257305	.270350	.273373	.276373	.279350	.282305	.285236
.8	.288145	.291030	.293892	.296731	.299546	.302337	.305105	.307850	.310570	.313267
.9	.315940	.318589	.321214	.323814	.326391	.328944	.331472	.333977	.336457	.338913
1.0	.341345	.343752	.346136	.348495	.350830	.353141	.355428	.357690	.359929	.362143
1.1	.364334	.366500	.368643	.370762	.372857	.374928	.376976	.379000	.381000	.382977
1.2	.384930	.386861	.388768	.390651	.392512	.393350	.396165	.397958	.399727	.401475
1.3	.403200	.404902	.406582	.408241	.409877	.411492	.413085	.414657	.416207	.417736
1.4	.419243	.420730	.422196	.423641	.425066	.426471	.427855	.429219	.430563	.431888
1.5	.433193	.434476	.435745	.436992	.438220	.439429	.440620	.441792	.442947	.444083
1.6	.445201	.446301	.447384	.448449	.449497	.450529	.451543	.452540	.453521	.454486
1.7	.455435	.456367	.457284	.458185	.459070	.459941	.460796	.461636	.462462	.463273
1.8	.464070	.464852	.465620	.466375	.467116	.467843	.468557	.469258	.469946	.470621
1.9	.471283	.471933	.472571	.473197	.473610	.474412	.475002	.475581	.476148	.476705
2.0	.477250	.477784	.478308	.478822	.479325	.479818	.480301	.480774	.481237	.481691
2.1	.482136	.482571	.482997	.483414	.483823	.484222	.484614	.484997	.485371	.485738
2.2	.486097	.486447	.486791	.487126	.487455	.487776	.488089	.488396	.488696	.488989
2.3	.489276	.489556	.489830	.490097	.490358	.490613	.490863	.491106	.491344	.491576
2.4	.491802	.492024	.492240	.492451	.492656	.492857	.493053	.493244	.493431	.493613
2.5	.493790	.493963	.494132	.494297	.494457	.494614	.494766	.494915	.495060	.495201
2.6	.495339	.495473	.495604	.495731	.495855	.495975	.496093	.496207	.496319	.496427
2.7	.496533	.496636	.496736	.496833	.496928	.497020	.497110	.497197	.497282	.497365
2.8	.497445	.497523	.497599	.497673	.497744	.497814	.497882	.497948	.498012	.498074
2.9	.498134	.498193	.498250	.498305	.498359	.498411	.498462	.498511	.498559	.498605
3.0	.498650	.498694	.498736	.498777	.498817	.498856	.598893	.498930	.498965	.498999
3.1	.499032	.499065	.499096	.499126	.499155	.499184	.499211	.499238	.499264	.499289
3.2	.499313	.499336	.499359	.499381	.499402	.499423	.499443	.499462	.499481	.499499
3.3	.499517	.499534	.499550	.499566	.499581	.499596	.499610	.499624	.499638	.499651
3.4	.499663	.499675	.499687	.499698	.499709	.499720	.499730	.499740	.499749	.499758
3.5	.499767	.499776	.499784	.499792	.499800	.499807	.499815	.499822	.499828	.499835
3.6	.499841	.499847	.499853	.499858	.499864	.499869	.499874	.499879	.499883	.499888
3.7	.499892	.499896	.499900	.400904	.499908	.499912	.499915	.499918	.499922	.499925
3.8	.499928	.499931	.499933	.499936	.499938	.499941	.499943	.499946	.499948	.499950
3.9	.499952	.499954	.499956	.499958	.499959	.499961	.499963	.499964	.499966	.499967

The normal distribution for the safety margin is

$$f(\rho) = \frac{1}{\sigma_\rho \sqrt{2\pi}} \exp\left[-\frac{1}{2}\left(\frac{\rho - \bar{\rho}}{\sigma_\rho}\right)^2 \right] \qquad (12.42)$$

in which σ_ρ is the standard deviation of the safety margin and $\bar{\rho}$ is the mean of the safety margin. From Eq. 12.34

$$\sigma_\rho = \sqrt{\sigma_s^2 + \sigma_\tau^2} \tag{12.43}$$

in which σ_s is the standard deviation of shear strength and σ_τ is the standard deviation of shear stress. From Eq. 12.41

$$\bar{\rho} = \bar{s} - \bar{\tau} \tag{12.44}$$

in which $\bar{s}$ is the mean shear strength and $\bar{\tau}$ is the mean shear stress.

As failure occurs when $\rho \leq 0$, the probability of failure becomes

$$p_f = \frac{1}{2} - \psi \left(\frac{\bar{\rho}}{\sigma_\rho} \right) \tag{12.45}$$

in which ψ is the cumulative probability function of the standard normal variate as given in Table 12.2.

Example 1: Given a slope with $\bar{s} = 1.5$ tsf (143.7 kPa), the coefficient of variation with respect to strength $C_v[s]$ is 40 percent, $\bar{\tau}$ is 0.5 tsf (47.9 kPa), and $C_v[\tau]$ is 20 percent. Determine the factor of safety by the deterministic method and the probability of failure by the probabilistic method assuming the safety margin to be normally distributed.

In the conventional deterministic method, the mean shear strength and the mean shear stress are used, so the factor of safety is $F = \bar{s}/\bar{\tau} = 1.5/0.5 = 3$. If a lower shear strength and a higher shear stress are used

$$F = \frac{\bar{s}\,(1 - C_v[s])}{\bar{\tau}\,(1 - C_v[\tau]} = \frac{1.5 \times 0.6}{0.5 \times 1.2} = 1.5$$

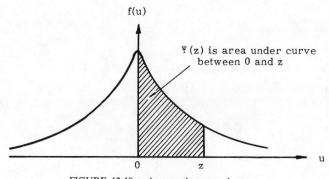

FIGURE 12.10.   Area under normal curve.

In the probabilistic method, $\sigma_s = \bar{s}C_v[s] = 1.5 \times 0.4 = 0.6$ tsf (57.5 kPa); $\sigma_\tau = \bar{\tau}C_v[\tau] = 0.5 \times 0.2 = 0.1$ tsf (9.6 kPa); from Eq. 12.43, $\sigma_\rho = \sqrt{(0.6)^2 + (0.1)^2} = 0.608$ tsf (58.2 kPa); From Eq. 12.44, $\bar{\rho} = 1.5 - 0.5 = 1.0$ tsf (95.8 kPa); so $\bar{\rho}/\sigma_\rho = 1.64$. From Eq. 12.45 and Table 12.2, the probability of failure is $p_f = 0.5 - \psi(1.64) = 0.5 - 0.4495 = 0.0505$, or approximately 5 per hundred.

Example 2: A failure surface has the following mean parameters: $c = 100$ psf (4.8 kPa), $\phi = 20°$, and length of failure surface $L = 30$ ft (9.1 m). The sum of normal and tangential components of the weight are: $\Sigma N = 12,460$ lb (55.4 kN) and $\Sigma T = 5440$ lb (24.2 kN). Calculate the conventional factor of safety. If $C_v[\tan \phi]$ is 5 percent and $C_v[c]$ is 35 percent, determine the probability of failure assuming the shear resistance to be normal variates.

The conventional factor of safety is

$$F = \frac{cL + \tan \phi \, \Sigma N}{\Sigma T} \qquad (12.46)$$

or $F = (100 \times 30 + \tan 20° \times 12,460)/5440 = 7535/5440 = 1.39$.

In the probabilistic method, $\bar{\rho} = 7535 - 5440 = 2095$ lb (9.3 kN). The variance of c is $V[c] = (100 \times 0.35)^2 = 1225$, and $V[\tan \phi] = (\tan 20° \times 0.05)^2 = 3.31 \times 10^{-4}$. From Eq. 12.34, $V[cL + \tan \phi \, \Sigma N] = L^2 V[c] + (\Sigma N)^2 V[\tan \phi] = (30)^2 \times 1225 + (12,460)^2 \times 3.31 \times 10^{-4} = 1.154 \times 10^6$, or $\sigma_\rho = 1074$ lb (4.8 kN). From Eq. 12.45 and Table 12.2, the probability of failure is $p_f = 0.5 - \psi (2095/1074) = 0.5 - \psi (1.95) = 0.5 - 0.4744 = 0.0256$ or 2.56 per hundred.

Appendix I

References

Algermissen, S. T., 1969. "Seismic Risk Studies in the United States," *Proceedings, Fourth World Conference on Earthquake Engineering*, Vol. 1, Santiago, Chile, pp. A1-14 to A1-27.

ASTM, 1981. "Standard Method for Field Vane Shear Test in Cohesive Soil, D-2573-72," *1981 Annual Book of ASTM Standards*, The American Society for Testing and Materials, Philadelphia.

Bailey, W. A., and J. T. Christian, 1969. *ICES-LEASE-I, A Problem Oriented Language for Slope Stability Analysis*, MIT Soil Mechanics Publication No. 235, Massachusetts Institute of Technology, Cambridge, MA.

Beles, A. A., and I. I. Stanculescu, 1958. "Thermal Treatment as A Means of Improving the Stability of Earth Masses," *Geotechnique*, Vol. 8, No. 4, pp. 158–165.

Bishop, A. W., 1955. "The Use of Slip Circle in the Stability Analysis of Slopes," *Geotechnique*, Vol. 5, No. 1, pp. 7–17.

Bishop, A. W., and D. J. Henkel, 1957. *The Measurement of Soil Properties in the Triaxial Test*, Edward Arnold, London.

Bishop, A. W., and N. Morgenstern, 1960. "Stability Coefficients for Earth Slopes," *Geotechnique*, Vol. 10, No. 4 pp. 129–150.

Bjerrum, L., and N. E. Simons, 1960. "Comparison of Shear Strength Characteristics of Normally Consolidated Clays," *Proceedings, Specialty Conference on Shear Strength of Cohesive Soils*, ASCE, pp. 711–726.

Bjerrum, L. 1972. "Embankments on Soft Ground," *Proceedings, Specialty Conference on Performance of Earth and Earth Supported Structures*, ASCE, Vol. 2, pp. 1–54.

Boutrup, E., 1977. *Computerized Slope Stability Analysis for Indiana Highways*, Technical Report, Joint Highway Research Project, No. 77–25, Purdue University, Lafayette, IN.

Bowles, J. E., 1979. *Physical and Geotechnical Properties of Soils*, McGraw-Hill, New York.

Bureau of Reclamation, 1973. *Design of Small Dams*, 2nd Ed., United States Government Printing Office, Washington, DC.

Casagrande, A. 1937. "Seepage Through Dams," *Contributions to Soil Mechanics*,

BSCE, 1925–1940 (paper first published in *Journal of New England Water Works Association*, June 1937), pp. 295–336.

Casagrande, L., R. W. Loughney and M. A. J. Matich, 1961. "Electro-Osmotic Stabilization of a High Slope in Loose Saturated Silt," *Proceedings, 5th International Conference on Soil Mechanics and Foundation Engineering*, Paris, Vol. 2, pp. 555–561.

Cedergren, H. R., 1977. *Seepage, Drainage, and Flow Nets*, John Wiley & Sons, New York.

Chowdhury, R. N., 1980. "Landslides as Natural Hazards—Mechanisms and Uncertainties," *Geotechnical Engineering*, Southeast Asian Society of Soil Engineering, Vol. 11, No. 2, pp. 135–180.

Chugh, A. K., 1981. *User Information Manual, Slope Stability Analysis Program SSTAB2*, Bureau of Reclamation, Department of the Interior, Denver, CO.

D'Appolonia Consulting Engineers, Inc., 1975. *Engineering and Design Manual–Coal Refuse Disposal Facilities*, Mining Enforcement and Safety Administration, U.S. Department of the Interior.

Dennis, T. H., and R. J. Allan, 1941. "Slide Problem: Storms Do Costly Damages on State Highways Yearly," *California Highways and Public Works*, Vol. 20, July, pp. 1–3.

Department of Navy, 1971. *Design Manual, Soil Mechanics, Foundations, and Earth Structures*, NAVFAC DM-7, Naval Facilities Engineering Command, Philadelphia.

Desai, C. S., and J. F. Abel, 1972. *Introduction to the Finite Element Method*, Van Nostrand Reinhold, New York.

Drnevich, V. P., 1972. *Generalized Sliding Wedge Method for Slope Stability and Earth Pressure Analysis*, Soil Mechanics Series No. 13, University of Kentucky, Lexington, KY.

Drnevich, V. P., C. T. Gorman, and T. C. Hopkins, 1974. "Shear Strength of Cohesive Soils and Friction Sleeve Resistance," *European Symposium on Penetration Testing*, Stockholm, Sweden, Vol. 2:2, pp. 129–132.

Federal Register, 1977. *Part II, Title 30–Mineral Resources*, Office of Surface Mining Reclamation and Enforcement, U.S. Department of the Interior, Chapter VII, Part 715, December.

Fellenius, W., 1936. "Calculation of the Stability of Earth Dams," *Transactions of 2nd Congress on Large Dams*, Washington, DC, Vol. 4, pp. 445–462.

Gilboy, G., 1933. "Hydraulic-Fill Dams," *Proceedings, International Commission on Large Dams*, World Power Conference, Stockholm, Sweden.

Handy, R. L., and W. W. Williams, 1967. "Chemical Stabilization of an Active Landslide," *Civil Engineering*, Vol. 37, No. 8, pp. 62–65.

Hansen, W. R., 1965. "Effects of the Earthquake of March 27, 1964 at Anchorage, Alaska," *USGS Professional Paper 542-A*, 68 pp.

Harr, M. E., 1962. *Groundwater and Seepage*, McGraw-Hill, New York.

Harr, M. E., 1977. *Mechanics of Particulate Media, A Probabilistic Approach*, McGraw-Hill, New York.

Haugen, J. J., and A. F. DiMillio, 1974. "A History of Recent Shale Problems in Indiana," *Highway Focus*, Vol. 6, No. 3, pp. 15–21.

Henkel, D. J. 1967. "Local Geology and the Stability of Natural Slopes," *Journal of the Soil Mechanics and Foundation Division*, ASCE, Vol. 93, No. SM4, pp. 437–446.

Hill, R. A., 1934. "Clay Stratum Dried Out to Prevent Landslides," *Civil Engineering*, Vol. 4, No. 8, pp. 403–407.

Hopkins, T. C., D. L. Allen and R. C. Deen, 1975. *Effect of Water on Slope Stability*, Research Report 435, Division of Research, Kentucky Department of Transportation.

Huang, Y. H., 1975. "Stability Charts for Earth Embankments," *Transportation Research Record 548*, Transportation Research Board, Washington, DC, pp. 1–12.

Huang, Y. H. and M. C. Avery, 1976. "Stability of Slopes by the Logarithmic-Spiral Method," *Journal of the Geotechnical Engineering Division*, ASCE, Vol. 102, No. GT1, pp. 41–49.

Huang, Y. H., 1977a. "Stability of Mine Spoil Banks and Hollow Fills," *Proceedings of the Conference on Geotechnical Practice for Disposal of Solid Waste Materials*, Specialty Conference of the Geotechnical Engineering Division, ASCE, Ann Arbor, MI, pp. 407–427.

Huang, Y. H., 1977b. "Stability Coefficients for Sidehill Benches," *Journal of the Geotechnical Engineering Division*, ASCE, Vol. 103, No. GT5, pp. 467–481.

Huang, Y. H., 1978a. "Stability Charts for Sidehill Fills," *Journal of the Geotechnical Engineering Division*, ASCE, Vol. 104, No. GT5, pp. 659–663.

Huang, Y. H., 1978b. *Stability of Spoil Banks and Hollow Fills Created by Surface Mining*, Research report IMMR34-RRR1-78, Institute for Mining and Minerals Research, University of Kentucky, Lexington, KY.

Huang, Y. H., 1979. "Stability charts for Refuse Dams," *Proceedings of the 5th Kentucky Coal Refuse Disposal and Utilization Seminar and Stability Analysis of Refuse Dam Workshop*, University of Kentucky, Lexington, KY, pp. 57–65.

Huang, Y. H., 1980. "Stability Charts for Effective Stress Analysis of Nonhomogeneous Embankments," *Transportation Research Record 749*, Transportation Research Board, Washington, DC, pp. 72–74.

Huang, Y. H., 1981a. "Line of Seepage in Earth Dams on Inclined Ledge," *Journal of the Geotechnical Engineering Division*, ASCE, Vol. 107, No. GT5, pp. 662–667.

Huang, Y. H. 1981b. *User's Manual, REAME, A Computer Program for the Stability Analysis of Slopes*, Institute for Mining and Minerals Research, University of Kentucky, Lexington, KY.

Huang, Y. H., 1982. *User's Manual, REAMES, A Simplified Version of REAME in Both BASIC and FORTRAN for the Stability Analysis of Slopes*, Institute for Mining and Minerals Research, University of Kentucky, Lexington, KY.

Hunter, J. H. and R. L. Schuster 1971. "Chart Solutions for Stability Analysis of Earth Slopes," *Highway Research Record 345*, Highway Research Board, Washington, DC, pp. 77–89.

International Business Machines Corporation, 1970. *System/360 Scientific Subroutine Package, Version III, Programmer's Manual*.

Janbu, N., 1954. "Application of Composite Slip Surface for Stability Analysis," *European Conference on Stability of Earth Slopes*, Stockholm, Sweden.

Janbu, N., 1973. "Slope Stability Computation," *Embankment-Dam Engineering, Casagrande Volume*, edited by R. C. Hirschfeld and S. J. Poulos, John Wiley & Sons, New York, pp. 47–86.

Karlsson, R., and L. Viberg, 1967. "Ratio *c/p'* in Relation to Liquid Limit and Plasticity Index, with Special Reference to Swedish Clays," *Proceedings, Geotechnical Conference*, Oslo, Norway, Vol. 1, pp. 43–47.

240 APPENDIX I

Kenney, T. C., 1959. Discussion, *Journal of the Soil Mechanics and Foundation Division*, ASCE, Vol. 85, No. SM3, pp. 67–79.

Ladd, C. C., and R. Foott, 1974. "New Design Procedure for Stability of Soft Clays," *Journal of the Geotechnical Engineering Division*, ASCE, Vol. 100, No. GT7, pp. 763–786.

Lambe, T. W., and R. V. Whitman, 1969. *Soil Mechanics*, John Wiley & Sons, New York.

Mendez, C., 1971. *Computerized Slope Stability, the Sliding Block Problem*, Technical Report No. 21, Purdue University Water Resources Research Center, Lafayette, IN.

Mines Branch, Canada, 1972. *Tentative Design Guide for Mine Waste Embankments in Canada*, Department of Energy, Mines, and Resources, Canada.

Morgenstern, N., 1963. "Stability Charts for Earth Slopes During Rapid Drawdown," *Geotechnique*, Vol. 13, No. 2, pp. 121–131.

Morgenstern, N., and V. E. Price, 1965. "The Analysis of the Stability of General Slip Surfaces," *Geotechique*, Vol. 15, No. 1, pp. 79–93.

National Coal Board, 1970. *Spoil Heaps and Lagoons*, Technical Handbook, London, England.

Neumann, F., 1954. *Earthquake Intensity and Related Ground Motion*, University of Washington Press, Seattle, WA.

O'Colman, E., and R. J. Trigo, 1970. "Design and Construction of Tied-Back Sheet Pile Wall," *Highway Focus*, Vol. 2, No. 5, pp. 63–71.

Peck, R. B., and H. O. Ireland, 1953. "Investigation of Stability Problems," *Bulletin 507*, American Railway Engineering Association, Chicago, pp. 1116–1128.

Peck, R. B., 1967. "Stability of Natural Slopes," *Journal of the Soil Mechanics and Foundation Division*, ASCE, Vol. 93, No. SM4, pp. 403–417.

Root, A. W., 1958. "Prevention of Landslides," *Landslides and Engineering Practice* (E. B. Eckel, editor), Special Report 29, Highway Research Board, pp. 113–149.

Royster, D. L., 1966. "Construction of a Reinforced Earth Fill Along I-40 in Tennessee," *Proceedings, 25th Highway Geology Symposium*, pp. 76–93.

Schiffman, R. L., 1972. *A Computer Program to Analyze the Stability of Slopes by Morgenstern's Method*, Report No. 72–18, University of Colorado, Boulder, CO.

Schmertmann, J. 1975. "Measurement of Insitu Shear Strength," *Proceedings of Conference on Insitu Measurement of Soil Properties*, Specialty Conference of the Geotechnical Engineering Division, ASCE, Raleigh, NC, pp. 57–138.

Schuster, R. L. and R. J. Krizek, 1978. *Landslides Analysis and Control*, Special Report 176, Transportation Research Board, Washington, DC.

Seed, H. B., and H. A. Sultan, 1967. "Stability Analysis for a Sloping Core Embankment," *Journal of the Soil Mechanics and Foundations Division*, ASCE, Vol. 93, No. SM4, pp. 69–84.

Seed, H. B., K. L. Lee, I. M. Idriss, and F. I. Makdisi, 1975a. "The Slides in the San Fernando Dams during the Earthquake of February 9, 1971," *Journal of the Geotechnical Engineering Division*, ASCE, Vol. 101, No. GT7, pp. 651–688.

Seed, H. B., I. M. Idriss, K. L. Lee, and F. I. Makdisi, 1975b, "Dynamic Analysis of the Slide in the Lower San Fernando Dam during the Earthquake of February 9, 1971," *Journal of the Geotechnical Engineering Division*, ASCE, Vol. 101, No. GT9, pp. 889–911.

Shannon and Wilson, Inc., 1968. *Slope Stability Investigation, Vicinity of Prospect Park, Minneapolis-St. Paul*, Shannon and Wilson, Seattle, WA.

Siegel, R. A., 1975. *STABL User Manual*, Technical Report 75-9, Joint Highway Research Project, Purdue University, Lafayette, IN.

Skempton, A. W., 1954. "The Pore-Pressure Coefficients A and B," *Geotechnique*, Vol. 4, No. 4, pp. 143–147.

Skempton, A. W., 1964. "Long Term Stability of Clay Slopes," *Geotechnique*, Vol. 14, No. 2, pp. 77–101.

Sowers, G. F., 1979. *Introductory Soil Mechanics and Foundations: Geotechnical Engineering*, 4th Ed., Macmillan, New York.

Smith, T. W., and H. R. Cedergren, 1962. "Cut Slope Design in Field Testing of Soils and Landslides," *Special Technical Publication 322*, ASTM, pp. 135–158.

Smith, T. W., and R. A. Forsyth, 1971. "Potrero Hill Slide and Correction," *Journal of Soil Mechanics and Foundations Division*, ASCE, Vol. 97, No. SM3, pp. 541–564.

Spencer, E., 1967. "A Method of Analysis of the Stability of Embankments Assuming Parallel Inter-slice Forces," *Geotechnique*, Vol. 17, No. 1, pp. 11–26.

Spencer, E., 1973. "Thrust Line Criterion in Embankment Stability Analysis," *Geotechnique*, Vol. 23, No. 1, pp. 85–100.

Spencer, E., 1981. "Slip Circles and Critical Shear Planes," *Journal of the Geotechnical Engineering Division*, ASCE, Vol. 107, No. GT7, pp. 929–942.

Taylor, D. W., 1937. "Stability of Earth Slopes," *Journal of Boston Society of Civil Engineers*, Vol. 24, pp. 197–246.

Taylor, D. W., 1948. *Fundamentals of Soil Mechanics*, John Wiley & Sons, New York.

Terzaghi, K., and R. B. Peck, 1967. *Soil Mechanics in Engineering Practice*, John Wiley & Sons, New York.

Trofimenknov, J. G., 1974. "Penetration Testing in the USSR," *European Symposium on Penetration Testing*, Stockholm, Sweden, Vol. 1, pp. 147–154.

Varnes, D. J., 1978. "Slope Movement Types and Processes," *Landslides Analysis and Control* (Schuster, R. L. and R. J. Krizek, editors), Special Report 176, Transportation Research Board, Washington, DC.

Wang, F. D., M. C. Sun, and D. M. Ropchan, 1972. *Computer Program for Pit Slope Stability Analysis by the Finite Element Stress Analysis and Limiting Equilibrium Method*, RI 7685, Bureau of Mines.

Whitman, R. V., and P. J. Moore, 1963. "Thoughts Concerning the Mechanics of Slopes Stability Analysis," *Proceedings of the Second Panamerican Conference on Soil Mechanics and Foundation Engineering*, São Paulo, Brazil, Vol. 1, pp. 391–411.

Whitman, R. V., an W. A. Bailey, 1967. "Use of Computers for Slope Stability Analysis," *Journal of the Soil Mechanics and Foundation Division*, ASCE, Vol. 93, No. SM4, pp. 475–498.

Wright, S. G., 1969. "A Study of Slope Stability and the Undrained Shear Strength of Clay Shales," Ph.D. dissertation, University of California, Berkeley, CA.

Wright, S. G., 1974. *SSTAB1—A General Computer Program for Slope Stability Analyses*, Research Report No. GE74-1, University of Texas at Austin, TX.

Zaruba, Q., and V. Mencl, 1969. *Landslides and Their Control*, American Elsevier, New York.

Zienkiewicz, O. C., 1971. *The Finite Element Method in Engineering Science*, McGraw-Hill, New York.

Appendix II

Symbols

a	=	y-coordinate of the exit point on line of seepage; or width of a free body in infinite slope; or a constant
$\bar{a}$	=	y-intercept of K_f-line
a_1 to a_5	=	coefficient for determining factor of safety of plane failure
a_i	=	moment arm of a seismic force on slice i
A	=	area between an arc and a chord of a failure circle; or pore pressure coefficient
b	=	y-coordinate of the midpoint on a line of seepage; or a constant
b_i	=	width of slice i
B	=	ratio between the base width and height of a fill; or pore pressure coefficient
$\bar{B}$	=	soil parameter
c	=	cohesion of soil
$\bar{c}$	=	effective cohesion of soil
c_0	=	compacted cohesion
c_1	=	constant of integration; or cohesion for soil 1
c_2 and c_3	=	cohesion for soils 2 and 3, respectively
$\bar{c}_1$	=	effective cohesion for soil 1
$\bar{c}_2$	=	effective cohesion for soil 2
c_d	=	developed cohesion
c_f	=	correction factor for the factor of safety; or correction factor for the exit point on a line of seepage
c_{sat}	=	cohesion after saturation
c_u	=	undrained cohesion
C	=	resultant of cohesion resistance along a failure surface
C_d	=	resultant of developed cohesion along a failure surface
C_s	=	seismic coefficient
C_v	=	coefficient of variation

d	=	horizontal distance between the toe and farthest point on a line of seepage; or depth of infinite slope
D	=	depth ratio of a failure circle; or driving force
D_A	=	driving force at an active wedge
D_p	=	driving force at a passive wedge
E	=	normal force on the vertical side of a slice
$f(x)$	=	a function of x in Morgenstern and Price's method; or probability density function of x
F	=	factor of safety
F_l	=	lower or left bound for a factor of safety
F_r	=	upper or right bound for a factor of safety
F_c	=	factor of safety with respect to cohesion
F_m	=	factor of safety at mth iteration
F_{m+1}	=	factor of safety at $(m+1)$th iteration
F_ϕ	=	factor of safety with respect to the friction angle
h	=	vertical distance between toe of a dam and the pool elevation; or depth of a point in a soil mass below the soil surface
h_1	=	vertical distance from a line of thrust to the base on one side of a slice
h_2	=	vertical distance from a line of thrust to the base on other side of slice
h_i	=	height of slice i
h_{iw}	=	vertical distance between a phreatic surface and the bottom of slice i
h_t	=	vertical distance from a line of thrust to the base of a slice
H	=	height of fill
H_A	=	height of an active wedge
H_e	=	equivalent height
H_p	=	height of a passive wedge
i	=	subscript for slice i
k	=	permeability of soil
K	=	coefficient of the friction circle
K_f-line	=	line through $\bar{p}$ versus q at the time of failure
l	=	distance along a line of seepage; or distance along downstream slope between a line of seepage and the origin; or length of a failure plane at the middle block
l_1	=	length of a failure plane at the lower block
l_2	=	length of a failure plane at the upper block
l_A	=	length of a failure plane at an active wedge
l_P	=	length of a failure plane at a passive wedge
L	=	arc length of a failure circle; or length of plane failure; or height of rapid drawdown
L_1	=	length of a failure plane at the lower block; or length of a failure arc in soil 1
L_2	=	length of a failure plane at the upper block; or length of a failure arc in soil 2
L_c	=	chord length of a failure circle
L_f	=	length factor, or fraction of a failure arc in soil 2
L_i	=	length of a failure arc in slice i

m	=	a/h; or stability coefficient in Bishop and Morgenstern's method
m_1	=	stability coefficient for the lower depth ratio
m_2	=	stability coefficient for the higher depth ratio
M_c	=	moment about the origin of a logarithmic spiral due to cohesion
M_w	=	moment about the origin of a logarithmic spiral due to weight
M'_w	=	moment about the origin of a logarithmic spiral due to additional weight
MM	=	modified Mercalli intensity scale
n	=	number of slices or samples; or horizontal distance between a failure circle and the toe in terms of embankment height; or stability coefficient in Bishop and Morgenstern's method
n_1	=	stability coefficient for the lower depth ratio
n_2	=	stability coefficient for the higher depth ratio
N	=	total force normal to a failure plane; or standard penetration blow count
$\bar{N}$	=	effective force normal to a failure plane; or effective force normal to a failure plane at the middle block
N_1	=	total force normal to a failure plane at the lower block
N_2	=	total force normal to a failure plane at the upper block
$\bar{N}_1$	=	effective force normal to a failure plane at the lower block
$\bar{N}_2$	=	effective force normal to a failure plane at the upper block
N_e	=	earthquake number
N_f	=	friction number
N_{f1}	=	friction number for soil 1
N_{f2}	=	friction number for soil 2
N_i	=	total force normal to the failure surface at slice i
$\bar{N}_i$	=	effective force normal to the failure surface at slice i
N_s	=	stability number
$\bar{p}$	=	$(\bar{\sigma}_1 + \bar{\sigma}_3)/2$; or effective overburden pressure
p_f	=	probability of failure
P	=	total force between two blocks; or resultant of normal and frictional forces along a failure arc
P_1	=	total force between the lower and middle blocks
P_2	=	total force between the upper and middle blocks
P_A	=	total active earth pressure
P_{A1}	=	total active earth pressure in soil 1
P_{A2}	=	total active earth pressure in soil 2
P_c	=	percent of cohesion resistance
P_p	=	total passive earth pressure
P_{p1}	=	total passive earth pressure in soil 1
P_{p2}	=	total passive earth pressure in soil 2
q	=	discharge; or maximum shear stress, $(\sigma_1 - \sigma_3)/2$
q_c	=	bearing stress on a Dutch cone
q_u	=	unconfined compressive strength
r	=	radius from the origin to the logarithmic spiral
r_0	=	initial radius

r_u	=	pore pressure ratio
r_{ue}	=	equal pore pressure ratio
r_{u1}	=	pore pressure ratio in soil 1
r_{u2}	=	pore pressure ratio in soil 2
R	=	radius of a failure circle; or resultant of the normal and frictional forces
R_A	=	resultant of the normal and frictional forces at an active wedge
R_p	=	resultant of the normal and frictional forces at a passive wedge
s	=	shear strength of soil
$\bar{s}$	=	mean shear strength
s_u	=	undrained shear strength
S	=	slope in terms of horizontal to vertical; or shear force at the side of a slice
t	=	angle used to locate a logarithmic spiral
T	=	shear force on a failure surface; or shear force on failure plane at the middle block
T_1	=	shear force on a failure plane at the lower block
T_2	=	shear force on a failure plane at the upper block
T_i	=	shear force at the bottom of slice i
u	=	pore water pressure; or normal deviate
U	=	neutral force
V	=	variance
W	=	total weight of soil above a failure surface; or weight of the middle block
W_1	=	weight of the lower block
W_2	=	weight of the upper block
W_A	=	weight of soil above a failure plane in an active wedge
W_f	=	width of a fill bench
W_i	=	weight of slice i
W_{i1}	=	weight of slice i with a failure arc in soil 1
W_{i2}	=	weight of slice i with a failure arc in soil 2
W_p	=	weight of soil above a failure plane in a passive wedge
x	=	horizontal coordinate; or a variable
x_i	=	ith value of variable x
X	=	ratio between x-coordinate of a circle and height of fill
$\bar{X}$	=	sample mean
y	=	vertical coordinate; or y-coordinate of a failure surface
y_0	=	y-coordinate of a line of seepage when $x = 0$
y_t	=	y-coordinate for the line of thrust
Y	=	ratio between y-coordinate of a circle and height of fill
z	=	distance below ground surface; or y-coordinate of a slope surface; or integration limit of normal distribution
Z_1	=	resultant of normal and shear forces on one side of a slice
Z_2	=	resultant of normal and shear forces on other side of a slice
α	=	degree of natural slope
$\bar{\alpha}$	=	slope of K_f-line
α_A	=	angle of inclination of active earth pressure
α_P	=	angle of inclination of passive earth pressure

β	=	angle of outslope
γ	=	unit weight of soil
γ_1, γ_2, and γ_3	=	unit weight of soils 1, 2, and 3, respectively
γ_w	=	unit weight of water
δ	=	angle of inclination of interslice force
Δ	=	horizontal distance between a point on upstream slope at pool elevation and the toe; or infinitesimal quantity
ΔE	=	increment of thrust across a slice
ΔS	=	increment of shear force across a slice
Δl	=	distance along the downstream slope between a basic parabola and a line of seepage; or a failure arc of infinitesimal length
θ	=	angle of inclination of a given chord in a failure circle; or angle of inclination of a failure plane at the middle block; or angle between the initial radius, r_0, of logarithmic spiral and radius r.
θ_1 and θ_2	=	angle of inclination of a failure plane at the top and bottom blocks, respectively
θ_A	=	angle of inclination of a failure plane in an active wedge
θ_i	=	angle of inclination of a failure surface at the bottom of slice i
θ_P	=	angle of inclination of a failure plane in a passive wedge
λ	=	a constant used in Morgenstern and Price's method
μ	=	true or population mean
ρ	=	safety margin
$\bar{\rho}$	=	mean safety margin
σ	=	standard deviation
σ_1	=	major total principal stress
$\bar{\sigma}_1$	=	major effective principal stress
σ_3	=	minor total principal stress
$\bar{\sigma}_3$	=	minor effective principal stress
σ_n	=	total normal stress on a failure plane
$\bar{\sigma}_n$	=	effective normal stress on a failure plane
σ_s	=	standard deviation of shear strength
σ_x	=	stress in x-direction
σ_y	=	stress in y-direction
σ_ρ	=	standard deviation of the safety margin
σ_τ	=	standard deviation of shear stress
τ	=	shear stress on a failure plane
$\bar{\tau}$	=	mean shear stress
τ_{xy}	=	shear stress on the xy-plane
ϕ	=	angle of internal friction of soil
ϕ_1, ϕ_2, and ϕ_3	=	angle of internal friction for soils 1, 2, and 3, respectively
$\bar{\phi}$	=	effective angle of internal friction of soil
$\bar{\phi}_1$	=	effective friction angle for soil 1
$\bar{\phi}_2$	=	effective friction angle for soil 2
$\bar{\phi}_c$	=	corrected effective friction angle
ϕ_d	=	developed friction angle after taking a factor of safety into account $= \tan^{-1} (\tan \phi / F)$
ϕ_u	=	undrained friction angle
ψ	=	cumulative distribution function

Appendix III

List of SWASE in FORTRAN

```
C                                                                        MAIN0001
C      SWASE (SLIDING WEDGE ANALYSIS OF SIDEHILL EMBANKMENTS)            MAIN0002
C                                                                        MAIN0003
       REAL LNS,LNS2                                                     MAIN0004
       EXTERNAL FCT                                                      MAIN0005
       COMMON/A1/AL1,C,CA,CD,CW,DSTPR,DSBMR,LNS2,PHIMAX,RU,SA,SD,SEIC,SW,MAIN0006
      1TP,W,W1,W2                                                        MAIN0007
       COMMON/A2/LNS,T,T1,T2,XLI                                         MAIN0008
       DIMENSION C(3),PHI(3),PHIR(3),TITLE(20),TP(3)                     MAIN0009
       DATA IEND,EPS,XL ,XR ,DF/ 20,1E-4,0.1,10.,0.5/                    MAIN0010
       L=0                                                               MAIN0011
C                                                                        MAIN0012
C   READ AND WRITE INPUT DATA                                           MAIN0013
C                                                                        MAIN0014
       READ(5,10) TITLE                                                  MAIN0015
C      ****************************************************************** MAIN0016
   10 FORMAT(20A4)                                                      MAIN0017
       READ(5,20)NSET,IPBW,IPUW                                          MAIN0018
C      ****************************************************************** MAIN0019
   20 FORMAT(3I5)                                                       MAIN0020
       WRITE (6,30) TITLE                                                MAIN0021
   30 FORMAT(1H1,///10X,20A4,//)                                        MAIN0022
       WRITE(6,40)NSET,IPBW,IPUW                                         MAIN0023
   40 FORMAT(//10X,15HPROGRAM OPTIONS,//10X,16HNO. OF DATA SETS,I4,/10X, MAIN0024
      126HSTARTING FROM DATA SET NO.,I2,46H USE TABULAR PRINTOUT WITH BLOMAIN0025
      2CK WEIGHTS GIVEN,/10X,26HSTARTING FROM DATA SET NO.,I2,45H USE TABMAIN0026
      3ULAR PRINTOUT WITH UNIT WEIGHTS GIVEN)                            MAIN0027
       READ (5,50) (C(I),PHI(I),I=1,3),PHIMA                            MAIN0028
C      ****************************************************************** MAIN0029
   50 FORMAT (8F10.3)                                                   MAIN0030
       WRITE(6,60) (C(I),PHI(I),I=1,3),PHIMA                            MAIN0031
   60 FORMAT (//,10X,24HCOHESION AT BOTTOM BLOCK,F10.3,5X,30HFRICTION ANMAIN0032
      1GLE AT BOTTOM BLOCK,F10.3,/,10X,21HCOHESION AT TOP BLOCK,F10.3,5X,MAIN0033
      227HFRICTION ANGLE AT TOP BLOCK,F10.3,/,10X,24HCOHESION AT MIDDLE BMAIN0034
      3LOCK,F10.3,5X,30HFRICTION ANGLE AT BOTTOM BLOCK,F10.3,/,10X,      MAIN0035
      434HANGLE OF INTERNAL FRICTION OF FILL,F10.3)                      MAIN0036
   70 L=L+1                                                             MAIN0037
       READ (5,50) DSBM,DSTP,DSME,LNS2,LNS,RU,SEIC,GAMMA                MAIN0038
C      ****************************************************************** MAIN0039
```

```
      IF(GAMMA) 80,100,80                                           MAIN0040
   80 READ (5,50) ANOUT,BW                                          MAIN0041
C     ************************************************************  MAIN0042
      B=BW                                                          MAIN0043
      IF(IPUW .EQ. L) WRITE(6,90)                                   MAIN0044
   90 FORMAT(//4X,3HSET,4X,4HDSBM,5X,4HDSTP,5X,4HDSME,4X,4HLNS2,6X, MAIN0045
     13HLNS,5X,2HRU,4X,4HSEIC,4X,5HGAMMA,4X,5HANOUT,6X,2HBW,7X,1HF) MAIN0046
      GO TO 130                                                     MAIN0047
  100 READ(5,110) W1,W2,W,AL1                                       MAIN0048
  110 FORMAT (3F10.0,F10.3)                                         MAIN0049
C     ************************************************************  MAIN0050
      IF(IPBW .EQ. L) WRITE(6,120)                                  MAIN0051
  120 FORMAT(//4X,3HSET,4X,4HDSBM,5X,4HDSTP,5X,4HDSME,4X,4HLNS2,6X, MAIN0052
     13HLNS,5X,2HRU,4X,4HSEIC,7X,2HW1,12X,2HW2,16X,1HW,6X,3HAL1,7X,1HF) MAIN0053
  130 IRC=0                                                         MAIN0054
      ND=1                                                          MAIN0055
      IER=3                                                         MAIN0056
      VAL=99999999                                                  MAIN0057
      F=9999.99                                                     MAIN0058
      IRC=0                                                         MAIN0059
      XRI=XR                                                        MAIN0060
      XLI=XL                                                        MAIN0061
      IF(IPBW .EQ. 0 .AND. IPUW .EQ. 0 .OR. IPBW .EQ. 0 .AND. L .LT. MAIN0062
     1 IPUW .OR. IPUW .EQ. 0 .AND. L .LT. IPBW .OR. L .LT. IPBW .AND. MAIN0063
     2 L .LT. IPUW) GO TO 140                                       MAIN0064
      GO TO 200                                                     MAIN0065
  140 WRITE (6,150) L                                               MAIN0066
  150 FORMAT(////10X,8HDATA SET,2X,I4)                              MAIN0067
      IF(GAMMA) 160,180,160                                         MAIN0068
  160 WRITE(6,170) DSBM,DSTP,DSME,LNS2,LNS,RU,SEIC,GAMMA,ANOUT,B    MAIN0069
  170 FORMAT (//10X,25HDEGREE OF SLOPE AT BOTTOM,F10.3,//10X,22HDEGREE OMAIN0070
     1F SLOPE AT TOP,F10.3,//10X,25HDEGREE OF SLOPE AT MIDDLE,F10.3,//10MAIN0071
     2X,30HLENGTH OF NATURAL SLOPE AT TOP,F10.3,//10X,33HLENGTH OF NATURMAIN0072
     3AL SLOPE AT MIDDLE,F10.3,//10X,19HPORE PRESSURE RATIO,F10.3,//10X,MAIN0073
     419HSEISMIC COEFFICIENT,F10.3,//10X,11HUNIT WEIGHT,F10.3,//10X,    MAIN0074
     517HANGLE OF OUTSLOPE,F10.3,//10X,11HBENCH WIDTH,F10.3)          MAIN0075
      GO TO 200                                                     MAIN0076
  180 WRITE(6,190)  DSBM,DSTP,DSME,LNS2,LNS,RU,SEIC,W1,W2,W,AL1     MAIN0077
  190 FORMAT (//10X,25HDEGREE OF SLOPE AT BOTTOM,F10.3,//10X,22HDEGREE OMAIN0078
     1F SLOPE AT TOP,F10.3,//10X,25HDEGREE OF SLOPE AT MIDDLE,F10.3,//10MAIN0079
     2X,30HLENGTH OF NATURAL SLOPE AT TOP,F10.3,//10X,33HLENGTH OF NATURMAIN0080
     3AL SLOPE AT MIDDLE,F10.3,//10X,19HPORE PRESSURE RATIO,F10.3,//10X,MAIN0081
     419HSEISMIC COEFFICIENT,F10.3,//10X,22HWEIGHT OF BOTTOM BLOCK,F15.3MAIN0082
     5,//10X,19HWEIGHT OF TOP BLOCK,F15.3,//10X,22HWEIGHT OF MIDDLE BLOCMAIN0083
     6K,F15.3,//10X,33HLENGTH OF NATURAL SLOPE AT BOTTOM,F10.3)      MAIN0084
C                                                                   MAIN0085
C   CONVERT DEGREES TO RADIANS                                      MAIN0086
C                                                                   MAIN0087
  200 PHIMAX=PHIMA *3.1416/180.                                     MAIN0088
      DO 210 I=1,3                                                  MAIN0089
  210 PHIR(I)=PHI(I)*3.1416/180.                                    MAIN0090
      IF (GAMMA .NE. 0.) ANOUTR=ANOUT*3.1416/180.                   MAIN0091
      DSTPR=DSTP*3.1416/180.                                        MAIN0092
      DSBMR=DSBM*3.1416/180.                                        MAIN0093
      DSMER=DSME*3.1416/180.                                        MAIN0094
      IF(GAMMA .EQ. 0) GO TO 260                                    MAIN0095
C                                                                   MAIN0096
C   DETERMINE THE DIMENSION AND WEIGHT OF DIFFERENT WEDGES          MAIN0097
C                                                                   MAIN0098
      X2=LNS2*COS(DSTPR)                                            MAIN0099
      IF(B-X2.GT.0.) GO TO 230                                      MAIN0100
C                                                                   MAIN0101
C   WHEN B .LE. HORIZONTAL PROJECTION OF LNS2                       MAIN0102
```

```
C                                                          MAIN0103
      SX=X2-B                                              MAIN0104
      AL3=SIN(DSTPR)*LNS2-SX*TAN(ANOUTR)                   MAIN0105
      W2=(SIN(2.*DSTPR)*LNS2**2/4.-SX*SX*TAN(ANOUTR)/2.)*GAMMA  MAIN0106
      IF(LNS .EQ. 0.) GO TO 220                            MAIN0107
      X1=LNS*COS(DSMER)                                    MAIN0108
      Z1=X1*TAN(ANOUTR)                                    MAIN0109
      Z2=AL3+LNS *SIN(DSMER)-Z1                            MAIN0110
      W=(AL3+Z2)*X1*GAMMA/2.                               MAIN0111
  220 IF(LNS .EQ. 0.) Z2=AL3                               MAIN0112
      AL1= Z2*COS(ANOUTR)/SIN(ANOUTR-DSBMR)                MAIN0113
      W1=COS(DSBMR)*AL1* Z2*GAMMA/2.                       MAIN0114
      GO TO 260                                            MAIN0115
C                                                          MAIN0116
C   WHEN B .GT. HORIZONTAL PROJECTION OF LNS2              MAIN0117
C                                                          MAIN0118
  230 W2 =SIN(2*DSTPR)*LNS2**2*GAMMA/4.                    MAIN0119
      IF(LNS .EQ. 0.) GO TO 250                            MAIN0120
      X1=LNS*COS(DSMER)                                    MAIN0121
      IF(B-X2-X1.GT.0.) GO TO 240                          MAIN0122
C                                                          MAIN0123
C   WHEN B .LE. HORIZONTAL PROJECTION OF LNS2+LNS          MAIN0124
C                                                          MAIN0125
      AL3=LNS2*SIN(DSTPR)                                  MAIN0126
      Z3=LNS*SIN(DSMER)                                    MAIN0127
      SX1=X1-(B-X2)                                        MAIN0128
      W=(X1*AL3-SX1*SX1*TAN(ANOUTR)/2.+X1*Z3/2.)*GAMMA     MAIN0129
      AL1=(AL3+Z3-(X1+X2-B)*TAN(ANOUTR))*COS(ANOUTR)/SIN(ANOUTR-DSBMR)  MAIN0130
      W1=(AL3+Z3-(X1+X2-B)*TAN(ANOUTR))*AL1*COS(DSBMR)*GAMMA/2.  MAIN0131
      GO TO 260                                            MAIN0132
C                                                          MAIN0133
C   WHEN B .GT. HORIZONTAL PROJECTION OF LNS2+LNS          MAIN0134
C                                                          MAIN0135
  240 W=(X1*(LNS2*SIN(DSTPR))+(SIN(2.*DSMER)*LNS*LNS/4.))*GAMMA  MAIN0136
      Z3 =SIN(DSMER)*LNS                                   MAIN0137
  250 IF(LNS .EQ. 0.) X1=0                                 MAIN0138
      IF(LNS .EQ. 0.) Z3=0                                 MAIN0139
      X=B-X1-X2                                            MAIN0140
      Y=SIN(DSTPR)*LNS2+Z3                                 MAIN0141
      AL3=SQRT(X*X+Y*Y)                                    MAIN0142
      W4=X*Y/2.*GAMMA                                      MAIN0143
      SINAX=Y/AL3                                          MAIN0144
      AX=ARSIN(SINAX)                                      MAIN0145
      ANGX=3.1416-(ANOUTR+AX)                              MAIN0146
      AL1=AL3*SIN(ANGX)/SIN(ANOUTR-DSBMR)                  MAIN0147
      ANGY=3.1416-(ANGX+ANOUTR-DSBMR)                      MAIN0148
      W1=AL3*AL1*SIN(ANGY)*GAMMA/2.+W4                     MAIN0149
  260 SA=SIN(DSTPR)                                        MAIN0150
      SD=SIN(DSBMR)                                        MAIN0151
      CA=COS(DSTPR)                                        MAIN0152
      CD=COS(DSBMR)                                        MAIN0153
      DO 270 I=1,3                                         MAIN0154
  270 TP(I)=TAN(PHIR(I))                                   MAIN0155
      SW=SIN(DSMER)                                        MAIN0156
      CW=COS(DSMER)                                        MAIN0157
C                                                          MAIN0158
C   CALL FIXLI TO DETERMINE THE INITIAL LEFT BOUND OF F    MAIN0159
C                                                          MAIN0160
      CALL FIXLI(DF,IRC)                                   MAIN0161
      IF(IRC.EQ.1) GO TO 290                               MAIN0162
C                                                          MAIN0163
C   CALL RTMI TO DETERMINE THE FACTOR OF SAFETY            MAIN0164
C                                                          MAIN0165
```

```
    280 CALL RTMI (F,VAL,FCT,XLI,XRI,EPS,IEND,IER)                    MAIN0166
        IF(IER.NE.2) GO TO 290                                       MAIN0167
        ND=2*ND                                                      MAIN0168
        IF (ND.GT.1024) GO TO 290                                    MAIN0169
        XLI=XLI-DF/ND                                                MAIN0170
        GO TO 280                                                    MAIN0171
    290 IF(IPBW .EQ. 0 .AND. IPUW .EQ. 0 .OR. IPBW .EQ. 0 .AND. L .LT. MAIN0172
       1 IPUW .OR. IPUW .EQ. 0 .AND. L .LT. IPBW .OR. L .LT. IPBW .AND. MAIN0173
       2 L .LT. IPUW) GO TO 300                                      MAIN0174
        GO TO 310                                                    MAIN0175
    300 IF(IRC .EQ. 1) GO TO 360                                     MAIN0176
        WRITE (6,370) F                                              MAIN0177
        GO TO 360                                                    MAIN0178
    310 IF(GAMMA) 320,340,320                                        MAIN0179
    320 WRITE(6,330) L,DSBM,DSTP,DSME,LNS2,LNS,RU,SEIC,GAMMA,ANOUT,B,F MAIN0180
    330 FORMAT (/,4X,I2,1X,5F9.3,2F7.3,4F9.3)                        MAIN0181
        GO TO 360                                                    MAIN0182
    340 WRITE(6,350)L,DSBM,DSTP,DSME,LNS2,LNS,RU,SEIC,W1,W2,W,AL1,F  MAIN0183
    350 FORMAT(/,4X,I2,1X,5F9.3,2F7.3,3F14.3,2F9.3)                  MAIN0184
    360 NSET=NSET-1                                                  MAIN0185
        IF(NSET.GT.0) GO TO 70                                       MAIN0186
        STOP                                                         MAIN0187
    370 FORMAT (//,10X,20HTHE FACTOR OF SAFETY,2X,F10.3,//)          MAIN0188
        END                                                          MAIN0189
        SUBROUTINE FIXLI(DF,IRC)                                     FILI0001
C                                                                    FILI0002
C   THIS SUBROUTINE IS USED TO DETERMINE THE INITIAL LEFT BOUND OF THE FILI0003
C   ROOT F.  THE VALUE OF F MUST SATISFY THE BASIC REQUIREMENT THAT THE FILI0004
C   SHEAR RESISTANCES ALONG THE BASE OF ALL SLIDING WEDGES ARE POSITIVE. FILI0005
C                                                                    FILI0006
        REAL LNS                                                     FILI0007
        COMMON/A2/LNS,T,T1,T2,XLI                                    FILI0008
     10 VAL=FCT(XLI)                                                 FILI0009
        IF(LNS .EQ. 0.) GO TO 20                                     FILI0010
        IF(T.GT.0..AND.T1.GT.0..AND.T2.GT.0.) RETURN                 FILI0011
        GO TO 30                                                     FILI0012
     20 IF(T1.GT.0..AND.T2.GT.0.) RETURN                            FILI0013
     30 XLI=XLI+DF                                                   FILI0014
        IF(XLI.LE.10.) GO TO 10                                      FILI0015
        WRITE(6,40)                                                  FILI0016
        IRC=1                                                        FILI0017
        RETURN                                                       FILI0018
     40 FORMAT(///10X,'** XLI IS GREATER THAN 10, SO THE FACTOR OF SAFETY FILI0019
       1CANNOT BE FOUND. PLEASE CHECK YOUR INPUT DATA')              FILI0020
        END                                                          FILI0021
        SUBROUTINE RTMI(X,F,FCT,XLI,XRI,EPS,IEND,IER)                RTMI0001
C                                                                    RTMI0002
C                                                                    RTMI0003
C   THIS SUBROUTINE, WHICH IS OBTAINED FROM THE IBM SUBROUTINE PACKAGE, RTMI0004
C   SOLVES THE GENERAL NONLINEAR EQUATION OF THE FORM FCT(X)=0 BY MEANS RTMI0005
C   OF MUELLER'S ITERATION METHOD.                                  RTMI0006
C                                                                    RTMI0007
        IER=0                                                        RTMI0008
        XL=XLI                                                       RTMI0009
        XR=XRI                                                       RTMI0010
        X=XL                                                         RTMI0011
        TOL=X                                                        RTMI0012
        F=FCT(TOL)                                                   RTMI0013
        IF(F)10,160,10                                               RTMI0014
     10 FL=F                                                         RTMI0015
        X=XR                                                         RTMI0016
        TOL=X                                                        RTMI0017
        F=FCT(TOL)                                                   RTMI0018
        IF(F)20,160,20                                               RTMI0019
```

```
 20 FR=F
    IF(SIGN(1.,FL)+SIGN(1.,FR))250,30,250              RTMI0021
 30 I=0                                                RTMI0022
    TOLF=100.*EPS                                      RTMI0023
 40 I=I+1                                              RTMI0024
    DO 130 K=1,IEND                                    RTMI0025
    X=.5*(XL+XR)                                       RTMI0026
    TOL=X                                              RTMI0027
    F=FCT(TOL)                                         RTMI0028
    IF(F)50,160,50                                     RTMI0029
 50 IF(SIGN(1.,F)+SIGN(1.,FR))70,60,70                 RTMI0030
 60 TOL=XL                                             RTMI0031
    XL=XR                                              RTMI0032
    XR=TOL                                             RTMI0033
    TOL=FL                                             RTMI0034
    FL=FR                                              RTMI0035
    FR=TOL                                             RTMI0036
 70 TOL=F-FL                                           RTMI0037
    A=F*TOL                                            RTMI0038
    A=A+A                                              RTMI0039
    IF(ABS(FR) .GT. 1.0E30) GO TO 90                   RTMI0040
    IF(A-FR*(FR-FL))80,90,90                           RTMI0041
 80 IF(I-IEND)170,170,90                               RTMI0042
 90 XR=X                                               RTMI0043
    FR=F                                               RTMI0044
    TOL=EPS                                            RTMI0045
    A=ABS(XR)                                          RTMI0046
    IF(A-1.)110,110,100                                RTMI0047
100 TOL=TOL*A                                          RTMI0048
110 IF(ABS(XR-XL)-TOL)120,120,130                      RTMI0049
120 IF(ABS(FR-FL)-TOLF)140,140,130                     RTMI0050
130 CONTINUE                                           RTMI0051
    IER=1                                              RTMI0052
140 IF(ABS(FR)-ABS(FL))160,160,150                     RTMI0053
150 X=XL                                               RTMI0054
    F=FL                                               RTMI0055
160 RETURN                                             RTMI0056
170 A=FR-F                                             RTMI0057
    DX=(X-XL)*FL*(1.+F*(A-TOL)/(A*(FR-FL)))/TOL        RTMI0058
    XM=X                                               RTMI0059
    FM=F                                               RTMI0060
    X=XL-DX                                            RTMI0061
    TOL=X                                              RTMI0062
    F=FCT(TOL)                                         RTMI0063
    IF(F)180,160,180                                   RTMI0064
180 TOL=EPS                                            RTMI0065
    A=ABS(X)                                           RTMI0066
    IF(A-1.)200,200,190                                RTMI0067
190 TOL=TOL*A                                          RTMI0068
200 IF(ABS(DX)-TOL)210,210,220                         RTMI0069
210 IF(ABS(F)-TOLF)160,160,220                         RTMI0070
220 IF(SIGN(1.,F)+SIGN(1.,FL))240,230,240              RTMI0071
230 XR=X                                               RTMI0072
    FR=F                                               RTMI0073
    GO TO 40                                           RTMI0074
240 XL=X                                               RTMI0075
    FL=F                                               RTMI0076
    XR=XM                                              RTMI0077
    FR=FM                                              RTMI0078
    GO TO 40                                           RTMI0079
250 IER=2                                              RTMI0080
    RETURN                                             RTMI0081
    END                                                RTMI0082
    FUNCTION FCT(F)                                    FUNC0001
```

The first line RTMI numbers:
```
 20 FR=F                                               RTMI0020
```

```
C                                                                      FUNC0002
C  THE EQUILIBRIUM OF SLIDING WEDGES REQUIRES FCT(F)=0, WHERE FCT(F) IS FUNC0003
C  GENERATED BY THIS FUNCTION SUBPROGRAM.                              FUNC0004
C                                                                      FUNC0005
      REAL LNS,LNS2,N,N1,N2                                            FUNC0006
      COMMON/A1/AL1,C,CA,CD,CW,DSTPR,DSBMR,LNS2,PHIMAX,RU,SA,SD,SEIC,SW,FUNC0007
     1TP,W,W1,W2                                                       FUNC0008
      COMMON/A2/LNS,T,T1,T2,XLI                                        FUNC0009
      DIMENSION C(3),TP(3)                                             FUNC0010
      AP=ATAN(TAN(PHIMAX)/F)                                           FUNC0011
      SAP=SIN(AP)                                                      FUNC0012
      CAP=COS(AP)                                                      FUNC0013
      N2=(W2*(CAP-RU*CA*COS(AP-DSTPR)-SEIC*SAP)+C(2)*LNS2*SIN(AP-DSTPR)/FUNC0014
     1F)/(COS(AP-DSTPR)-TP(2)*SIN(AP-DSTPR)/F)                         FUNC0015
      T2=(C(2)*LNS2+N2*TP(2))/F                                        FUNC0016
      IF(LNS .EQ. 0.) GO TO 10                                         FUNC0017
      P2 = (N2*SA-T2*CA+RU*W2*CA*SA+SEIC*W2)/CAP                       FUNC0018
      N1=(W1*(CAP-RU*CD*COS(AP-DSBMR)-SEIC*SAP)+C(1)*AL1*SIN(AP-DSBMR)/ FUNC0019
     1F)/(COS(AP-DSBMR)-TP(1)*SIN(AP-DSBMR)/F)                         FUNC0020
      T1=(C(1)*AL1+N1*TP(1))/F                                         FUNC0021
      P1=(T1*CD-N1*SD-RU*W1*CD*SD-SEIC*W1)/CAP                         FUNC0022
      N = (W*(1.-RU*CW*CW)+P2*SAP-P1*SAP-C(3)*LNS*SW/F)/(CW+SW*TP(3)/F) FUNC0023
      T=(C(3)*LNS+N*TP(3))/F                                           FUNC0024
      FCT=CAP*(P1-P2)-N*SW+T*CW-RU*W*CW*SW-SEIC*W                      FUNC0025
      RETURN                                                           FUNC0026
   10 P=(N2*SA-T2*CA+RU*W2*CA*SA+SEIC*W2)/CAP                          FUNC0027
      N1 = (W1*(1.-RU*CD*CD)+P*SAP-C(1)*AL1*SD/F)/(CD+TP(1)*SD/F)      FUNC0028
      T1=(C(1)*AL1+N1*TP(1))/F                                         FUNC0029
      FCT=P*CAP+N1*SD-T1*CD+RU*W1*CD*SD+SEIC*W1                        FUNC0030
      RETURN                                                           FUNC0031
      END                                                             FUNC0032
```

Appendix IV

List of SWASE in BASIC

```
10 REM SWASE (SLIDING WEDGE ANALYSIS OF SIDEHILL EMBANKMENTS)
20 DIM C(3),F4(30),P1(3),P2(3),T3(3),T$(72)
30 L=0
40 REM INPUT DATA
50 PRINT "INPUT TITLE";
60 INPUT T$
70 PRINT
80 PRINT "INPUT NO. OF DATA SETS TO BE RUN ",
90 INPUT N
100 PRINT
110 PRINT
120 PRINT
130 PRINT "COHESION ALONG FAILURE PLANE AT BOTTOM BLOCK";
140 INPUT C(1)
150 PRINT
160 PRINT "FRIC. ANGLE ALONG FAILURE PLANE AT BOTTOM BLOCK";
170 INPUT P1(1)
180 PRINT
190 PRINT "COHESION ALONG FAILURE PLANE AT TOP BLOCK";
200 INPUT C(2)
210 PRINT
220 PRINT "FRIC. ANGLE ALONG FAILURE PLANE AT TOP BLOCK";
230 INPUT P1(2)
240 PRINT
250 PRINT "COHESION ALONG FAILURE PLANE AT MIDDLE BLOCK";
260 INPUT C(3)
270 PRINT
280 PRINT "FRIC. ANGLE ALONG FAILURE PLANE AT MIDDLE BLOCK";
290 INPUT P1(3)
300 PRINT
310 PRINT
320 PRINT "ANGLE OF INTERNAL FRICTION OF FILL ",
330 INPUT P4
340 PRINT
350 PRINT
360 L=L+1
370 I=20
380 E=1E-4
390 X1=0.1
```

255

```
400 X2=10
410 D2=0.5
420 PRINT "DATA SET ",L
430 PRINT
440 PRINT
450 PRINT "DEGREE OF SLOPE AT BOTTOM ",
460 INPUT D3
470 PRINT
480 PRINT "DEGREE OF SLOPE AT TOP ",
490 INPUT D4
500 PRINT
510 PRINT "DEGREE OF SLOPE AT MIDDLE ",
520 INPUT D5
530 PRINT
540 PRINT "LENGTH OF NATURAL SLOPE AT TOP ",
550 INPUT L1
560 PRINT
570 PRINT "LENGTH OF NATURAL SLOPE AT MIDDLE ",
580 INPUT L2
590 PRINT
600 PRINT "PORE PRESSURE RATIO ",
610 INPUT R
620 PRINT
630 PRINT "SEISMIC COEFFICIENT ",
640 INPUT S2
650 PRINT
660 PRINT "UNIT WEIGHT OF FILL ",
670 INPUT G
680 PRINT
690 IF G=0 THEN 780
700 PRINT "ANGLE OF OUTSLOPE ",
710 INPUT A
720 PRINT
730 PRINT "BENCH WIDTH ",
740 INPUT B1
750 PRINT
760 B=B1
770 GO TO 900
780 PRINT "WEIGHT OF BOTTOM BLOCK ",
790 INPUT W1
800 PRINT
810 PRINT "WEIGHT OF TOP BLOCK ",
820 INPUT W2
830 PRINT
840 PRINT "WEIGHT OF MIDDLE BLOCK ",
850 INPUT W
860 PRINT
870 PRINT "LENGTH OF NATURAL SLOPE AT BOTTOM ",
880 INPUT A3
890 PRINT
900 I3=0
910 N1=1
920 I4=3
930 V=99999999
940 F=9999.99
950 X3=X2
960 X=X1
970 REM CONVERT DEGREES TO RADIANS
980 P3=P4*3.1416/180
990 FOR K=1 TO 3
1000 P2(K)=P1(K)*3.1416/180
1010 NEXT K
1020 IF G=0 THEN 1040
1030 A5=A*3.1416/180
```

```
1040 D=D4*3.1416/180
1050 D1=D3*3.1416/180
1060 D6=D5*3.1416/180
1070 IF G=0 THEN 1580
1080 REM DETERMINE DIMENSION AND WEIGHT OF DIFFERENT WEDGES
1090 X4= L1*COS(D)
1100 IF (B-X4)>0 THEN 1260
1110 REM WHEN B<=HORIZONTAL PROJECTION OF LNS2
1120 S4=X4-B
1130 A6=SIN(D)*L1-S4*TAN(A5)
1140 W2=(SIN(2*D)*L1**2/4-S4*S4*TAN(A5)/2)*G
1150 IF L2=0 THEN 1200
1160 X5=L2*COS(D6)
1170 Z1=X5*TAN(A5)
1180 Z2=A6+L2*SIN(D6)-Z1
1190 W=(A6+Z2)*X5*G/2
1200 IF L2<>0 THEN 1220
1210 Z2=A6
1220 A3=Z2*COS(A5)/SIN(A5-D1)
1230 W1=COS(D1)*A3*Z2*G/2
1240 GO TO 1580
1250 REM WHEN B > HORIZONTAL PROJECTION OF LNS2
1260 W2=SIN(2*D)*L1**2*G/4
1270 IF L2=0 THEN 1410
1280 X5=L2*COS(D6)
1290 IF (B-X4-X5)>0 THEN 1390
1300 REM WHEN B<= HORIZONTAL PROJECTION OF LNS2 +LNS
1310 A6=L1*SIN(D)
1320 Z3=L2*SIN(D6)
1330 S5=X5-(B-X4)
1340 W=(X5*A6-S5*S5*TAN(A5)/2+X5*Z3/2)*G
1350 A3=(A6+Z3-(X5+X4-B)*TAN(A5))*COS(A5)/SIN(A5-D1)
1360 W1=(A6+Z3-(X5+X4-B)*TAN(A5))*A3*COS(D1)*G/2
1370 GO TO 1580
1380 REM WHEN B> HORIZONTAL PROJECTION OF LNS2 + LNS
1390 W=(X5*(L1*SIN(D))+(SIN(2*D6)*L2*L2/4))*G
1400 Z3=SIN(D6)*L2
1410 IF L2<>0 THEN 1430
1420 X5=0
1430 IF L2<>0 THEN 1450
1440 Z3=0
1450 X6=B-X5-X4
1460 Y=SIN(D)*L1+Z3
1470 A6=SQR(X6*X6+Y*Y)
1480 W3=X6*Y/2*G
1490 S6=Y/A6
1500 IF ABS(S6)<0.99 THEN 1530
1510 A7=1.5708
1520 GO TO 1540
1530 A7=ATN(S6/SQR(1-S6**2))
1540 A8=3.1416-(A5+A7)
1550 A3=A6*SIN(A8)/SIN(A5-D1)
1560 A9=3.1416-(A8+A5-D1)
1570 W1=A6*A3*SIN(A9)*G/2+W3
1580 S=SIN(D)
1590 S1=SIN(D1)
1600 C1=COS(D)
1610 C2=COS(D1)
1620 FOR K=1 TO 3
1630 T3(K)=TAN(P2(K))
1640 NEXT K
1650 S3= SIN (D6)
1660 C3=COS(D6)
1670 REM CALL FIXLI TO DETERMINE INITIAL LEFT BOUND OF F
```

```
1680 GOSUB 2070
1690 IF I3=0   THEN 1730
1700 F4(L)=10
1710 GO TO 1830
1720 REM CALL RTM1 TO DETERMINE FACTOR OF SAFETY
1730 GOSUB 2320
1740 IF I4<>2 THEN 1790
1750 N1=2*N1
1760 IF N1>1024 THEN 1790
1770 X=X-D2/N1
1780 GO TO 1730
1790 PRINT "THE FACTOR OF SAFETY ",F
1800 PRINT
1810 PRINT
1820 F4(L)=F
1830 N=N-1
1840 IF N>0 THEN 360
1850 PRINT
1860 PRINT
1870 PRINT "    SET           FACTOR OF SAFETY  "
1880 PRINT
1890 L0=0
1900 F0=100
1910 FOR K=1 TO L
1920 PRINT "     ";K;"              ";F4(K)
1930 PRINT
1940 IF F0<=F4(K) THEN 1970
1950 F0=F4(K)
1960 L0=K
1970 NEXT K
1980 PRINT
1990 PRINT
2000 PRINT "   MINIMUM FACTOR OF SAFETY = ";F0;" AT SET ";L0
2010 GO TO 3290
2020 REM SUBROUTINE FIXLI
2030 REM THIS SUBROUTINE IS USED TO DETERMINE THE INITIAL LEFT BOUND OF
2040 REM THE ROOT OF F.THE VALUE OF F MUST SATIAFY THE BASIC REQUIREMENT
2050 REM THAT THE SHEAR RESISTANCES ALONG THE BASE OF ALL SLIDING WEDGES
2060 REM ARE POSITIVE
2070 T9=X
2080 GOSUB 3060
2090 T9=X
2100 IF L2=0 THEN 2170
2110 IF T>0 THEN 2130
2120 GO TO 2160
2130 IF T1>0 THEN 2150
2140 GO TO 2160
2150 IF T2>0 THEN   2280
2160 GO TO 2200
2170 IF T1>0 THEN 2190
2180 GO TO  2200
2190 IF T2>0 THEN 2280
2200 X=X+D2
2210 IF X<=10 THEN 2070
2220 PRINT
2230 PRINT
2240 PRINT
2250 PRINT "XLI IS GREATER THAN 10, SO THE FACTOR OF SAFETY CANNOT"
2260 PRINT "BE FOUND. PLEASE CHECK YOUR INPUT DATA."
2270 I3=1
2280 RETURN
2290 REM THIS SUBROUTINE WHICH IS OBTAINED FROM THE IBM PACKAGE,
2300 REM SOLVES THE GENERAL NONLINEAR EQUATION OF THE FORM FCT(F)=0
2310 REM BY MEANS OF MUELLER'S ITERATION METHOD
```

```
2320 I4=0
2330 X1=X
2340 X2=X3
2350 F=X1
2360 T9=F
2370 GOSUB 3060
2380 IF V=0 THEN 2800
2390 F1=V
2400 F=X2
2410 T9=F
2420 GOSUB 3060
2430 IF V=0 THEN 2800
2440 F2=V
2450 IF SGN(F1)=SGN(F2) THEN 3040
2460 I5=0
2470 T8=100*E
2480 I5=I5+1
2490 FOR J=1 TO I
2500 F=0.5*(X1+X2)
2510 T9=F
2520 GOSUB 3060
2530 IF V=0 THEN 2800
2540 IF SGN(V)=SGN(F2) THEN 2610
2550 T9= X1
2560 X1=X2
2570 X2=T9
2580 T9=F1
2590 F1=F2
2600 F2=T9
2610 T9=V-F1
2620 C4=V*T9
2630 C4=C4+C4
2640 IF ABS(F2)>1E30 THEN 2670
2650 IF (C4-F2*(F2-F1))>=0 THEN 2670
2660 IF (I5-I)<=0 THEN 2810
2670 X2=F
2680 F2=V
2690 T9=E
2700 C4=ABS(X2)
2710 IF (C4-1)<=0 THEN 2730
2720 T9=T9*C4
2730 IF (ABS(X2-X1)-T9)>0 THEN 2750
2740 IF (ABS(F2-F1)-T8)<=0 THEN 2770
2750 NEXT J
2760 I4=1
2770 IF (ABS(F2)-ABS(F1))<=0 THEN 2800
2780 F=X1
2790 V=F1
2800 RETURN
2810 C4=F2-V
2820 D7=(F-X1)*F1*(1+V*(C4-T9)/(C4*(F2-F1)))/T9
2830 X7=F
2840 F3=V
2850 F=X1-D7
2860 T9=F
2870 GOSUB 3060
2880 IF V=0 THEN 2800
2890 T9=E
2900 C4=ABS(F)
2910 IF (C4-1)<=0 THEN 2930
2920 T9=T9*C4
2930 IF (ABS(D7)-T9)>0 THEN 2950
2940 IF (ABS(V)-T8)<=0 THEN 2800
2950 IF SGN(V)=SGN(F1) THEN 2990
```

```
2960 X2=F
2970 F2=V
2980 GO TO 2480
2990 X1=F
3000 F1=V
3010 X2=X7
3020 F2=F3
3030 GO TO 2480
3040 I4=2
3050 RETURN
3060 REM THIS SUBROUTINE IS USED TO GENERATE THE FUNCTION FCT
3070 REM SO THAT FCT(F)=0
3080 Y1=ATN(TAN(P3)/T9)
3090 S7=SIN(Y1)
3100 Y2=COS(Y1)
3110 Y3=(W2*(Y2-R*C1*COS(Y1-D)-S2*S7)+C(2)*L1*SIN(Y1-D)/T9)
3120 Y3=Y3/(COS(Y1-D)-T3(2)*SIN(Y1-D)/T9)
3130 T2=(C(2)*L1+Y3*T3(2))/T9
3140 IF L2=0 THEN 3240
3150 P5=(Y3*S-T2*C1+R*W2*C1*S+S2*W2)/Y2
3160 Y4=(W1*(Y2-R*C2*COS(Y1-D1)-S2*S7)+C(1)*A3*SIN(Y1-D1)/T9)
3170 Y4=Y4/(COS(Y1-D1)-T3(1)*SIN(Y1-D1)/T9)
3180 T1=(C(1)*A3+Y4*T3(1))/T9
3190 Y5=(T1*C2-Y4*S1-R*W1*C2*S1-S2*W1)/Y2
3200 Y6=(W*(1-R*C3*C3)+P5*S7-Y5*S7-C(3)*L2*S3/T9)/(C3+S3*T3(3)/T9)
3210 T=(C(3)*L2+Y6*T3(3))/T9
3220 V=Y2*(Y5-P5)-Y6*S3+T*C3-R*W*C3*S3-S2*W
3230 GO TO 3280
3240 Y7= (Y3*S-T2*C1+R*W2*C1*S+S2*W2)/Y2
3250 Y4=(W1*(1-R*C2*C2)+Y7*S7-C(1)*A3*S1/T9)/(C2+T3(1)*S1/T9)
3260 T1=(C(1)*A3+Y4*T3(1))/T9
3270 V=Y7*Y2+Y4*S1-T1*C2+R*W1*C2*S1+S2*W1
3280 RETURN
3290 END
```

Appendix V

List of REAME in FORTRAN

```
C     REAME(ROTATIONAL EQUILIBRIUM ANALYSIS OF MULTILAYERED EMBANKMENTS)MAIN0001
C                                                                      MAIN0002
C     ****************************************************************MAIN0003
C     THIS PROGRAM WAS DEVELOPED BY DR. Y. H. HUANG, PROFESSOR OF CIVIL MAIN0004
C     ENGINEERING, UNIVERSITY OF KENTUCKY, LEXINGTON, KY. 40506.       MAIN0005
C     ****************************************************************MAIN0006
      REAL LX,LY                                                        MAIN0007
      COMMON/A1/C,DMIN,G,GW,NSLI   ,NSRCH,RL,RS,RU,SEIC,SLM,TANPHI,XO,YO MAIN0008
      COMMON/A4/NBL,NPBL,NPWT,NSPG,NWT,SLOPE,XBL,XWT,YBL,YINT,YWT        MAIN0009
      COMMON/A5/LM,XMINX,YMINY,XMINR                                    MAIN0010
      COMMON/A2/NK,RR,FF                                                MAIN0011
      COMMON/A3/KR,KRA,NQ,R,ROO                                         MAIN0012
      COMMON/A6/BFS,IC,ISTP,LBSHP,METHOD,NSTP                           MAIN0013
      DIMENSION C(19),EFF(40,5),FS(5),FSM(5),G(19),LINO(9,11),NBP(9,11),MAIN0014
     * NEP(9,11),NCIR(10),NPBL(20),PHID(19),RDEC(10),RL(11),RRM(5),RU(5)MAIN0015
     * ,SLOPE(49,20),TANPHI(19),X(3),XBL(50,20),XMINFS(5),XMINR(5),     MAIN0016
     * XMINX(5),XWT(50,5),Y(3),YBL(50,20),YINT(49,20),YMINY(5),YWT(50,5)MAIN0017
      DIMENSION FF(90,5),INFC(10),NPWT(5),NOL(11),RR(90,5)              MAIN0018
      DIMENSION BFS(5),ISTP(5),KRA(5),TITLE(20)                         MAIN0019
      READ (5,10) TITLE                                                 MAIN0020
C     ****************************************************************MAIN0021
   10 FORMAT (20A4)                                                     MAIN0022
      WRITE(6,20) TITLE                                                 MAIN0023
   20 FORMAT (1H1,///,10X,20A4,//)                                      MAIN0024
      READ (5,30) NCASE                                                 MAIN0025
C     ****************************************************************MAIN0026
   30 FORMAT (16I5)                                                     MAIN0027
      WRITE(6,40) NCASE                                                 MAIN0028
   40 FORMAT (//,10X,30HNUMBER OF CASES TO BE ANALYZED,I5,//)           MAIN0029
      DO 1580 NC=1,NCASE                                                MAIN0030
      WRITE (6,50) NC                                                   MAIN0031
   50 FORMAT (//,10X,11HCASE NUMBER,I5,///)                             MAIN0032
      READ(5,30) NBL,(NPBL(I),I=1,NBL)                                  MAIN0033
C     ****************************************************************MAIN0034
      WRITE(6,60)NBL,(NPBL(I),I=1,NBL)                                  MAIN0035
   60 FORMAT ( 5X,25HNUMBER OF BOUNDARY LINES=,I5,/,5X,39HNUMBER OF POIMAIN0036
     *NTS ON BOUNDARY LINES ARE:,8(5X,I5))                             MAIN0037
      DO 70 J=1,NBL                                                     MAIN0038
      NPBLJ=NPBL(J)                                                     MAIN0039
      READ (5,80) (XBL(I,J),YBL(I,J),I=1,NPBLJ)                         MAIN0040
```

```
C     ***************************************************************MAIN0041
   70 WRITE (6,90) J,(I,XBL(I,J),YBL(I,J),I=1,NPBLJ)                 MAIN0042
   80 FORMAT (8F10.3)                                               MAIN0043
   90 FORMAT (/,5X,20HON BOUNDARY LINE NO.,I2,31H,POINT NO. AND COORDINAMAIN0044
     *TES ARE:,/,5(I5,2F10.3))                                      MAIN0045
C DETERMINE SLOPE OF EACH LINE SEGMENT AND INTERCEPT AT Y AXIS.     MAIN0046
      WRITE (6,100)                                                 MAIN0047
  100 FORMAT (/,5X,39HLINE NO. AND SLOPE OF EACH SEGMENT ARE:)       MAIN0048
      DO 130 J=1,NBL                                                MAIN0049
      NPBL1=NPBL(J)-1                                               MAIN0050
      DO 120 I=1,NPBL1                                              MAIN0051
      IF(XBL(I+1,J) .EQ. XBL(I,J)) GO TO 110                        MAIN0052
      SLOPE(I,J)=(YBL(I+1,J)-YBL(I,J))/(XBL(I+1,J)-XBL(I,J))        MAIN0053
      GO TO 120                                                     MAIN0054
  110 SLOPE(I,J)=99999.                                             MAIN0055
  120 YINT(I,J)=YBL(I,J)- SLOPE(I,J)*XBL(I,J)                       MAIN0056
      WRITE(6,140) J,(SLOPE(I,J),I=1,NPBL1)                         MAIN0057
  130 CONTINUE                                                      MAIN0058
  140 FORMAT (7X,I5,(12F10.3))                                      MAIN0059
      READ (5,30) NRCZ,NPLOT,NQ                                     MAIN0060
C     ***************************************************************MAIN0061
      IF(NQ .EQ. 0) NQ=1                                            MAIN0062
      WRITE (6,150) NRCZ,NPLOT,NQ                                   MAIN0063
  150 FORMAT (/,5X,28HNO. OF RADIUS CONTROL ZONES=,I3,5X,16HPLOT OR NO PMAIN0064
     *LOT=,I3,5X,21HNO. OF SEEPAGE CASES=,I3)                       MAIN0065
      NQQ=NQ                                                        MAIN0066
C NGS IS INITIALIZED 0 BUT WILL CHANGE TO 1 WHEN SEARCH FOLLOWS GRID.MAIN0067
      NGS=0                                                         MAIN0068
      KRR=0                                                         MAIN0069
      IF(NRCZ .EQ. 0) GO TO 250                                     MAIN0070
      READ (5,30) (NOL(I),I=1,NRCZ)                                 MAIN0071
C     ***************************************************************MAIN0072
      WRITE (6,160) (NOL(I),I=1,NRCZ)                               MAIN0073
  160 FORMAT (/,5X,57HTOTAL NO. OF LINES AT BOTTOM OF RADIUS CONTROL ZONMAIN0074
     *ES ARE:,10I5)                                                 MAIN0075
      READ(5,80) (RDEC(I),I=1,NRCZ)                                 MAIN0076
C     ***************************************************************MAIN0077
      DO 170 I=1,NRCZ                                               MAIN0078
      NOLI=NOL(I)                                                   MAIN0079
      READ (5,30) NCIR(I),INFC(I), (LINO(J,I),NBP(J,I),NEP(J,I)     MAIN0080
     *, J=1,NOLI)                                                   MAIN0081
C     ***************************************************************MAIN0082
      WRITE (6,180) I,RDEC(I),NCIR(I),INFC(I)                       MAIN0083
      DO 170 J=1,NOLI                                               MAIN0084
  170 WRITE (6,190) LINO(J,I),NBP(J,I),NEP(J,I)                     MAIN0085
  180 FORMAT (/,5X,23HFOR RAD. CONT. ZONE NO.,I3,5X,17HRADIUS DECREMENT=MAIN0086
     *,F10.3,5X,15HNO. OF CIRCLES=,I3,5X,24HID NO. FOR FIRST CIRCLE=,I3)MAIN0087
  190 FORMAT (8X,9HLINE NO.=,I3,5X,14HBEGIN PT. NO.=,I3,5X,12HEND PT. NOMAIN0088
     *.=,I3)                                                        MAIN0089
      NOL1=NOL(1)                                                   MAIN0090
      XLE=XBL(NBP(1,1),LINO(1,1))                                   MAIN0091
      XRE=XBL(NEP(1,1),LINO(1,1))                                   MAIN0092
      IF(NOL1 .EQ. 1) GO TO 210                                     MAIN0093
      DO 200 I=2,NOL1                                               MAIN0094
      IF(XBL(NBP(I,1),LINO(I,1)) .LT. XLE) XLE=XBL(NBP(I,1),LINO(I,1)) MAIN0095
      IF(XBL(NEP(I,1),LINO(I,1)) .GT. XRE) XRE=XBL(NEP(I,1),LINO(I,1)) MAIN0096
  200 CONTINUE                                                      MAIN0097
  210 IF(XLE .EQ. XBL(1,NBL) .AND. XRE .EQ. XBL(NPBL(NBL),NBL)) GO TO230MAIN0098
      IF(NRCZ .NE. 1) GO TO 230                                     MAIN0099
      WRITE(6,220)                                                  MAIN0100
  220 FORMAT (/,5X,89HEND POINT OF LOWEST BOUNDARY LINE IS NOT EXTENDED MAIN0101
     *TO THE SAME X COORDINATE AS GROUND LINE,/,5X,52HOR GROUND LINE IS MAIN0102
     *TOO LONG AND INCLUDES ROCK SURFACE)                           MAIN0103
      STOP                                                          MAIN0104
```

```
C  KRB IS THE NO. ASSIGNED TO THE FIRST ADDITIONAL CIRCLE AFTER ALL THE MAIN0105
C  REGULAR NCIR(NRCZ) CIRCLES ARE NUMBERED.                               MAIN0106
   230 DO 240 I=1,NRCZ                                                    MAIN0107
   240 KRR=KRR+NCIR(I)                                                    MAIN0108
       KRB=KRR+1                                                         MAIN0109
   250 NOL(NRCZ+1)=1                                                      MAIN0110
C  USE EVERY SEGMENT OF GROUND LINE TO DETERMINE MINIMUM RADIUS           MAIN0111
       LINO(1,NRCZ+1)=NBL                                                MAIN0112
       NBP(1,NRCZ+1)=1                                                   MAIN0113
       NEP(1,NRCZ+1)=NPBL(NBL)                                           MAIN0114
C  NSOIL IS NO. OF SOIL LAYERS,WHICH IS EQUAL TO NO. OF SOIL BOUNDARY-1 MAIN0115
       NSOIL=NBL-1                                                       MAIN0116
       DO 260 I=1,NSOIL                                                  MAIN0117
       READ (5,80) C(I),PHID(I),G(I)                                     MAIN0118
C  **************************************************************MAIN0119
       IF(G(I) .GT. 900.) G(I)=G(I)*0.00981                              MAIN0120
   260 CONTINUE                                                          MAIN0121
       WRITE(6,270)                                                      MAIN0122
   270 FORMAT (/,5X,8HSOIL NO.,2X,8HCOHESION,5X,8HF. ANGLE,5X,8HUNIT WT.)MAIN0123
       IF(G(1) .GT. 900.) WRITE (6,280)                                  MAIN0124
   280 FORMAT (/,5X,60HALL UNIT WEIGHTS HAVE BEEN CONVERTED FROM KG/CU M MAIN0125
      *TO KN/CU M)                                                       MAIN0126
       DO 290 I=1,NSOIL                                                  MAIN0127
   290 WRITE (6,300) I,C(I),PHID(I),G(I)                                 MAIN0128
   300 FORMAT (5X,I5,3(3X,F10.3))                                        MAIN0129
       DO 310 I=1,NSOIL                                                  MAIN0130
   310 TANPHI(I)=TAN(PHID(I)*3.141593/180.)                              MAIN0131
       READ (5,30) METHOD,NSPG,NSRCH,NSLI,NK                             MAIN0132
C  **************************************************************MAIN0133
       READ (5,80) SEIC,DMIN,GW                                          MAIN0134
C  **************************************************************MAIN0135
       IF(GW .GT. 900.) GW=GW*0.00981                                    MAIN0136
       WRITE (6,320) SEIC,DMIN,GW                                        MAIN0137
   320 FORMAT (/,5X,20HSEISMIC COEFFICIENT=,F10.3,5X,28HMIN. DEPTH OF TALMAIN0138
      *LEST SLICE=,F10.3,5X,21HUNIT WEIGHT OF WATER=,F10.3)              MAIN0139
       IF (METHOD) 330,330,350                                           MAIN0140
   330 WRITE(6,340)                                                      MAIN0141
   340 FORMAT (/,5X, 57HTHE FACTORS OF SAFETY ARE DETERMINED BY THE NORMAMAIN0142
      *L METHOD)                                                         MAIN0143
       GO TO 370                                                         MAIN0144
   350 WRITE (6,360)                                                     MAIN0145
   360 FORMAT (/,5X, 68HTHE FACTORS OF SAFETY ARE DETERMINED BY THE SIMPLMAIN0146
      *IFIED BISHOP METHOD)                                              MAIN0147
   370 WRITE (6,380) NSPG,NSRCH,NSLI,NK                                  MAIN0148
   380 FORMAT (/,5X,5HNSPG=,I2,2X,6HNSRCH=,I2,2X,14HNO. OF SLICES=,I3,    MAIN0149
      * 2X,18HNO. OF ADD. RADII=,I2)                                     MAIN0150
       IF(NRCZ .EQ. 0 .AND. NSRCH .NE. 2) GO TO 1540                     MAIN0151
       XINC=0.                                                           MAIN0152
       YINC=0.                                                           MAIN0153
       IF(NSPG .EQ. 2) GO TO 430                                         MAIN0154
       IF(NSPG .EQ. 0) GO TO 450                                         MAIN0155
       READ (5,30) (NPWT(I),I=1,NQ)                                      MAIN0156
C  **************************************************************MAIN0157
       WRITE (6,390) (NPWT(I),I=1,NQ)                                    MAIN0158
   390 FORMAT (/,5X,43HNO. OF POINTS ON WATER TABLE FOR EACH CASE=,5I4)  MAIN0159
       DO 410 J=1,NQ                                                     MAIN0160
       NPWTJ=NPWT(J)                                                     MAIN0161
       READ (5,80) (XWT(I,J),YWT(I,J),I=1,NPWTJ)                         MAIN0162
C  **************************************************************MAIN0163
       WRITE (6,420) J,(I,XWT(I,J),YWT(I,J),I=1,NPWTJ)                   MAIN0164
       IF(XWT(1,J) .LE. XBL(1,NBL) .AND. XWT(NPWT(J),J) .GE. XBL(NPBL    MAIN0165
      * (NBL),NBL)) GO TO 410                                            MAIN0166
       WRITE(6,400) J                                                    MAIN0167
   400 FORMAT (/,5X,20HPIEZOMETRIC LINE NO.,I2,89H IS NOT EXTENDED AS FARMAIN0168
```

```
       * OUT AS GROUND LINE. PLEASE CHANGE DATA AND RUN THE PROGRAM AGAIN.MAIN0169
       * )                                                                 MAIN0170
         GO TO 1580                                                        MAIN0171
   410 CONTINUE                                                            MAIN0172
   420 FORMAT(/,5X,23HUNDER SEEPAGE CONDITION,I2,46H,POINT NO. AND COORDIMAIN0173
       *NATES OF WATER TABLE ARE:,/,5(I5,2F10.3))                         MAIN0174
         GO TO 450                                                         MAIN0175
   430 READ (5,80) (RU(I),I=1,NQ)                                          MAIN0176
C        ****************************************************************MAIN0177
         WRITE(6,440) (RU(I),I=1,NQ)                                       MAIN0178
   440 FORMAT(/,5X,20HPORE PRESSURE RATIO=,5F10.3)                        MAIN0179
   450 IF(NSRCH .NE. 0) GO TO 530                                         MAIN0180
C  USE GRID FOR DETERMINING FACTORS OF SAFETY AT VARIOUS POINTS           MAIN0181
         READ (5,80) (X(I),Y(I),I=1,3),XINC,YINC                          MAIN0182
C        ****************************************************************MAIN0183
         READ (5,30) NJ,NI                                                 MAIN0184
C        ****************************************************************MAIN0185
         WRITE(6,460)(X(I),Y(I),I=1,3),NJ,NI                              MAIN0186
   460 FORMAT(/,9H POINT1=(,F10.3,1H,,F10.3,1H),9H POINT2=(,F10.3,1H,,F10MAIN0187
       *.3,1H),9H POINT3=(,F10.3,1H,,F10.3,1H),4H NJ=,I5,4H NI=,I5)       MAIN0188
         IF(XINC .NE. 0. .OR. YINC .NE. 0.) WRITE (6,470) XINC,YINC       MAIN0189
   470 FORMAT (5X,50HAUTOMATIC SEARCH WILL FOLLOW AFTER GRID WITH XINC=, MAIN0190
       * F10.3,10H AND YINC=,F10.3)                                        MAIN0191
         DO 480 I=1,NQ                                                     MAIN0192
         ISTP(I)=0                                                         MAIN0193
   480 XMINR(I)=1.0E06                                                     MAIN0194
         ROO=0.                                                            MAIN0195
         NP=0                                                              MAIN0196
         IF (NI .NE. 0) GO TO 490                                         MAIN0197
         LX=0.                                                             MAIN0198
         LY=0.                                                             MAIN0199
         GO TO 500                                                         MAIN0200
   490 LX=(X(3)-X(2))/NI                                                   MAIN0201
         LY=(Y(3)-Y(2))/NI                                                 MAIN0202
   500 IF (NJ .NE. 0) GO TO 510                                           MAIN0203
         DX=0.                                                             MAIN0204
         DY=0.                                                             MAIN0205
         GO TO 520                                                         MAIN0206
   510 DX=(X(2)-X(1))/NJ                                                   MAIN0207
         DY=(Y(2)-Y(1))/NJ                                                 MAIN0208
   520 XV=X(1)                                                             MAIN0209
         YV=Y(1)                                                           MAIN0210
         XH=X(1)                                                           MAIN0211
         YH=Y(1)                                                           MAIN0212
         GO TO 650                                                         MAIN0213
C  DO NOT USE GRID                                                         MAIN0214
   530 READ(5,30) NP                                                       MAIN0215
C        ****************************************************************MAIN0216
         WRITE(6,540) NP                                                   MAIN0217
   540 FORMAT (/,5X,30HNO. OF CENTERS TO BE ANALYZED=,I5)                 MAIN0218
   550 IF(NSRCH .NE. 1) GO TO 560                                         MAIN0219
         NQ=1                                                              MAIN0220
         ISTP(1)=0                                                         MAIN0221
         XMINR(1)=1.0E06                                                   MAIN0222
         GO TO 580                                                         MAIN0223
   560 READ (5,80) XV,YV,R                                                 MAIN0224
C        ****************************************************************MAIN0225
         IF(NRCZ .EQ. 0 .AND. R .EQ.0.) GO TO 1520                        MAIN0226
         DO 570 I=1,NQ                                                     MAIN0227
         ISTP(I)=0                                                         MAIN0228
   570 XMINR(I)=1.0E06                                                     MAIN0229
         GO TO 630                                                         MAIN0230
   580 IF(NGS .EQ. 1) GO TO 600                                           MAIN0231
         READ (5,80) XV,YV,XINC,YINC                                      MAIN0232
```

```
C     *****************************************************************MAIN0233
      WRITE(6,590)XV,YV,XINC,YINC                                      MAIN0234
590   FORMAT(/T2,'POINT=(',F8.3,',',F8.3,')',' XINC=',F8.3,' YINC=',F8.3MAIN0235
     * )                                                               MAIN0236
      GO TO 620                                                        MAIN0237
600   LM=NQQ-NP+1                                                      MAIN0238
      IF(ISTP(LM) .EQ. 0) GO TO 610                                    MAIN0239
      LM=LM+1                                                          MAIN0240
      NP=NP-1                                                          MAIN0241
610   XV=XMINX(LM)                                                     MAIN0242
      YV=YMINY(LM)                                                     MAIN0243
620   R=0.                                                            MAIN0244
630   XOO=XV                                                          MAIN0245
      YOO=YV                                                          MAIN0246
      ROO=R                                                           MAIN0247
      NP=NP-1                                                          MAIN0248
      SLM=0.                                                          MAIN0249
       NI = 0                                                          MAIN0250
      NJ=0                                                            MAIN0251
      IF(NSRCH .EQ. 1) GO TO 640                                      MAIN0252
      DX=0                                                            MAIN0253
      DY=0                                                            MAIN0254
      GO TO 650                                                       MAIN0255
640   DX=XINC                                                         MAIN0256
      DY=YINC                                                         MAIN0257
      NCT=0                                                           MAIN0258
      NFD=0                                                           MAIN0259
      NSW=0                                                           MAIN0260
      NHV=0                                                           MAIN0261
      MOM=2                                                           MAIN0262
C  NAGAIN IS ASSIGNED 1 WHEN THE FULL INCREMENT OF SEARCH IS COMPLETED MAIN0263
C  AND THE QUARTER SEARCH INCREMENT BEGINS.                           MAIN0264
      NAGAIN=0                                                         MAIN0265
C  CHOOSE THE STARTING POINT FROM THE GRID WHICH IS DETERMINED BY      MAIN0266
C  THE GIVEN THREE POINTS,(X(1),Y(1)),(X(2),Y(2)),(X(3),Y(3))         MAIN0267
650   NSTP=0                                                          MAIN0268
      IF(NGS .EQ. 0) GO TO 660                                        MAIN0269
      ISTP(LM)=0                                                       MAIN0270
      XMINR(LM)=1.0E06                                                 MAIN0271
660   NI=NI+1                                                         MAIN0272
      NJ=NJ+1                                                         MAIN0273
      LBSHP=0                                                         MAIN0274
C  NSTATE IS ASSIGNED 2 WHEN A PREVIOUS SEARCH IS COMPLETED AND A SEARCHMAIN0275
C  FROM A NEW INITIAL TRIAL CENTER BEGINS.                            MAIN0276
      NSTATE=0                                                         MAIN0277
      M=0                                                             MAIN0278
C  BEGIN CENTER LOOP                                                  MAIN0279
      DO 1500 II=1,NI                                                  MAIN0280
      DO 1490 JJ=1,NJ                                                  MAIN0281
      IF(NGS .EQ. 0) LM=0                                             MAIN0282
      IF(NSPG .EQ. 0)LM=1                                             MAIN0283
      IF(NSRCH .EQ. 1 .AND. NSTATE .EQ. 0 .AND. NGS .EQ. 0) LM=1       MAIN0284
C  KR IS THE COUNTER FOR NO. OF NCIR CIRCLES BEING COMPUTED.          MAIN0285
670   KR=0                                                           MAIN0286
C  KRA IS THE COUNTER FOR NO. OF ADDITIONAL RADII BEING COMPUTED.     MAIN0287
      IF(NGS .EQ. 1) GO TO 690                                        MAIN0288
      DO 680 I=1,NQ                                                    MAIN0289
680   KRA(I)=KRR                                                      MAIN0290
690   IF(NSRCH .EQ. 1 .AND. NSTATE .EQ. 2) GO TO 700                  MAIN0291
      GO TO 710                                                       MAIN0292
C  RETURN TO FULL INCREMENT FOR SEARCH AFTER COMPLETING ONE QUARTER    MAIN0293
C  INCREMENT FOR PREVIOUS SEARCH POINT.                               MAIN0294
700   IF(NGS .EQ. 0) LM=LM+1                                          MAIN0295
      M=0                                                             MAIN0296
```

```
         NCT=0                                                        MAIN0297
         NFD=0                                                        MAIN0298
         NSW=0                                                        MAIN0299
         NHV=0                                                        MAIN0300
         MOM=2                                                        MAIN0301
         NAGAIN=0                                                     MAIN0302
         XV=XOO                                                       MAIN0303
         YV=YOO                                                       MAIN0304
         DX=XINC                                                      MAIN0305
         DY=YINC                                                      MAIN0306
C  LB IS A PARAMETER INDICATING THE CASE NO. FOR SEEPAGE CONDITIONS;  MAIN0307
C  ASSIGN 1 FOR ONE SEEPAGE CONDITION AND 0 FOR SEVERAL SEEPAGE       MAIN0308
C  CONDITIONS TO BE COMPUTED AT THE SAME TIME.                        MAIN0309
     710 LB=LM                                                        MAIN0310
         IF(NGS .EQ. 1) KRA(LB)=KRR                                   MAIN0311
         XO=XV                                                        MAIN0312
         YO=YV                                                        MAIN0313
         R=ROO                                                        MAIN0314
         IF(NSRCH .NE. 1) GO TO 730                                   MAIN0315
C  CHECK DISTANCE FROM CENTER TO INITIAL TRIAL CENTER                 MAIN0316
         IF(ABS(XO-XOO).LT. 20.*XINC .OR. ABS(YO-YOO) .LT. 20.*YINC)  MAIN0317
       * GO TO 730                                                    MAIN0318
         WRITE (6,720)                                                MAIN0319
     720 FORMAT(/,5X,106HTHE INCREMENTS USED FOR SEARCH ARE TOO SMALL,OR EQMAIN0320
       *UAL TO 0,SO THE MINIMUM FACTOR OF SAFETY CANNOT BE FOUND)     MAIN0321
         GO TO 1560                                                   MAIN0322
C  RLL IS DISTANCE FROM CENTER TO THE NEAREST END POINT OF GROUND LINE. MAIN0323
     730 RLL=SQRT((XO-XBL(1,NBL))**2+(YO-YBL(1,NBL))**2)              MAIN0324
         RLR=SQRT((XO-XBL(NPBL(NBL),NBL))**2+ (YO-YBL(NPBL(NBL),NBL))**2) MAIN0325
         IF(RLR .LT. RLL) RLL=RLR                                     MAIN0326
         GO TO 830                                                    MAIN0327
C  DETERMINE FACTOR OF SAFETY WHEN RADIUS IS SPECIFIED.               MAIN0328
     740 IF(R .GT. RS) GO TO 760                                      MAIN0329
         WRITE (6,750) XO,YO,R,RS                                     MAIN0330
     750 FORMAT(/,11H AT POINT (,F10.3,1H,,F10.3,1H),16H WITH RADIUS OF ,  MAIN0331
       * F8.3,/,45H RADIUS IS SMALLER THAN THE MINIMUM RADIUS OF, F10.3,   MAIN0332
       * 43H SO THE CIRCLE DOES NOT INTERSECT THE SLOPE)              MAIN0333
         GO TO 1560                                                   MAIN0334
     760 IF(NRCZ .EQ. 0) GO TO 780                                    MAIN0335
         IF(R .LE. RL(1)) GO TO 800                                   MAIN0336
         WRITE (6,770) XO,YO,R,RL(1)                                  MAIN0337
     770 FORMAT(/,11H AT POINT (,F10.3,1H,,F10.3,1H),16H WITH RADIUS OF ,  MAIN0338
       * F10.3,/,69H THE CIRCLE WILL CUT INTO LOWEST BOUNDARY,SO THE RADIUMAIN0339
       *S IS REDUCED TO,F10.3)                                       MAIN0340
         R=RL(1)                                                      MAIN0341
         GO TO 800                                                    MAIN0342
     780 IF(R .LE. RLL) GO TO 800                                     MAIN0343
         WRITE (6,790) XO,YO,R,RLL                                    MAIN0344
     790 FORMAT(/,11H AT POINT (,F10.3,1H,,F10.3,1H),16H WITH RADIUS OF ,  MAIN0345
       * F8.3,/,41H THE RADIUS IS GREATER THAN THE RADIUS OF, F10.3,  MAIN0346
       * 46H AS DETERMINED BY THE END POINT OF GROUND LINE)          MAIN0347
         GO TO 1560                                                   MAIN0348
     800 SLM=0.                                                       MAIN0349
         CALL FSAFTY(FS,LB,XANBL,XBNBL)                               MAIN0350
         IF(NSTP .EQ. 1) GO TO 1560                                   MAIN0351
         DO 810 I=1,NQ                                                MAIN0352
     810 WRITE(6,820) XO,YO,R,FS(I),I                                 MAIN0353
     820 FORMAT(/,11H AT POINT (,F10.3,1H,,F10.3,1H),16H WITH RADIUS OF ,  MAIN0354
       * F10.3,21H FACTOR OF SAFETY IS ,F10.3,15H UNDER SEEPAGE ,I5)  MAIN0355
         GO TO 1560                                                   MAIN0356
C  DETERMINE MAXIMUM RADIUS IN EACH CONTROL ZONE.                     MAIN0357
     830 NRCZ1=NRCZ+1                                                 MAIN0358
         DO 900 I=1,NRCZ1                                             MAIN0359
         PRL=99999.                                                   MAIN0360
```

```
           NOLI=NOL(I)                                                   MAIN0361
           DO 880 J=1,NOLI                                               MAIN0362
           NBPJI=NBP(J,I)                                                MAIN0363
           NEPJI=NEP(J,I)-1                                              MAIN0364
           DO 880 K=NBPJI,NEPJI                                          MAIN0365
           IF(XBL(K,LINO(J,I)) .EQ. XBL(K+1,LINO(J,I))) GO TO 870        MAIN0366
           XI=(XO+SLOPE(K,LINO(J,I))*(YO-YINT(K,LINO(J,I))))/(SLOPE(K,   MAIN0367
         * LINO(J,I))**2+1.)                                             MAIN0368
           IF(XI .LT. XBL(K,LINO(J,I))) GO TO 840                        MAIN0369
           GO TO 850                                                     MAIN0370
       840 RLI=SQRT((XO-XBL(K,LINO(J,I)))**2+ (YO-YBL(K,LINO(J,I)))**2)  MAIN0371
           GO TO 880                                                     MAIN0372
       850 IF(XI .GT. XBL(K+1,LINO(J,I))) GO TO 860                      MAIN0373
           RLI=SQRT((XO-XI)**2+ (YO-SLOPE(K,LINO(J,I))  *XI- YINT(K,LINO(J,I)MAIN0374
         * ))**2)                                                        MAIN0375
           GO TO 880                                                     MAIN0376
       860 RLI=SQRT((XO-XBL(K+1,LINO(J,I)))**2+ (YO-YBL(K+1,LINO(J,I)))**2) MAIN0377
           GO TO 880                                                     MAIN0378
       870 IF(YO .LT. YBL(K,LINO(J,I))) GO TO 840                        MAIN0379
           IF(YO .GT. YBL(K+1,LINO(J,I))) GO TO 860                      MAIN0380
           RLI=ABS(XO-XBL(K,LINO(I,J)))                                  MAIN0381
       880 IF (RLI .LT. PRL) PRL=RLI                                     MAIN0382
           IF(PRL .LE. RLL) GO TO 900                                    MAIN0383
           WRITE (6,890) I                                               MAIN0384
       890 FORMAT (/,58H ****WARNING AT NEXT CENTER**** AT RADIUS CONTROL ZONMAIN0385
         *E NO.,I2,58H,MAXIMUM RADIUS IS LIMITED BY THE END POINT OF GROUND MAIN0386
         *LINE)                                                          MAIN0387
           PRL=RLL                                                       MAIN0388
       900 RL(I)=PRL                                                     MAIN0389
           RS=RL(NRCZ1)                                                  MAIN0390
           IF(R .NE. 0.) GO TO 740                                       MAIN0391
C   COMPUTE FACTOR OF SAFETY FOR EACH RADIUS AND COMPARE WITH PREVIOUS   MAIN0392
C   FACTOR OF SAFETY                                                     MAIN0393
           DO 1050 I=1,NRCZ                                              MAIN0394
           IF(NCIR(I) .EQ. 0) GO TO 1050                                 MAIN0395
           R=RL(I)                                                       MAIN0396
           IF(RDEC(I)) 920,910,920                                       MAIN0397
       910 DR=(RL(I)-RL(I+1))/NCIR(I)                                    MAIN0398
           GO TO 930                                                     MAIN0399
       920 DR=RDEC(I)                                                    MAIN0400
       930 NCIRI=NCIR(I)                                                 MAIN0401
           R=R-(INFC(I)-1)*DR                                           MAIN0402
           DO 1040 J=1,NCIRI                                             MAIN0403
           SLM=0.                                                        MAIN0404
           IF(R .GE. RS) GO TO 960                                       MAIN0405
           IF(KR .NE. 0) GO TO 1130                                      MAIN0406
C   IF CIRCLE DOES NOT INTERCEPT EMBANKMENT, ASSIGN A LARGE F. S.        MAIN0407
           IF(LB .NE. 0) GO TO 950                                       MAIN0408
           DO 940 K=1,NQ                                                 MAIN0409
       940 FSM(K)=1.0E06                                                 MAIN0410
           GO TO 1130                                                    MAIN0411
       950 FSM(LB)=1.0E06                                                MAIN0412
           GO TO 1130                                                    MAIN0413
       960 IF(R .LT. RL(I+1))GO TO 1050                                  MAIN0414
           CALL FSAFTY(FS,LB,XANBL,XBNBL)                                MAIN0415
           IF(NSTP .EQ. 1) GO TO 1560                                    MAIN0416
           IF(SLM .GE. DMIN) GO TO 990                                   MAIN0417
           IF(KR .NE. 0) GO TO 1060                                      MAIN0418
           IF(LB .NE. 0) GO TO 980                                       MAIN0419
           DO 970 K=1,NQ                                                 MAIN0420
       970 FSM(K)=1.0E06                                                 MAIN0421
           RRM(K)=R                                                      MAIN0422
           GO TO 1060                                                    MAIN0423
       980 FSM(LB)=1.0E06                                                MAIN0424
```

```
      RRM(LB)=R                                                       MAIN0425
      GO TO 1060                                                      MAIN0426
  990 CALL SAVE(FS,RRM,FSM,LB,XANBL,XBNBL,XASM,XBSM)                  MAIN0427
      IF(NGS-1) 1010,1000,1010                                        MAIN0428
 1000 IF(FS(LB) .LT. 100) GO TO 1030                                  MAIN0429
      GO TO 1130                                                      MAIN0430
 1010 DO 1020 K=1,NQ                                                  MAIN0431
      IF(FS(K) .LT. 100.) GO TO 1030                                  MAIN0432
 1020 CONTINUE                                                        MAIN0433
      GO TO 1130                                                      MAIN0434
 1030 IF(DR .LE. 0.) GO TO 1050                                       MAIN0435
 1040 R=R-DR                                                          MAIN0436
 1050 CONTINUE                                                        MAIN0437
      R=R+DR                                                          MAIN0438
C   ADD NK MORE CIRCLES IF THE FACTOR OF SAFETY FOR THE SMALLEST CIRCLE MAIN0439
C   IS SMALLER THAN THAT OF THE PRECEDING CIRCLE.                     MAIN0440
 1060 IF(KR .EQ. 0) GO TO 1130                                        MAIN0441
      SLM2=SLM                                                        MAIN0442
      R2=R                                                            MAIN0443
      DO 1120 IQ=1,NQ                                                 MAIN0444
      I=IQ                                                            MAIN0445
      IF(LB .GT. 1) I=LB                                              MAIN0446
      IF(KR .EQ. 0 .OR. KR .EQ. 1 .AND. SLM2 .GT. DMIN) GO TO 1120   MAIN0447
      IF(KR .EQ. 1) GO TO 1070                                        MAIN0448
      IF(FF(KR,I) .GE. FF(KR-1,I)) GO TO 1120                         MAIN0449
 1070 RI=RR(KR,I)                                                     MAIN0450
      IF(SLM2-DMIN) 1090,1120,1080                                    MAIN0451
 1080 DELTA2=(RI-RS)/(NK+1)                                           MAIN0452
      GO TO 1100                                                      MAIN0453
 1090 DELTA2=(RI-R2)/(NK+1)                                           MAIN0454
 1100 DO 1110 N=1,NK                                                  MAIN0455
      RI=RI-DELTA2                                                    MAIN0456
      R=RI                                                            MAIN0457
      SLM=0.                                                          MAIN0458
      CALL FSAFTY(FS,LB,XANBL,XBNBL)                                  MAIN0459
      IF(SLM .LT. DMIN) GO TO 1120                                    MAIN0460
      KRA(I)=KRA(I)+1                                                 MAIN0461
      RR(KRA(I),I)=R                                                  MAIN0462
      FF(KRA(I),I)=FS(I)                                              MAIN0463
      IF(FS(I) .GE. FSM(I)) GO TO 1110                                MAIN0464
      FSM(I)=FS(I)                                                    MAIN0465
      RRM(I)=R                                                        MAIN0466
      XASM=XANBL                                                      MAIN0467
      XBSM=XBNBL                                                      MAIN0468
 1110 CONTINUE                                                        MAIN0469
 1120 CONTINUE                                                        MAIN0470
 1130 M=M+1                                                           MAIN0471
      IF(NSRCH .NE. 1) GO TO 1140                                     MAIN0472
      IF(NSTATE-1)1160,1160,1150                                      MAIN0473
 1140 IF(NSPG .GE. 1 .AND. NGS .EQ. 0) LM=LM+1                        MAIN0474
 1150 NSTATE=1                                                        MAIN0475
C   WRITE OUT FACTORS OF SAFETY FOR ALL CIRCLES AND THE LOWEST F. S.  MAIN0476
 1160 IF(KR .NE. 0) GO TO 1200                                        MAIN0477
      IF(SLM .EQ. 0.) GO TO 1180                                      MAIN0478
      WRITE (6,1170) XO,YO,LM                                         MAIN0479
 1170 FORMAT(/,11H AT POINT (,F10.3,1H,,F10.3,1H),14H UNDER SEEPAGE,I5, MAIN0480
     * 45H THE DEPTH OF TALLEST SLICE IS LESS THAN DMIN)              MAIN0481
      GO TO 1250                                                      MAIN0482
 1180 WRITE (6,1190) XO,YO,LM                                         MAIN0483
 1190 FORMAT(/,11H AT POINT (,F10.3,1H,,F10.3,1H),14H UNDER SEEPAGE,I5, MAIN0484
     * 40H THE CIRCLE DOES NOT INTERCEPT THE SLOPE)                   MAIN0485
      GO TO 1250                                                      MAIN0486
 1200 IF(ISTP(LM) .EQ. 1) GO TO 1270                                  MAIN0487
      WRITE (6,1210) XO,YO,LM                                         MAIN0488
```

```
 1210 FORMAT(/,11H AT POINT (,F10.3,1H,,F10.3,1H),15H UNDER SEEPAGE ,I5,MAIN0489
      * 55H,THE RADIUS AND THE CORRESPONDING FACTOR OF SAFETY ARE:)    MAIN0490
      WRITE(6,1220) (RR(KK,LM),FF(KK,LM),KK=1,KR)                      MAIN0491
      IF(KRA(LM) .LT. KRB) GO TO 1230                                  MAIN0492
      KRE=KRA(LM)                                                      MAIN0493
      WRITE(6,1220) (RR(KK,LM),FF(KK,LM),KK=KRB,KRE)                   MAIN0494
 1220 FORMAT(5(5X,2F10.3))                                            MAIN0495
 1230 WRITE(6,1240) FSM(LM),RRM(LM)                                    MAIN0496
 1240 FORMAT (5X,24HLOWEST FACTOR OF SAFETY=,F10.3,24H AND OCCURS AT RADMAIN0497
      *IUS = ,F10.3,/)                                                 MAIN0498
C                                                                      MAIN0499
C  DETERMINE THE MINIMUM FACTOR OF SAFETY FOR ALL POINTS ON GRID.      MAIN0500
C                                                                      MAIN0501
 1250 IF(NSRCH .EQ. 1) GO TO 1280                                      MAIN0502
      IF(M .EQ. 1) GO TO 1260                                          MAIN0503
      IF(FSM(LM) .GE. XMINFS(LM)) GO TO 1270                           MAIN0504
 1260 XMINFS(LM)=FSM(LM)                                               MAIN0505
      XMINR(LM)=RRM(LM)                                                MAIN0506
      XMINX(LM)=XO                                                     MAIN0507
      YMINY(LM)=YO                                                     MAIN0508
 1270 IF(LM .LT. NQ) GO TO 1140                                        MAIN0509
      IF(NSRCH .EQ. 2) GO TO 1560                                      MAIN0510
      XV=XV+DX                                                         MAIN0511
      YV=YV+DY                                                         MAIN0512
      GO TO 1490                                                       MAIN0513
C  AUTOMATIC SEARCH                                                    MAIN0514
 1280 IF (M .EQ. 1) GO TO 1300                                         MAIN0515
      NCT=NCT+1                                                        MAIN0516
      IF(FSM(LM)-XMINFS(LM)) 1300,1290,1290                            MAIN0517
 1290 IF(NHV .EQ. 1) GO TO 1310                                        MAIN0518
      IF(NSW .NE. 0) GO TO 1390                                        MAIN0519
      IF(NCT .NE. 1) GO TO 1380                                        MAIN0520
C  REVERSE X-DIRECTION                                                 MAIN0521
      NCT=0                                                            MAIN0522
      XV=XV-MOM*DX                                                     MAIN0523
      NSW=1                                                            MAIN0524
      GO TO 1400                                                       MAIN0525
C                                                                      MAIN0526
C  SAVE MIN. FS                                                        MAIN0527
 1300 XMINFS(LM)=FSM(LM)                                               MAIN0528
      XMINR(LM)=RRM(LM)                                                MAIN0529
      XMINX(LM)=XO                                                     MAIN0530
      YMINY(LM)=YO                                                     MAIN0531
      XAMIN=XASM                                                       MAIN0532
      XBMIN=XBSM                                                       MAIN0533
      GO TO 1340                                                       MAIN0534
 1310 IF(NSW .NE. 0) GO TO 1320                                        MAIN0535
      IF (NCT .NE. 1) GO TO 1330                                       MAIN0536
C  REVERSE Y-DIRECTION                                                 MAIN0537
      NSW=1                                                            MAIN0538
      YV=YV-MOM*DY                                                     MAIN0539
      NCT=0                                                            MAIN0540
      GO TO 1400                                                       MAIN0541
 1320 IF (NCT .NE. 1) GO TO 1330                                       MAIN0542
      IF(NFD .NE. 1) GO TO 1330                                        MAIN0543
      NFD=2                                                            MAIN0544
      GO TO 1400                                                       MAIN0545
C  HORIZONTAL DIRECTION                                                MAIN0546
 1330 NCT=0                                                            MAIN0547
      NFD=1                                                            MAIN0548
      M=1                                                              MAIN0549
      NSW=0                                                            MAIN0550
      NHV=0                                                            MAIN0551
      MOM=2                                                            MAIN0552
```

```
            XV=XMINX(LM)+DX                                                    MAIN0553
            YV=YMINY(LM)                                                       MAIN0554
            IF (DX .EQ. 0.) NFD=2                                             MAIN0555
            GO TO 1400                                                         MAIN0556
      1340 IF(NHV .EQ. 1) GO TO 1360                                          MAIN0557
            IF(NSW .EQ. 0) GO TO 1350                                         MAIN0558
            XV=XV-DX                                                           MAIN0559
            GO TO 1400                                                         MAIN0560
      1350 XV=XV+DX                                                            MAIN0561
            IF(DX .EQ. 0.) GO TO 1380                                         MAIN0562
            GO TO 1400                                                         MAIN0563
      1360 IF(NSW .EQ. 0) GO TO 1370                                          MAIN0564
            YV=YV-DY                                                           MAIN0565
            GO TO 1400                                                         MAIN0566
      1370 YV=YV+DY                                                            MAIN0567
            GO TO 1400                                                         MAIN0568
C     VERTICAL DIRECTION                                                       MAIN0569
      1380 M=1                                                                 MAIN0570
            NFD=1                                                              MAIN0571
            NCT=0                                                              MAIN0572
            NHV=1                                                              MAIN0573
            MOM=2                                                              MAIN0574
            NSW=0                                                              MAIN0575
            XV=XMINX(LM)                                                       MAIN0576
            YV=YMINY(LM)+DY                                                    MAIN0577
            IF(DY .EQ. 0.) NFD=2                                              MAIN0578
            GO TO 1400                                                         MAIN0579
      1390 IF(NCT .NE. 1) GO TO 1380                                          MAIN0580
            IF(NFD .NE . 1) GO TO 1380                                        MAIN0581
            NFD=2                                                              MAIN0582
      1400 R=0                                                                 MAIN0583
            IF(NFD .NE. 2) GO TO 670                                          MAIN0584
            NAGAIN=NAGAIN+1                                                    MAIN0585
            IF(NAGAIN .NE. 1) GO TO 1430                                      MAIN0586
C                                                                             MAIN0587
C  DO THE SEARCH AGAIN BY USING ONE FOURTH OF THE INCREMENT IN X AND Y        MAIN0588
C  DIRECTIONS                                                                  MAIN0589
C                                                                             MAIN0590
            IF(XMINFS(LM) .LT. 100.) GO TO 1420                              MAIN0591
            WRITE (6,1410)                                                     MAIN0592
      1410 FORMAT (/,5X,34HIMPROPER CENTER IS USED FOR SEARCH)               MAIN0593
            NSTATE=2                                                          MAIN0594
            GO TO 1560                                                         MAIN0595
      1420 DX=XINC/4.                                                         MAIN0596
            DY=YINC/4.                                                         MAIN0597
            NFD=0                                                              MAIN0598
            NCT=0                                                              MAIN0599
            NSW=0                                                              MAIN0600
            NHV=0                                                              MAIN0601
            MOM=2                                                              MAIN0602
            M=1                                                                MAIN0603
            NAGAIN=NAGAIN+1                                                    MAIN0604
            XV=XMINX(LM)+DX                                                    MAIN0605
            YV=YMINY(LM)                                                       MAIN0606
            GO TO 670                                                          MAIN0607
      1430 IF(NSPG .EQ. 1) WRITE (6,1440) LM                                 MAIN0608
      1440 FORMAT (///,25H FOR PIEZOMETRIC LINE NO.,I5)                      MAIN0609
            IF(NSPG .EQ. 2) WRITE(6,1450)RU(LM)                             MAIN0610
      1450 FORMAT (//,24H FOR PORE PRESSURE RATIO,F10.3)                    MAIN0611
            WRITE(6,1460)XMINX(LM),YMINY(LM),XMINR(LM),XMINFS(LM)           MAIN0612
      1460 FORMAT(/,11H AT POINT (,F10.3,1H,,F10.3,1H),7H,RADIUS,F10.3,/,    MAIN0613
          * 5X,31HTHE MINIMUM FACTOR OF SAFETY IS ,F10.3,//)               MAIN0614
            IF(METHOD .EQ. 1) GO TO 1480                                    MAIN0615
            LBSHP=1                                                          MAIN0616
```

```
          XO=XMINX(LM)                                                  MAIN0617
          YO=YMINY(LM)                                                  MAIN0618
          R=XMINR(LM)                                                   MAIN0619
          CALL FSAFTY(FS,LM,XAMIN,XBMIN)                                MAIN0620
          WRITE(6,1470) BFS(LM),IC                                      MAIN0621
 1470 FORMAT (4X,'THE CORRESPONDING FACTOR OF SAFETY BY THE SIMPLIFIED',MAIN0622
     *' BISHOP METHOD=',F10.3,5X,'THE NUMBER OF ITERATIONS=',I5)        MAIN0623
 1480 CONTINUE                                                          MAIN0624
          IF(NPLOT .EQ. 1) CALL XYPLOT(XMINFS(LM),XAMIN,XBMIN)          MAIN0625
          IF(NSRCH .EQ. 1) GO TO 1560                                   MAIN0626
          NSTATE=2                                                      MAIN0627
          IF(NP .NE. 0) GO TO 550                                       MAIN0628
 1490 CONTINUE                                                          MAIN0629
          XH=XH+LX                                                      MAIN0630
          YH=YH+LY                                                      MAIN0631
          XV=XH                                                         MAIN0632
          YV=YH                                                         MAIN0633
 1500 CONTINUE                                                          MAIN0634
C  END CENTER LOOP                                                      MAIN0635
          DO 1510 LM=1,NQ                                               MAIN0636
          IF(ISTP(LM) .EQ. 1) GO TO 1510                                MAIN0637
          IF(NSPG .EQ. 1)WRITE(6,1440) LM                               MAIN0638
          IF(NSPG .EQ. 2) WRITE(6,1450)RU(LM)                           MAIN0639
          WRITE (6,1460) XMINX(LM),YMINY(LM),XMINR(LM),XMINFS(LM)       MAIN0640
 1510 CONTINUE                                                          MAIN0641
          GO TO 1560                                                    MAIN0642
 1520 WRITE(6,1530) XV,YV,R                                             MAIN0643
 1530 FORMAT(/,11H AT POINT (,F10.3,1H,,F10.3,1H),7H,RADIUS,F10.3,/,    MAIN0644
     * 5X,78HWHEN THE NUMBER OF RADIUS CONTROL ZONE IS 0,RADIUS SHOULD NMAIN0645
     *OT BE ASSIGNED ZERO)                                             MAIN0646
          NP=NP-1                                                       MAIN0647
          GO TO 1560                                                    MAIN0648
 1540 WRITE (6,1550)                                                    MAIN0649
 1550 FORMAT(/,5X,102HRADIUS CONTROL ZONE IS NOT DEFINED SO THE PROGRAM MAIN0650
     *IS STOPPED. UNLESS NSRCH=2, NCRZ SHOULD NOT BE ZERO.)            MAIN0651
          GO TO 1580                                                    MAIN0652
 1560 IF(NP .NE. 0) GO TO 550                                           MAIN0653
          IF(NSRCH .EQ. 0 .AND. (XINC .NE. 0. .OR. YINC .NE. 0.) .AND. NSTP MAIN0654
     * .EQ. 0) GO TO 1570                                              MAIN0655
          GO TO 1580                                                    MAIN0656
 1570 NSRCH=1                                                           MAIN0657
          NP=NQ                                                         MAIN0658
          NGS=1                                                         MAIN0659
          NQ=1                                                          MAIN0660
          GO TO 600                                                     MAIN0661
 1580 CONTINUE                                                          MAIN0662
          STOP                                                          MAIN0663
          END                                                           MAIN0664
C                                                                       FSTY0001
C                                                                       FSTY0002
      SUBROUTINE FSAFTY(FS,LM,XANBL,XBNBL)                              FSTY0003
C                                                                       FSTY0004
C  THIS SUBROUTINE IS USED TO DETERMINE THE FACTOR OF SAFETY FOR EACH   FSTY0005
C  TRIAL CIRCLE.  TO SAVE THE COMPUTER TIME, SEVERAL SEEPAGE CONDITIONS FSTY0006
C  ARE COMPUTED AT THE SAME TIME IF THE AUTOMATIC SEARCH IS NOT USED.   FSTY0007
C  LM IS THE CASE NO. FOR SEEPAGE CONDITIONS;ASSIGN 0 FOR SEVERAL       FSTY0008
C  SEEPAGE CONDITIONS TO BE COMPUTED AT THE SAME TIME.                  FSTY0009
C                                                                       FSTY0010
      COMMON/A1/C,DMIN,G,GW,NSLI   ,NSRCH,RL,RS,RU,SEIC,SLM,TANPHI,XO,YO FSTY0011
      COMMON/A4/NBL,NPBL,NPWT,NSPG,NWT,SLOPE,XBL,XWT,YBL,YINT,YWT        FSTY0012
      COMMON/A3/KR,KRA,NQ,R,ROO                                         FSTY0013
      COMMON/A6/BFS,IC,ISTP,LBSHP,METHOD,NSTP                           FSTY0014
      DIMENSION COSA(40),EFF(40,5),FS(5),LC(40),SINA(40),SLOPE(49,20),  FSTY0015
     * W(40),WW(40),XC(40),XBL(50,20), YB(40,20),YBL(50,20),YC(40),     FSTY0016
```

```
      *  YINT(49,20),YT(40,5),NPBL(20),XWT(50,5),YWT(50,5),ISTP(5)          FSTY0017
         DIMENSION BFS(5),KRA(5),XA(20),XAPT(40),XB(20),BB(40)              FSTY0018
         DIMENSION C(19),G(19),NPWT(5),RL(11),RU(5),SL(40),TANPHI(19)       FSTY0019
         GO TO 40                                                           FSTY0020
   10 IF(LM .NE. 0) GO TO 30                                                FSTY0021
         DO 20 L=1,NQ                                                       FSTY0022
   20 FS(L)=1.0E06                                                          FSTY0023
         GO TO 960                                                          FSTY0024
   30 FS(LM)=1.0E06                                                         FSTY0025
         GO TO 960                                                          FSTY0026
C  COMPUTE POINTS OF INTERSECTION BETWEEN CIRCLE AND BOUNDARY LINES.        FSTY0027
   40 DO 100 J=1,NBL                                                        FSTY0028
         XA(J)=-99999.                                                      FSTY0029
         XB(J)= 99999.                                                      FSTY0030
         NPBL1=NPBL(J)-1                                                    FSTY0031
         DO 90 I=1,NPBL1                                                    FSTY0032
         IF(SLOPE(I,J) .EQ. 99999.) GO TO 90                               FSTY0033
         A=1.+SLOPE(I,J)**2                                                 FSTY0034
         B=SLOPE(I,J)*(YINT(I,J)-YO)-XO                                     FSTY0035
         D=(YINT(I,J)-YO)**2+ XO**2-R**2                                    FSTY0036
         TEST=B**2-A*D                                                      FSTY0037
         IF(ABS(TEST) .LT. 1.) GO TO 100                                   FSTY0038
         IF(TEST .GT. 0.) GO TO 50                                         FSTY0039
         GO TO 90                                                           FSTY0040
   50 XM=(-B-SQRT(TEST))/A                                                  FSTY0041
         XP=(-B+SQRT(TEST))/A                                               FSTY0042
         IF(XM.GE.XBL(I, J )-0.01 .AND. XM.LE.XBL(I+1, J )+0.01) GO TO 60   FSTY0043
         GO TO 70                                                           FSTY0044
   60 XA(J)=XM                                                              FSTY0045
         IF(J .EQ. NBL) YA=SLOPE(I,NBL)*XA(NBL)+ YINT(I,NBL)               FSTY0046
   70 IF(XP.GE.XBL(I, J )-0.01 .AND. XP.LE.XBL(I+1, J )+0.01) GO TO 80      FSTY0047
         GO TO 90                                                           FSTY0048
   80 XB(J)=XP                                                              FSTY0049
         IF(J .EQ. NBL) YBB=SLOPE(I,NBL)*XB(NBL)+ YINT(I,NBL)             FSTY0050
   90 CONTINUE                                                              FSTY0051
  100 CONTINUE                                                              FSTY0052
         IF(XA(NBL) .NE. -99999. .AND. XB(NBL) .NE. 99999.) GO TO 130      FSTY0053
  110 WRITE(6,120) R                                                        FSTY0054
  120 FORMAT (/,46H ****WARNING AT NEXT CENTER**** WHEN RADIUS IS,F10.3,    FSTY0055
      */119H CENTER OF CIRCLE LIES BELOW GROUND LINE OR CIRCLE DOES NOT IFSTY0056
      *NTERCEPT GROUND LINE PROPERLY,OR THE CIRCLE CUTS THE SLOPE,/,        FSTY0057
      *56H VERY SLIGHTLY, SO A LARGE FACTOR OF SAFETY IS ASSIGNED.)         FSTY0058
         GO TO 10                                                           FSTY0059
C  REPLACE THE ARC AT THE END BY A VERTICAL LINE WHEN CENTER OF CIRCLE      FSTY0060
C  LIES BELOW POINT A OR B.                                                 FSTY0061
  130 IF (YO .GE. YA .OR. YO .GE. YBB) GO TO 150                           FSTY0062
         WRITE(6,140) R                                                     FSTY0063
  140 FORMAT (/,46H ****WARNING AT NEXT CENTER**** WHEN RADIUS IS,F10.3,    FSTY0064
      * /,122H THE CENTER OF CIRCLE IS LOCATED BELOW THE INTERSECTION OF    FSTY0065
      *CIRCLE WITH GROUND LINE,SO A LARGE FACTOR OF SAFETY IS ASSIGNED)     FSTY0066
         GO TO 10                                                           FSTY0067
  150 RMI=0.                                                                FSTY0068
         IF(YO .GT. YA) GO TO 250                                          FSTY0069
         XA(NBL)=XO-R                                                       FSTY0070
         XX=XA(NBL)                                                         FSTY0071
  160 DO 240 K=1,NBL                                                        FSTY0072
         KK=NBL+1-K                                                         FSTY0073
         NPBL1=NPBL(KK)-1                                                   FSTY0074
         YB(1,KK)=-9999.                                                    FSTY0075
         DO 170 I=1,NPBL1                                                   FSTY0076
         IF(XBL(I,KK) .EQ. XBL(I+1,KK)) GO TO 170                          FSTY0077
         M=I                                                                FSTY0078
         IF(XX .GT. XBL(I,KK) .AND. XX .LE. XBL(I+1,KK)) GO TO 180         FSTY0079
  170 CONTINUE                                                              FSTY0080
```

```
        GO TO 190                                              FSTY0081
 180 YB(1,KK)=SLOPE(M,KK)*XX+YINT(M,KK)                        FSTY0082
        IF(K .EQ. 1) GO TO 230                                 FSTY0083
 190 IF(YB(1,KK)+9999.) 200,240,200                            FSTY0084
 200 IF(YO-YB(1,KK)) 210,210,220                               FSTY0085
 210 RMI=RMI+(YBJK-YB(1,KK))*C(KK)                             FSTY0086
        GO TO 230                                              FSTY0087
 220 RMI=RMI+ (YBJK-YO)*C(KK)                                  FSTY0088
        GO TO 260                                              FSTY0089
 230 YBJK=YB(1,KK)                                             FSTY0090
 240 CONTINUE                                                  FSTY0091
 250 IF(YO .GT. YBB) GO TO 260                                 FSTY0092
        XB(NBL)=XO+R                                           FSTY0093
        XX=XB(NBL)                                             FSTY0094
        GO TO 160                                              FSTY0095
C  COMPUTE WIDTH OF SLICE,BB,AND SUBDIVIDE INTO SMALLER SLICES IF NEEDEDFSTY0096
 260 NBLM=NBL-1                                                FSTY0097
        NPBLM=NPBL(NBL)-1                                      FSTY0098
        RB=(XB(NBL)-XA(NBL))/NSLI                              FSTY0099
        XANBL=XA(NBL)                                          FSTY0100
        XBNBL=XB(NBL)                                          FSTY0101
C  NSLICE IS THE ACTUAL NO. OF SLICES, WHICH IS GENERALLY GREATER THAN  FSTY0102
C  NSLI.                                                       FSTY0103
        NSLICE=0                                               FSTY0104
        XDIS=XA(NBL)                                           FSTY0105
        DO 370 I=1,NSLI                                        FSTY0106
        NAPT=0                                                 FSTY0107
        DO 270 J=2,NPBLM                                       FSTY0108
        IF(XBL(J,NBL) .LE. XDIS+0.01 .OR. XBL(J,NBL) .GE. XDIS+RB-0.01)  FSTY0109
     * GO TO 270                                               FSTY0110
        NAPT=NAPT+1                                            FSTY0111
        XAPT(NAPT)= XBL(J,NBL)                                 FSTY0112
 270 CONTINUE                                                  FSTY0113
        DO 290 J=2,NBLM                                        FSTY0114
        IF(XA(J) .LE. XDIS+0.01 .OR. XA(J) .GE. XDIS+RB-0.01) GO TO 280  FSTY0115
        NAPT=NAPT+1                                            FSTY0116
        XAPT(NAPT)=XA(J)                                       FSTY0117
        GO TO 290                                              FSTY0118
 280 IF(XB(J) .LE. XDIS+0.01 .OR. XB(J) .GE. XDIS+RB-0.01) GO TO 290     FSTY0119
        NAPT=NAPT+1                                            FSTY0120
        XAPT(NAPT)= XB(J)                                      FSTY0121
 290 CONTINUE                                                  FSTY0122
C  ARRANGE COORDINATES OF INTERMEDIATE POINTS BY INCREASING ORDER       FSTY0123
        IF(NAPT-1) 350,340,300                                 FSTY0124
 300 NAPT1=NAPT-1                                              FSTY0125
        NKOT=0                                                 FSTY0126
        DO 320 L=1,NAPT1                                       FSTY0127
        NKOT=NKOT+1                                            FSTY0128
        XSM=XAPT(NKOT)                                         FSTY0129
        NKOT1=NKOT+1                                           FSTY0130
        KID=NKOT                                               FSTY0131
        DO 310 K=NKOT1,NAPT                                    FSTY0132
        IF(XAPT(K) .GE. XSM) GO TO 310                         FSTY0133
        XSM=XAPT(K)                                            FSTY0134
        KID=K                                                  FSTY0135
 310 CONTINUE                                                  FSTY0136
        XAPT(KID)=XAPT(NKOT)                                   FSTY0137
        XAPT(NKOT)= XSM                                        FSTY0138
 320 CONTINUE                                                  FSTY0139
        BB(NSLICE+1)=XAPT(1)-XDIS                              FSTY0140
        DO 330 L=2,NAPT                                        FSTY0141
 330 BB(NSLICE+L)=XAPT(L)-XAPT(L-1)                            FSTY0142
        BB(NSLICE+NAPT+1)=XDIS+RB-XAPT(NAPT)                   FSTY0143
        GO TO 360                                              FSTY0144
```

```
  340 BB(NSLICE+1)= XAPT(1)-XDIS                                         FSTY0145
      BB(NSLICE+2)= XDIS+RB-XAPT(1)                                      FSTY0146
      GO TO 360                                                          FSTY0147
  350 BB(NSLICE+1)= RB                                                   FSTY0148
  360 NSLICE=NSLICE+NAPT+1                                               FSTY0149
      XDIS=XDIS+RB                                                       FSTY0150
  370 CONTINUE                                                           FSTY0151
      IJK=0                                                              FSTY0152
      DO 380 J=1,NSLICE                                                  FSTY0153
      IF(J .GT. NSLICE-IJK) GO TO 390                                    FSTY0154
      IF(BB(J) .EQ. 0.) IJK=IJK+1                                        FSTY0155
  380 BB(J)=BB(J+IJK)                                                    FSTY0156
  390 NSLICE=NSLICE-IJK                                                  FSTY0157
      IF(NSLICE .LE. 40) GO TO 410                                       FSTY0158
      WRITE(6,400) NSLICE                                                FSTY0159
  400 FORMAT (/,5X,23HACTUAL NO. OF SLICES IS,I3,75H,WHICH IS GREATER THFSTY0160
     *AN THE DECLARED DIMENSION,SO THE COMPUTATION IS STOPPED)           FSTY0161
      NSTP=1                                                             FSTY0162
      GO TO 960                                                          FSTY0163
C  COMPUTE X COORDINATES FOR CENTERLINE OF SLICE,XC                      FSTY0164
  410 XC(1)=XA(NBL)+BB(1)/2.                                             FSTY0165
      DO 420 J=2,NSLICE                                                  FSTY0166
  420 XC(J)=XC(J-1)+(BB(J)+BB(J-1))/2.                                   FSTY0167
C  COMPUTE INTERSECTION OF SLICE CENTERLINE WITH CIRLE,YC, AND INCLINA-  FSTY0168
C  TION OF FAILURE ARC AT BOTTOM OF SLICE.                               FSTY0169
      DO 430 J=1,NSLICE                                                  FSTY0170
      YC(J)=YO-SQRT(R**2-(XC(J)-XO)**2)                                  FSTY0171
      SINA(J)=(XO-XC(J))/R                                               FSTY0172
      IF(YBB.GT. YA) SINA(J)=-SINA(J)                                    FSTY0173
  430 COSA(J)=(YO-YC(J))/R                                               FSTY0174
C  COMPUTE INTERSECTION OF SLICE CENTERLINE WITH BOUNDARY LINE,YB.       FSTY0175
      DO 490 J=1,NSLICE                                                  FSTY0176
      IDS=0                                                              FSTY0177
      DO 480 K=1,NBL                                                     FSTY0178
      YB(J,K)=-9999.                                                     FSTY0179
      NPBL1=NPBL(K)-1                                                    FSTY0180
      DO 440 I=1,NPBL1                                                   FSTY0181
      IF(XBL(I,K) .EQ. XBL(I+1,K)) GO TO 440                            FSTY0182
      M=I                                                                FSTY0183
      IF(XC(J) .GT. XBL(I,K) .AND. XC(J) .LE. XBL(I+1,K)) GO TO 450      FSTY0184
  440 CONTINUE                                                           FSTY0185
      GO TO 480                                                          FSTY0186
  450 YB(J,K)=SLOPE(M,K)*XC(J)+YINT(M,K)                                 FSTY0187
      IF(IDS .EQ. 0) GO TO 470                                           FSTY0188
      IF(K .EQ. 1) GO TO 470                                             FSTY0189
      IF(YB(J,K) .GT. YBJK-0.1) GO TO 470                               FSTY0190
      WRITE(6,460) K                                                     FSTY0191
  460 FORMAT (/,5X,17HBOUNDARY LINE NO.,I3,71H IS OUT OF PLACE,PLEASE CHFSTY0192
     *ANGE THE INPUT DATA AND RUN THE PROGRAM AGAIN)                     FSTY0193
      NSTP=1                                                             FSTY0194
      GO TO 960                                                          FSTY0195
  470 YBJK=YB(J,K)                                                       FSTY0196
      IDS=1                                                              FSTY0197
  480 CONTINUE                                                           FSTY0198
      SL(J)=YB(J,NBL)-YC(J)                                              FSTY0199
      IF(SL(J) .GT. SLM) SLM=SL(J)                                       FSTY0200
  490 CONTINUE                                                           FSTY0201
      IF(SLM .LT. 0.1) GO TO 110                                         FSTY0202
      IF(SLM .LT. DMIN) GO TO 960                                        FSTY0203
C  COMPUTE WEIGHT OF SLICE,W,ACTUATING FORCE,WW,AND EARTHQUAKE FORCE.    FSTY0204
      ID=0                                                               FSTY0205
      DO 670 J=1,NSLICE                                                  FSTY0206
      NINT=0                                                             FSTY0207
      W(J)=0                                                             FSTY0208
```

```
      WW(J)=0.                                              FSTY0209
      DO 670 K=1,NBL                                        FSTY0210
      IF(YB(J,K)+9999.) 510,500,510                         FSTY0211
  500 ID=ID+1                                               FSTY0212
      GO TO 670                                             FSTY0213
  510 SL1=YB(J,NBL)-YB(J,K)                                 FSTY0214
      IF(SL1 .GE. SL(J)-0.01) GO TO 640                     FSTY0215
      IF(NINT .EQ. 1) GO TO 530                             FSTY0216
      WRITE (6,520) XO,YO,R                                 FSTY0217
  520 FORMAT (/,1H0,'AT POINT (',F8.3,',',F8.3,')',' WITH RADIUS OF ', FSTY0218
     * F8.3,/,1X,'THE CIRCLE CUTS INTO THE LOWEST BOUNDARY, SO THE COMPUFSTY0219
     *TATION IS STOPPED.  PLEASE CHECK THE RADIUS CONTROL ZONE')  FSTY0220
      NSTP=1                                                FSTY0221
      GO TO 960                                             FSTY0222
  530 IF(NFG) 630,540,630                                   FSTY0223
  540 YBJ=YC(J)                                             FSTY0224
      LC(J)=K-ID                                            FSTY0225
      NFG=1                                                 FSTY0226
      IF(C(LC(J)) .EQ. 0. .AND. TANPHI(LC(J)) .EQ. 0.) GO TO 580  FSTY0227
      IF(J .EQ. 1) GO TO 650                                FSTY0228
      IF(C(LC(J-1)) .EQ. 0. .AND. TANPHI(LC(J-1)) .EQ. 0.) GO TO 550  FSTY0229
      GO TO 650                                             FSTY0230
  550 LCJ=LC(J-1)                                           FSTY0231
      YBBJ=YB(J-1,NBL)                                      FSTY0232
      IF(YBB-YA) 560,570,570                                FSTY0233
  560 ISIGN=1                                               FSTY0234
      GO TO 620                                             FSTY0235
  570 ISIGN=-1                                              FSTY0236
      GO TO 620                                             FSTY0237
  580 IF( J .EQ. 1) GO TO 660                               FSTY0238
      IF(C(LC(J-1)) .NE. 0. .OR. TANPHI(LC(J-1)) .NE. 0.) GO TO 590  FSTY0239
      GO TO 660                                             FSTY0240
  590 LCJ=LC(J)                                             FSTY0241
      YBBJ=YB(J,NBL)                                        FSTY0242
      IF(YBB-YA) 600,610,610                                FSTY0243
  600 ISIGN=-1                                              FSTY0244
      GO TO 620                                             FSTY0245
  610 ISIGN=1                                               FSTY0246
  620 YCW=YO-SQRT(R**2-(XC(J)-BB(J)/2-XO)**2)               FSTY0247
      WD=YBBJ-YCW                                           FSTY0248
      WW(J)=ISIGN*0.5*WD**2*G(LCJ)*(YO-YCW-WD/3)/R          FSTY0249
      IF(C(LC(J-1)) .EQ. 0. .AND. TANPHI(LC(J-1)) .EQ. 0.) GO TO 650  FSTY0250
      GO TO 660                                             FSTY0251
  630 YBJ=YB(J,K-ID)                                        FSTY0252
      GO TO 650                                             FSTY0253
  640 NFG=0                                                 FSTY0254
      NINT=1                                                FSTY0255
      GO TO 660                                             FSTY0256
  650 WG=BB(J)*(YB(J,K)-YBJ)*G(K-ID)                        FSTY0257
      W(J)=W(J)+WG                                          FSTY0258
      WW(J)=WW(J)+ WG*SINA(J)                               FSTY0259
      IF(SEIC .EQ. 0.) GO TO 660                            FSTY0260
      IF(C(K-ID) .EQ. 0. .AND. TANPHI(K-ID) .EQ. 0. .AND. (YA .GE. YBB  FSTY0261
     * .AND. XC(J) .GT. XBL(2,NBL) .OR. YA .LE. YBB .AND. XC(J) .LT.  FSTY0262
     * XBL(2,NBL))) GO TO 660                               FSTY0263
      WW(J)=WW(J)+WG*SEIC*(YO-0.5*(YB(J,K)+YBJ))/R          FSTY0264
  660 ID=1                                                  FSTY0265
  670 CONTINUE                                              FSTY0266
      L=0                                                   FSTY0267
      IF(NSPG .NE. 0) GO TO 680                             FSTY0268
      L=1                                                   FSTY0269
      GO TO 830                                             FSTY0270
  680 L=L+1                                                 FSTY0271
      IF(ISTP(L) .EQ. 1) GO TO 950                          FSTY0272
```

```
      IF(NSPG .EQ. 2) GO TO 780                                       FSTY0273
      IF(LM) 700,690,700                                              FSTY0274
  690 LL=L                                                            FSTY0275
      GO TO 710                                                       FSTY0276
  700 LL=LM                                                           FSTY0277
      L=LM                                                            FSTY0278
  710 NPWT1=NPWT(LL)-1                                                FSTY0279
C COMPUTE INTERSECTION OF SLICE CENTERLINE WITH WATER TABLE,YT.       FSTY0280
      DO 740 J=1,NSLICE                                               FSTY0281
      DO 720 I=1,NPWT1                                                FSTY0282
      M=I                                                             FSTY0283
      IF(XC(J) .GE. XWT(I,LL) .AND. XC(J) .LE. XWT(I+1,LL)) GO TO 730 FSTY0284
  720 CONTINUE                                                        FSTY0285
      GO TO 740                                                       FSTY0286
  730 YT(J,LL)=((YWT(M+1,LL)-YWT(M,LL))/(XWT(M+1,LL)-XWT(M,LL)))*(XC(J)-FSTY0287
     * XWT(M,LL))+ YWT(M,LL)                                          FSTY0288
  740 CONTINUE                                                        FSTY0289
C COMPUTE EFFECTIVE WEIGHT OF SLICE,EFF.                              FSTY0290
      DO 770 J=1,NSLICE                                               FSTY0291
      IF(YT(J,LL)-YC(J)) 760,760,750                                  FSTY0292
  750 EFF(J,LL)=W(J)-(YT(J,LL)-YC(J))*GW*BB(J)                        FSTY0293
      GO TO 770                                                       FSTY0294
  760 EFF(J,LL)=W(J)                                                  FSTY0295
  770 CONTINUE                                                        FSTY0296
      GO TO 850                                                       FSTY0297
  780 IF(LM)800,790,800                                               FSTY0298
  790 LL=L                                                            FSTY0299
      GO TO 810                                                       FSTY0300
  800 LL=LM                                                           FSTY0301
      L=LM                                                            FSTY0302
  810 DO 820 J=1,NSLICE                                               FSTY0303
  820 EFF(J,LL)=W(J)*(1.-RU(LL))                                      FSTY0304
      GO TO 850                                                       FSTY0305
  830 DO 840 J=1,NSLICE                                               FSTY0306
  840 EFF(J,L )=W(J)                                                  FSTY0307
C COMPUTE RESISTING FORCE,RM,ACTUATING FORCE,OTM,AND FACTOR OF SAFETY FSTY0308
  850 RM=RMI                                                          FSTY0309
      OTM=0.                                                          FSTY0310
      DO 860 J=1,NSLICE                                               FSTY0311
      RM=RM+C(LC(J))*BB(J)/COSA(J)+ EFF(J,L)*COSA(J)*TANPHI(LC(J))    FSTY0312
  860 OTM=OTM+WW(J)                                                   FSTY0313
      IF(OTM .GT.1. .AND. RM .NE. 0.) GO TO 880                       FSTY0314
      WRITE (6,870) R                                                 FSTY0315
  870 FORMAT (/,46H ****WARNING AT NEXT CENTER**** WHEN RADIUS IS,F10.3,FSTY0316
     * /,92H EITHER THE OVERTURNING OR THE RESISTING MOMENT IS 0,SO A LAFSTY0317
     *RGE FACTOR OF SAFETY IS ASSIGNED)                              FSTY0318
      FS(L)=1.0E06                                                    FSTY0319
      GO TO 940                                                       FSTY0320
  880 FS(L)=RM/OTM                                                    FSTY0321
      IF(FS(L) .GT. 0.) GO TO 890                                     FSTY0322
      GO TO 920                                                       FSTY0323
  890 IF(METHOD .EQ. 0 .AND. LBSHP .EQ. 0) GO TO 940                  FSTY0324
      IF(FS(L) .GT. 100.) GO TO 940                                   FSTY0325
      BFS(L)=FS(L)                                                    FSTY0326
      IF(FS(L) .LT. 1.) BFS(L)=2.                                     FSTY0327
      IC=0                                                            FSTY0328
  900 IC=IC+1                                                         FSTY0329
      PFS=BFS(L)                                                      FSTY0330
      RA1=0.                                                          FSTY0331
      RA2=0.                                                          FSTY0332
      DO 910 J=1,NSLICE                                               FSTY0333
      CJBJ=C(LC(J))*BB(J)+ EFF(J,L)*TANPHI(LC(J))                     FSTY0334
      PFSC=PFS*COSA(J)+TANPHI(LC(J))*SINA(J)                          FSTY0335
      RA1=RA1+CJBJ/PFSC                                               FSTY0336
```

```
  910 RA2=RA2+CJBJ*TANPHI(LC(J))*SINA(J)/PFSC**2            FSTY0337
      BFS(L)= PFS*(1.+(RMI/PFS+RA1-OTM)/(OTM-RA2))          FSTY0338
      IF(ABS((BFS(L)-PFS)/PFS) .GT. 0.0001 .AND. IC .LT. 10) GO TO 900  FSTY0339
      FS(L)=BFS(L)                                          FSTY0340
      IF(FS(L) .GT. 0.) GO TO 940                           FSTY0341
  920 WRITE (6,930) XO,YO,R,L                               FSTY0342
  930 FORMAT(/,11H AT POINT (,F10.3,1H,,F10.3,1H),16H WITH RADIUS OF ,  FSTY0343
     * F10.3, 15H UNDER SEEPAGE ,I2,/, 95H THE FACTOR OF SAFETY IS LESS FSTY0344
     *THAN OR EQUAL TO 0. CHECK PIEZOMETRIC LINE OR PORE PRESSURE RATIO)FSTY0345
      ISTP(L)=1                                             FSTY0346
      IF(NQ .EQ. 1) STOP                                    FSTY0347
  940 IF(NSPG .EQ. 0) GO TO 960                             FSTY0348
      IF(LM .NE. 0) GO TO 960                               FSTY0349
  950 IF(L .LT. NQ) GO TO 680                               FSTY0350
  960 RETURN                                                FSTY0351
      END                                                   FSTY0352
C                                                           SAVE0001
C                                                           SAVE0002
      SUBROUTINE SAVE(FS,RRM,FSM,LB,XANBL,XBNBL,XASM,XBSM)  SAVE0003
C                                                           SAVE0004
C THIS SUBROUTINE IS USED TO SAVE THE FACTOR OF SAFETY FOR VARIOUS      SAVE0005
C RADII AND DETERMINE WHICH ONE IS THE LOWEST. IF A FACTOR OF SAFETY    SAVE0006
C SMALLER THAN THAT OF THE TWO ADJOINING RADII IS FOUND, ADDITIONAL     SAVE0007
C RADII WILL BE TRIED TO DETERMINE THE LOWEST FACTOR OF SAFETY          SAVE0008
C                                                           SAVE0009
      COMMON/A2/NK,RR,FF                                    SAVE0010
      COMMON/A3/KR,KRA,NQ,R,ROO                             SAVE0011
      DIMENSION FF(90,5),FS(5),FSM(5),KRA(5),RR(90,5),RRM(5)  SAVE0012
      KR=KR+1                                               SAVE0013
      DO 50 LP=1,NQ                                         SAVE0014
      L=LP                                                  SAVE0015
      IF(LB .NE. 0) L=LB                                    SAVE0016
      RR(KR,L)=R                                            SAVE0017
      FF(KR,L)=FS(L)                                        SAVE0018
      IF (KR.NE. 1) GO TO 10                                SAVE0019
      FSM(L)=FF(KR,L)                                       SAVE0020
      RRM(L)=RR(KR,L)                                       SAVE0021
      XASM=XANBL                                            SAVE0022
      XBSM=XBNBL                                            SAVE0023
   10 IF(FF(KR,L) .GE. FSM(L)) GO TO 20                     SAVE0024
      FSM(L)=FF(KR,L)                                       SAVE0025
      RRM(L)=RR(KR,L)                                       SAVE0026
      XASM=XANBL                                            SAVE0027
      XBSM=XBNBL                                            SAVE0028
   20 IF (KR.LT. 3 .OR. NK .EQ. 0) GO TO 50                 SAVE0029
      IF(FF(KR,L).GT.FF(KR-1,L) .AND.FF(KR-1,L) .LT.FF(KR-2,L)) GO TO 30SAVE0030
      GO TO 50                                              SAVE0031
C                                                           SAVE0032
C SUBDIVIDE THE MINIMUM RANGE                               SAVE0033
C                                                           SAVE0034
   30 DOT=R                                                 SAVE0035
      DO 40 I=1,2                                           SAVE0036
      RI=RR(KR-3+I,L)                                       SAVE0037
      DELTA=(RR(KR-2+I,L)-RR(KR-3+I,L))/(NK+1)              SAVE0038
      DO 40 N=1,NK                                          SAVE0039
      RI=RI+DELTA                                           SAVE0040
      R=RI                                                  SAVE0041
      CALL FSAFTY(FS,LB,XANBL,XBNBL)                        SAVE0042
      KRA(L)=KRA(L)+1                                       SAVE0043
      RR(KRA(L),L)=R                                        SAVE0044
      FF(KRA(L),L)=FS(L)                                    SAVE0045
      IF(FS(L) .GE. FSM(L)) GO TO 40                        SAVE0046
      FSM(L)=FS(L)                                          SAVE0047
      RRM(L)=R                                              SAVE0048
```

```
            XASM=XANBL                                                    SAVE0049
            XBSM=XBNBL                                                    SAVE0050
         40 CONTINUE                                                      SAVE0051
            R=DOT                                                         SAVE0052
            IF(LB .NE. 0) GO TO 60                                        SAVE0053
         50 CONTINUE                                                      SAVE0054
         60 RETURN                                                        SAVE0055
            END                                                           SAVE0056
      C                                                                   XYPL0001
      C                                                                   XYPL0002
            SUBROUTINE XYPLOT(FOS,XAMIN,XBMIN)                            XYPL0003
      C                                                                   XYPL0004
      C       PURPOSE                                                     XYPL0005
      C                                                                   XYPL0006
      C             TO PREPARE PLOTS OF SINGLE OR MULTIPLE SETS OF DATA POINT XYPL0007
      C             VECTORS.  REQUIRES NO SCALING OR SORTING OF DATA.     XYPL0008
      C                                                                   XYPL0009
            COMMON/A4/NBL,NPBL,NPWT,NSPG,NWT,SLOPE,XBL,XWT,YBL,YINT,YWT   XYPL0010
            COMMON/A5/LM,XMINX,YMINY,XMINR                                XYPL0011
            DIMENSION X(50,22),Y(50,22),IX(51,22),IY(51,22),L(22),IXSN(22) XYPL0012
            DIMENSION LINE(51),XLABEL(6),YLABEL(11),POINT(22),JA(22)      XYPL0013
            DIMENSION NPBL(20),SLOPE(49,20),XBL(50,20),XWT(50,5),YBL(50,20), XYPL0014
           *YINT(49,20),YWT(50,5),XMINX(5),YMINY(5),XMINR(5),NPWT(5)      XYPL0015
            INTEGER BLANK,PLUS,EXXE,POINT                                 XYPL0016
            DATABLANK/' ',PLUS/'+',EXXE/'X',POINT/'1','2','3','4','5','6',' XYPL0017
           *7','8','9','0','A','B','C','D','E','F','G','J','K','L','P','*'/  XYPL0018
         10 FORMAT (1H1//,5X,32HCROSS SECTION IN DISTORTED SCALE,/,5X,61HNUMER XYPL0019
           *ALS INDICATE BOUNDARY LINE NO. IF THERE ARE MORE THAN 10,/,5X,53HB XYPL0020
           *OUND. LINES,ALPHABETS WILL THEN BE USED. P INDICATES,/,5X,64HP1ZOM XYPL0021
           *ETRIC LINE. IF A PORTION OF PIEZOMETRIC LINE COINCIDES WITH,/,5X,6 XYPL0022
           *3HTHE GROUND OR ANOTHER BOUNDARY LINE,ONLY THE GROUND OR BOUNDARY, XYPL0023
           */,5X,60HLINE WILL BE SHOWN. X INDICATES INTERSECTION OF TWO BOUNDA XYPL0024
           *RY,/,5X,36HLINES.  * INDICATES FAILURE SURFACE.,/,5X,31HTHE MINIMU XYPL0025
           *M FACTOR OF SAFETY IS,F10.3')                                XYPL0026
         20 FORMAT (1H0,3X,1PE10.3,1X,51A1)                              XYPL0027
         30 FORMAT (1H0,14X,51A1)                                        XYPL0028
         40 FORMAT (1H0,11X,6(1PE9.2,1X),//)                             XYPL0029
      C ENTER COORDINATES OF BOUNDARY LINES AND WATER TABLE.             XYPL0030
            M=NBL                                                        XYPL0031
            DO 50 J=1,M                                                  XYPL0032
            L(J)=NPBL(J)                                                 XYPL0033
            LJ=L(J)                                                      XYPL0034
            DO 50 I=1,LJ                                                 XYPL0035
            X(I,J)=XBL(I,J)                                              XYPL0036
         50 Y(I,J)=YBL(I,J)                                              XYPL0037
            IF(NSPG .NE. 1) GO TO 70                                     XYPL0038
            M=M+1                                                        XYPL0039
            L(M)=NPWT(LM)                                                XYPL0040
            NPWTJ=NPWT(LM)                                               XYPL0041
            DO 60 I=1,NPWTJ                                              XYPL0042
            X(I,M)=XWT(I,LM)                                             XYPL0043
         60 Y(I,M)=YWT(I,LM)                                             XYPL0044
      C FIND MAXIMA AND MINIMA                                           XYPL0045
         70 XMIN=X(1,1)                                                  XYPL0046
            XMAX=X(1,1)                                                  XYPL0047
            YMIN=Y(1,1)                                                  XYPL0048
            YMAX=Y(1,1)                                                  XYPL0049
            DO 160 J=1,M                                                 XYPL0050
            LJ=L(J)                                                      XYPL0051
            DO 150 I=1,LJ                                                XYPL0052
            IF (XMIN-X(I,J)) 90,90,80                                    XYPL0053
         80 XMIN=X(I,J)                                                  XYPL0054
         90 IF (XMAX-X(I,J)) 100,110,110                                 XYPL0055
        100 XMAX=X(I,J)                                                  XYPL0056
```

```
      110 IF (YMIN-Y(I,J)) 130,130,120                    XYPL0057
      120 YMIN=Y(I,J)                                     XYPL0058
      130 IF (YMAX-Y(I,J)) 140,150,150                    XYPL0059
      140 YMAX=Y(I,J)                                     XYPL0060
      150 CONTINUE                                        XYPL0061
      160 JA(J)=1                                         XYPL0062
C FIND GRAPHICAL LIMIT FOR X VALUES                       XYPL0063
      170 DIFF=XMAX-XMIN                                  XYPL0064
          N=0                                             XYPL0065
          IF (DIFF) 190,180,190                           XYPL0066
C CORRECT FOR ZERO DIFFERENCE AND NORMALIZE               XYPL0067
      180 DIFF=(.20)*XMIN                                 XYPL0068
          IF (XMIN.EQ.0) DIFF=0.2                         XYPL0069
          XMIN=XMIN-(.50)*DIFF                            XYPL0070
          XMAX=XMAX+(.45)*DIFF                            XYPL0071
      190 DO 230 I=1,75                                   XYPL0072
          IF (DIFF-10) 200,200,220                        XYPL0073
      200 IF (DIFF-1) 210,210,240                         XYPL0074
      210 DIFF=DIFF*10                                    XYPL0075
          N=-I                                            XYPL0076
          GO TO 230                                       XYPL0077
      220 DIFF=DIFF/10                                    XYPL0078
          N=I                                             XYPL0079
      230 CONTINUE                                        XYPL0080
C COMPARE WITH STANDARD INTERVAL                          XYPL0081
      240 IF (DIFF-8) 260,250,250                         XYPL0082
      250 KINT=10                                         XYPL0083
          GO TO 330                                       XYPL0084
      260 IF (DIFF-5) 280,270,270                         XYPL0085
      270 KINT=8                                          XYPL0086
          GO TO 330                                       XYPL0087
      280 IF (DIFF-4) 300,290,290                         XYPL0088
      290 KINT=5                                          XYPL0089
          GO TO 330                                       XYPL0090
      300 IF (DIFF-2) 320,310,310                         XYPL0091
      310 KINT=4                                          XYPL0092
          GO TO 330                                       XYPL0093
      320 KINT=2                                          XYPL0094
C UNNORMALIZE AND FIND MINIMUM VALUES OF GRAPH            XYPL0095
      330 IF(N)340,340,350                                XYPL0096
      340 XINT=KINT                                       XYPL0097
          XINT=XINT/(10.**(1-N))                          XYPL0098
          GO TO 360                                       XYPL0099
      350 XINT=KINT*(10.**(N-1))                          XYPL0100
      360 XGMIN=0.0                                        XYPL0101
          IF (XMIN) 370,380,380                           XYPL0102
      370 XGMIN=XGMIN-XINT                                XYPL0103
          IF (XGMIN-XMIN) 390,390,370                     XYPL0104
      380 XGMIN=XGMIN+XINT                                XYPL0105
          IF (XGMIN-XMIN) 380,390,370                     XYPL0106
C FIND MAXIMUM VALUES OF GRAPH AND COMPARE WITH PLOT.     XYPL0107
      390 XGMAX=XINT*10+XGMIN                             XYPL0108
          IF (XGMAX-XMAX) 400,410,410                     XYPL0109
      400 XMIN=XGMIN-.001*XINT                            XYPL0110
C CENTER PLOT OF GRAPH.                                   XYPL0111
          GO TO 170                                       XYPL0112
      410 I=(XGMAX-XMAX)/(2*XINT)                         XYPL0113
          XGMIN=XGMIN-I*XINT                              XYPL0114
          XGMAX=XGMIN+10*XINT                             XYPL0115
C FIND GRAPHICAL LIMIT FOR Y VALUES.                      XYPL0116
      420 DIFF=YMAX-YMIN                                  XYPL0117
          N=0                                             XYPL0118
          IF (DIFF) 440,430,440                           XYPL0119
C CORRECT FOR ZERO DIFFERENCE AND NORMALIZE               XYPL0120
```

```
    430 DIFF=(.20)*YMIN                                    XYPL0121
        IF (YMIN.EQ.0) DIFF=0.2                            XYPL0122
        YMIN=YMIN-(.50)*DIFF                               XYPL0123
        YMAX=YMAX+(.45)*DIFF                               XYPL0124
    440 DO 480 I=1,75                                      XYPL0125
        IF (DIFF-10) 450,450,470                           XYPL0126
    450 IF (DIFF-1) 460,460,490                            XYPL0127
    460 DIFF=DIFF*10                                       XYPL0128
        N=-I                                               XYPL0129
        GO TO 480                                          XYPL0130
    470 DIFF=DIFF/10                                       XYPL0131
        N=I                                                XYPL0132
    480 CONTINUE                                           XYPL0133
C COMPARE WITH STANDARD INTERVAL                           XYPL0134
    490 IF (DIFF-8) 510,500,500                            XYPL0135
    500 KINT=10                                            XYPL0136
        GO TO 580                                          XYPL0137
    510 IF (DIFF-5) 530,520,520                            XYPL0138
    520 KINT=8                                             XYPL0139
        GO TO 580                                          XYPL0140
    530 IF (DIFF-4) 550,540,540                            XYPL0141
    540 KINT=5                                             XYPL0142
        GO TO 580                                          XYPL0143
    550 IF (DIFF-2) 570,560,560                            XYPL0144
    560 KINT=4                                             XYPL0145
        GO TO 580                                          XYPL0146
    570 KINT=2                                             XYPL0147
C UNNORMALIZE AND FIND MAXIMUM VALUES OF GRAPH.            XYPL0148
    580 IF(N)590,590,600                                   XYPL0149
    590 YIN =KINT                                          XYPL0150
        YIN =YIN /(10.**(1-N))                             XYPL0151
        GO TO 610                                          XYPL0152
    600 YIN =KINT*(10.**(N-1))                             XYPL0153
    610 YGMIN=0.0                                          XYPL0154
        IF (YMIN) 620,630,630                              XYPL0155
    620 YGMIN=YGMIN-YIN                                    XYPL0156
        IF (YGMIN-YMIN) 640,640,620                        XYPL0157
    630 YGMIN=YGMIN+YIN                                    XYPL0158
        IF (YGMIN-YMIN) 630,630,620                        XYPL0159
C FIND MAXIMUM VALUE OF GRAPH AND COMPARE WITH PLOT.       XYPL0160
    640 YGMAX=YIN *10+YGMIN                                XYPL0161
        IF (YGMAX-YMAX) 650,660,660                        XYPL0162
    650 YMIN=YGMIN-.001*YIN                                XYPL0163
        GO TO 420                                          XYPL0164
C CENTER PLOT OF GRAPH.                                    XYPL0165
    660 I=(YGMAX-YMAX)/(2*YIN )                            XYPL0166
        YGMIN=YGMIN-I*YIN                                  XYPL0167
        YGMAX=YGMIN+10*YIN                                 XYPL0168
C DETERMINE INTERSECTION OF CIRCLE WITH GROUND LINE        XYPL0169
        M=M+1                                              XYPL0170
        JA(M)=1                                            XYPL0171
        X(1,M)=XAMIN                                       XYPL0172
        X(2,M)=XBMIN                                       XYPL0173
        L(M)=2                                             XYPL0174
C INTERPOLATE RECTANGULAR POINTS AND FIND AXIS.            XYPL0175
        DO 680 J=1,M                                       XYPL0176
        LJ=L(J)                                            XYPL0177
        DO 670 I=1,LJ                                      XYPL0178
        IX(I,J)=(((X(I,J)-XGMIN)/(XGMAX-XGMIN))*50.)+1.5   XYPL0179
        IF(J .EQ. M) GO TO 670                             XYPL0180
        IY(I,J)=51.5-(((Y(I,J)-YGMIN)/(YGMAX-YGMIN))*50.0) XYPL0181
    670 CONTINUE                                           XYPL0182
    680 CONTINUE                                           XYPL0183
        KY=1                                               XYPL0184
```

```
        JX=51                                                   XYPL0185
        DO 730 I=1,10                                           XYPL0186
        IF(I .GT. 5) GO TO 710                                  XYPL0187
        IF(XGMIN+XINT*(I-1)*2) 690,700,710                      XYPL0188
690 KY=10*(I-1)+6                                               XYPL0189
        GO TO 710                                               XYPL0190
700 KY=10*(I-1)+1                                               XYPL0191
710 IF (YGMIN+(YIN *(I-1)*1.05)) 720,730,730                    XYPL0192
720 JX=51-5*I                                                   XYPL0193
730 CONTINUE                                                    XYPL0194
        DO 740 I=1,11                                           XYPL0195
        IF(I .GT. 6) GO TO 740                                  XYPL0196
        XLABEL(I)=XGMIN+XINT*(I-1)*2.                           XYPL0197
740 YLABEL(I)=YGMIN+YIN *(11-I)                                 XYPL0198
C SET UP THE GRAPH LINE BY ADDING POINTS AT EACH HORIZONTAL DIVISION. XYPL0199
        DO 820 N=1,M                                            XYPL0200
        IF(N .EQ. M) GO TO 770                                  XYPL0201
        NPBL1=L(N)-1                                            XYPL0202
        IXS=L(N)                                                XYPL0203
        DO 760 I1=1,NPBL1                                       XYPL0204
        IXD=IX(I1+1,N)-IX(I1,N)                                 XYPL0205
        IYD=IY(I1+1,N)-IY(I1,N)                                 XYPL0206
        IF(IXD .LT. 2) GO TO 760                                XYPL0207
        IXD1=IXD-1                                              XYPL0208
        DO 750 I2=1,IXD1                                        XYPL0209
        IX(IXS+I2,N)=IX(I1,N)+I2                                XYPL0210
750 IY(IXS+I2,N)=IY(I1,N)+I2*IYD/IXD                            XYPL0211
        IXS=IXS+IXD1                                            XYPL0212
760 CONTINUE                                                    XYPL0213
        GO TO 790                                               XYPL0214
770 IXS=IX(2,N)-IX(1,N)-1                                       XYPL0215
        IX1=IX(1,N)                                             XYPL0216
        DO 780 I=1,IXS                                          XYPL0217
        IX(I,N)=IX1+I                                           XYPL0218
        XI=(IX(I,N)-1.5)*(XGMAX-XGMIN)/50.+XGMIN                XYPL0219
        YI =YMINY(LM)-SQRT(XMINR(LM)**2-(XI -XMINX(LM))**2)     XYPL0220
780 IY(I,N)=51.5-(YI -YGMIN)/(YGMAX-YGMIN)*50.0                 XYPL0221
790 DO 810 I=1,IXS                                              XYPL0222
        DO 810 J=I,IXS                                          XYPL0223
        IF (IY(I,N)-IY(J,N)) 810,810,800                        XYPL0224
800 K=IY(I,N)                                                   XYPL0225
        IY(I,N)=IY(J,N)                                         XYPL0226
        IY(J,N)=K                                               XYPL0227
        K=IX(I,N)                                               XYPL0228
        IX(I,N)=IX(J,N)                                         XYPL0229
        IX(J,N)=K                                               XYPL0230
810 CONTINUE                                                    XYPL0231
        IXSN(N)=IXS                                             XYPL0232
820 CONTINUE                                                    XYPL0233
        WRITE (6,10) FOS                                        XYPL0234
        DO 980 I=1,51                                           XYPL0235
        DO 830 N=1,51                                           XYPL0236
830 LINE(N)=BLANK                                               XYPL0237
C RECTANGULAR SET.                                              XYPL0238
        LINE(1)=PLUS                                            XYPL0239
        LINE(KY)=PLUS                                           XYPL0240
        LINE(51)=PLUS                                           XYPL0241
        IF (MOD(I,5)-1) 860,840,860                             XYPL0242
840 DO 850 N=1,51                                               XYPL0243
850 IF (MOD(N,10).EQ.1) LINE(N)=PLUS                            XYPL0244
        LINE(1)=EXXE                                            XYPL0245
        LINE(KY)=EXXE                                           XYPL0246
        LINE(51)=EXXE                                           XYPL0247
860 IF (I.NE.1.AND.I.NE.JX.AND.I.NE.51) GO TO 890               XYPL0248
```

```
        DO 880 N=2,50                                              XYPL0249
        LINE(N)=PLUS                                               XYPL0250
        IF (MOD(N,10)-1) 880,870,880                               XYPL0251
 870 LINE(N)=EXXE                                                  XYPL0252
 880 CONTINUE                                                      XYPL0253
C       LOAD LINE                                                  XYPL0254
 890 DO 950 J=1,M                                                  XYPL0255
 900 IF (JA(J)-IXSN(J)) 910,910,950                                XYPL0256
 910 JJ=JA(J)                                                      XYPL0257
        IF (IY(JJ,J)-I) 950,920,950                                XYPL0258
 920 JB=IX(JJ,J)                                                   XYPL0259
        IF (LINE(JB).NE.BLANK.AND.LINE(JB).NE.PLUS) GO TO 930      XYPL0260
        JK=J                                                       XYPL0261
        IF(J .EQ. M) JK=22                                         XYPL0262
        IF(NSPG .EQ. 1 .AND. J .EQ. M-1) JK=21                     XYPL0263
        LINE(JB)=POINT(JK)                                         XYPL0264
        GO TO 940                                                  XYPL0265
 930 IF(NSPG .EQ. 1 .AND. J .GE. M-1 .OR. NSPG .NE. 1 .AND. J .EQ. M)  XYPL0266
     * GO TO 940                                                   XYPL0267
        IF (LINE(JB).NE.POINT(J)) LINE(JB)=EXXE                    XYPL0268
 940 JA(J)=JA(J)+1                                                 XYPL0269
        GO TO 900                                                  XYPL0270
 950 CONTINUE                                                      XYPL0271
C PRINT A LINE OF THE GRAPH.                                       XYPL0272
        IF (MOD(I,5)-1) 970,960,970                                XYPL0273
 960 K=((.2)*I+1.)                                                 XYPL0274
        WRITE(6,20) YLABEL(K),(LINE(N),N=1,51)                     XYPL0275
        GO TO 980                                                  XYPL0276
 970 WRITE(6,30) (LINE(N),N=1,51)                                  XYPL0277
 980 CONTINUE                                                      XYPL0278
        WRITE(6,40) (XLABEL(N),N=1,6)                              XYPL0279
        RETURN                                                     XYPL0280
        END                                                        XYPL0281
```

Appendix VI

List of REAME in BASIC

```
5 REM REAME(ROTATIONAL EQUILIBRIUM ANALYSIS OF MULTILAYED EMBANKMENTS)
10 REM INTERACTIVE OR BATCH MODE
15 DIM C(19),E8(80),F2(90),G(19),L2(9,11),L4(9,11),N0(10),L3(9,11)
20 DIM N6(20),P2(19),R4(11),R5(90),S4(49,20),T1(19),T6(10),T7(10)
25 DIM T8(11),T$(72),X(3),X1(50,20),X3(50),Y(3),Y8(50,20)
30 DIM A0(5),B9(5),F8(5),F9(5),S5(5),Y7(50),Y9(49,20)
35 PRINT "TITLE -";
40 MAT INPUT T$
45 PRINT
50 PRINT "FILE NAME";
55 INPUT F$
60 FILE #1,F$
65 PRINT "READ FROM FILE?(ENTER 1 WHEN READ FROM FILE & 0 WHEN NOT)";
70 INPUT B0
75 PRINT
80 IF B0=1 THEN 95
85 SCRATCH #1
90 GO TO 100
95 RESTORE #1
100 PRINT "NO. OF STATIC AND SEISMIC CASES-";
105 INPUT P6
110 FOR P5=1 TO P6
115 PRINT
120 PRINT "CASE NO.";P5,"SEISMIC COEFFICIENT=";
125 INPUT S5(P5)
130 PRINT
135 IF P5<>1 THEN 1195
140 IF B0=1 THEN 165
145 PRINT "NUMBER OF BOUNDARY LINES -";
150 INPUT N5
155 WRITE #1,N5
160 GO TO 175
165 READ #1,N5
170 PRINT "NO. OF BOUNDARY LINES=";N5
175 PRINT
180 FOR J=1 TO N5
185 IF B0=1 THEN 210
190 PRINT "NO. OF POINTS ON BOUNDARY LINE";J;"=";
195 INPUT N6(J)
200 WRITE #1,N6(J)
```

```
205 GO TO 220
210 READ #1,N6(J)
215 PRINT "NO. OF POINTS ON BOUNDARY LINE";J;"=";N6(J)
220 PRINT
225 PRINT "BOUNDARY LINE -";J
230 FOR I=1 TO N6(J)
235 IF B0=1 THEN 270
240 PRINT I;"X-COORDINATE=";
245 INPUT X1(I,J)
250 PRINT TAB(3);"Y-COORDINATE=";
255 INPUT Y8(I,J)
260 WRITE #1,X1(I,J),Y8(I,J)
265 GO TO 280
270 READ #1,X1(I,J),Y8(I,J)
275 PRINT I;"X COORD.=";X1(I,J),"Y COORD.=";Y8(I,J)
280 NEXT I
285 PRINT
290 NEXT J
295 PRINT
300 PRINT"LINE NO. AND SLOPE OF EACH SEGMENT ARE:"
305 FOR J=1 TO N5
310 N1=N6(J)-1
315 PRINT J,
320 FOR I=1 TO N1
325 IF X1(I+1,J)=X1(I,J) THEN 340
330 S4(I,J)=(Y8(I+1,J)-Y8(I,J))/(X1(I+1,J)-X1(I,J))
335 GO TO 345
340 S4(I,J)=99999
345 Y9(I,J)=Y8(I,J)-S4(I,J)*X1(I,J)
350 PRINT S4(I,J);
355 NEXT I
360 PRINT
365 NEXT J
370 PRINT
375 IF B0=1 THEN 415
380 PRINT "MIN. DEPTH OF TALLEST SLICE=";
385 INPUT D4
390 PRINT
395 PRINT "NO. OF RADIUS CONTROL ZONES-";
400 INPUT F6
405 WRITE #1,D4,F6
410 GO TO 430
415 READ #1,D4,F6
420 PRINT "MIN. DEPTH OF TALLEST SLICE=";D4
425 PRINT "NO. OF RADIUS CONTROL ZONES=";F6
430 FOR I=1 TO F6
435 PRINT
440 IF B0=1 THEN 510
445 PRINT "RADIUS DECREMENT FOR ZONE";I;"=";
450 INPUT T6(I)
455 PRINT
460 PRINT "NO. OF CIRCLE FOR ZONE";I;"=";
465 INPUT N0(I)
470 PRINT
475 PRINT "ID NO. FOR FIRST CIRCLE FOR ZONE";I;"=";
480 INPUT T7(I)
485 PRINT
490 PRINT "NO. OF BOTTOM LINES FOR ZONE";I;"=";
495 INPUT T8(I)
500 WRITE #1,T6(I),N0(I),T7(I),T8(I)
505 GO TO 535
510 READ #1,T6(I),N0(I),T7(I),T8(I)
515 PRINT "RADIUS DECREMENT FOR ZONE";I;"=";T6(I)
520 PRINT "NO. OF CIRCLES FOR ZONE";I;"=";N0(I)
```

```
525 PRINT "ID NO. FOR FIRST CIRCLE FOR ZONE";I;"=";T7(I)
530 PRINT "NO. OF BOTTOM LINES FOR ZONE";I;"=";T8(I)
535 PRINT
540 IF B0=1 THEN 555
545 PRINT "INPUT LINE NO.,BEGIN PT. NO.,AND END PT. NO. FOR ZONE";I
550 PRINT "EACH LINE ON ONE LINE & EACH ENTRY SEPARATED BY COMMA"
555 FOR J=1 TO T8(I)
560 IF B0=1 THEN 580
565 INPUT L2(J,I),L3(J,I),L4(J,I)
570 WRITE #1,L2(J,I),L3(J,I),L4(J,I)
575 GO TO 595
580 READ #1,L2(J,I),L3(J,I),L4(J,I)
585 PRINT "FOR ZONE";I,"LINE SEQUENCE";J
590 PRINT "LINE NO.=";L2(J,I),"BEG. NO.=";L3(J,I),"END NO.=";L4(J,I)
595 NEXT J
600 NEXT I
605 E7=X1(L3(1,1),L2(1,1))
610 G8=X1(L4(1,1),L2(1,1))
615 IF T8(1)=1 THEN 650
620 FOR I=2 TO T8(1)
625 IF X1(L3(I,1),L2(I,1))>=E7 THEN 635
630 E7=X1(L3(I,1),L2(I,1))
635 IF X1(L4(I,1),L2(I,1))<=G8 THEN 645
640 G8=X1(L4(I,1),L2(I,1))
645 NEXT I
650 IF E7<>X1(1,N5) THEN 665
655 IF G8<>X1(N6(N5),N5) THEN 665
660 GO TO 680
665 PRINT
670 PRINT "ROCK LINE IS TOO SHORT OR EXTENDS BEYOND GROUND LINE"
675 STOP
680 L2(1,F6+1)=N5
685 L3(1,F6+1)=1
690 L4(1,F6+1)=N6(N5)
695 T8(F6+1)=1
700 T9=N5-1
705 PRINT
710 IF B0=1 THEN 725
715 PRINT"INPUT COHESION, FRIC. ANGLE, UNIT WT. OF SOIL"
720 PRINT "EACH SOIL ON ONE LINE & EACH ENTRY SEPARATED BY COMMA"
725 FOR I=1 TO T9
730 IF B0=1 THEN 750
735 INPUT C(I),P2(I),G(I)
740 WRITE #1,C(I),P2(I),G(I)
745 GO TO 755
750 READ #1,C(I),P2(I),G(I)
755 NEXT I
760 IF B0<>1 THEN 785
765 PRINT "SOIL NO.","COHESION","FRIC. ANGLE","UNIT WEIGHT"
770 FOR I=1 TO T9
775 PRINT I,C(I),P2(I),G(I)
780 NEXT I
785 FOR I=1 TO T9
790 T1(I)=TAN(P2(I)*3.141593/180)
795 NEXT I
800 PRINT
805 IF B0=1 THEN 835
810 PRINT "ANY SEEPAGE? (ENTER 0 WITHOUT SEEPAGE, 1 WITH PHREATIC"
815 PRINT "SURFACE, AND 2 WITH PORE PRESSURE RATIO)";
820 INPUT N3
825 WRITE #1,N3
830 GO TO 875
835 READ #1,N3
840 IF N3=0 THEN 860
```

```
845 IF N3=1 THEN 870
850 PRINT "USE PORE PRESSURE RATIO"
855 GO TO 875
860 PRINT "NO SEEPAGE"
865 GO TO 875
870 PRINT "USE PHREATIC SURFACE"
875 IF B0=1 THEN 960
880 IF N3<>1 THEN 905
885 PRINT
890 PRINT "UNIT WEIGHT OF WATER=";
895 INPUT G5
900 WRITE #1,G5
905 PRINT
910 PRINT "ANY SEARCH?(ENTER 0 WITH GRID AND 1 WITH SEARCH)";
915 INPUT Z0
920 PRINT
925 PRINT "NO. OF SLICES=";
930 INPUT P4
935 PRINT
940 PRINT "NO. OF ADD. RADII=";
945 INPUT N7
950 WRITE #1,Z0,P4,N7
955 GO TO 1005
960 IF N3<>1 THEN 975
965 READ #1,G5
970 PRINT "UNIT WEIGHT OF WATER=";G5
975 READ #1,Z0,P4,N7
980 IF Z0=0 THEN 995
985 PRINT "USE SEARCH"
990 GO TO 1000
995 PRINT "USE GRID"
1000 PRINT "NO. OF SLICES=";P4,"NO. OF ADD. RADII=";N7
1005 P7=Z0
1010 IF N3=0 THEN 1195
1015 IF N3=2 THEN 1155
1020 PRINT
1025 IF B0=1 THEN 1055
1030 PRINT "NO. OF POINTS ON WATER TABLE =";
1035 INPUT N4
1040 PRINT
1045 WRITE #1,N4
1050 GO TO 1065
1055 READ #1,N4
1060 PRINT "NO. OF POINTS ON WATER TABLE=";N4
1065 FOR I=1 TO N4
1070 IF B0=1 THEN 1110
1075 PRINT I;"X-COORDINATE=";
1080 INPUT X3(I)
1085 PRINT TAB(3);"Y-COORDINATE=";
1090 INPUT Y7(I)
1095 PRINT
1100 WRITE #1,X3(I),Y7(I)
1105 GO TO 1120
1110 READ #1,X3(I),Y7(I)
1115 PRINT I;"X COORD.=";X3(I),"Y COORD.=";Y7(I)
1120 NEXT I
1125 IF X3(1)<=X1(1,N5) THEN 1135
1130 GO TO 1140
1135 IF X3(N4)>=X1(N6(N5),N5) THEN 1195
1140 PRINT
1145 PRINT"PIEZOMETRIC LINE IS NOT EXTENDED AS FAR OUT AS GROUND LINE"
1150 STOP
1155 PRINT
1160 IF B0=1 THEN 1185
```

```
1165 PRINT "PORE PRESSURE RATIO=";
1170 INPUT R8
1175 WRITE #1,R8
1180 GO TO 1195
1185 READ #1,R8
1190 PRINT "PORE PRESSURE RATIO=";R8
1195 IF P5=1 THEN 1215
1200 Z=F4
1205 Z3=F5
1210 Z0=P7
1215 IF Z0<>0 THEN 1510
1220 IF P5<>1 THEN 1415
1225 PRINT
1230 PRINT "INPUT COORD. OF GRID POINTS 1,2,AND 3"
1235 PRINT
1240 FOR I=1 TO 3
1245 IF B0=1 THEN 1285
1250 PRINT "POINT";I;"X-COORDINATE =";
1255 INPUT X(I)
1260 PRINT TAB(8);"Y-COORDINATE =";
1265 INPUT Y(I)
1270 PRINT
1275 WRITE #1,X(I),Y(I)
1280 GO TO 1295
1285 READ #1,X(I),Y(I)
1290 PRINT "POINT";I;"X COORD.=";X(I);"Y COORD.=";Y(I)
1295 NEXT I
1300 IF B0=1 THEN 1385
1305 PRINT "X INCREMENT=";
1310 INPUT Z1
1315 PRINT
1320 PRINT "Y INCREMENT=";
1325 INPUT Z2
1330 PRINT
1335 PRINT "NO. OF DIVISIONS BETWEEN POINTS 1 AND 2=";
1340 INPUT Z
1345 PRINT
1350 PRINT "NO. OF DIVISIONS BETWEEN POINTS 2 AND 3=";
1355 INPUT Z3
1360 WRITE #1,Z1,Z2,Z,Z3
1365 PRINT "CONTINUE?(ENTER 1 FOR CONTINUING AND 0 FOR STOP";
1370 INPUT G7
1375 IF G7=0 THEN 6260
1380 GO TO 1405
1385 READ #1,Z1,Z2,Z,Z3
1390 PRINT "X INCREMENT=";Z1,"Y INCREMENT=";Z2
1395 PRINT "NO. OF DIVISIONS BETWEEN POINTS 1 AND 2=";Z
1400 PRINT "NO. OF DIVISIONS BETWEEN POINTS 2 AND 3=";Z3
1405 F5=Z3
1410 F4=Z
1415 IF Z1<>0 THEN 1430
1420 IF Z2<>0 THEN 1430
1425 GO TO 1440
1430 PRINT
1435 PRINT "AUTOMATIC SEARCH WILL FOLLOW AFTER GRID"
1440 Z4=Z5=0
1445 IF Z3<>0 THEN 1460
1450 Z6=Z7=0
1455 GO TO 1470
1460 Z6=(X(3)-X(2))/Z3
1465 Z7=(Y(3)-Y(2))/Z3
1470 IF Z<>0 THEN 1485
1475 Z8=Z9=0
1480 GO TO 1495
```

```
1485 Z8=(X(2)-X(1))/Z
1490 Z9=(Y(2)-Y(1))/Z
1495 A1=A3=X(1)
1500 A2=A4=Y(1)
1505 GO TO 1810
1510 IF P5<>1 THEN 1685
1515 PRINT
1520 IF B0=1 THEN 1545
1525 PRINT"NO. OF CENTERS TO BE ANALYZED =";
1530 INPUT Z5
1535 WRITE #1,Z5
1540 GO TO 1555
1545 READ #1,Z5
1550 PRINT "NO. OF CENTERS TO BE ANLYZED=";Z5
1555 D3=Z5
1560 IF P5<>1 THEN 1690
1565 PRINT
1570 IF B0=1 THEN 1645
1575 PRINT "X COORDINATE OF TRIAL CENTER=";
1580 INPUT A1
1585 PRINT "Y COORDINATE OF TRIAL CENTER=";
1590 INPUT A2
1595 PRINT
1600 PRINT "X INCREMENT=";
1605 INPUT Z1
1610 PRINT "Y INCREMENT=";
1615 INPUT Z2
1620 WRITE #1,A1,A2,Z1,Z2
1625 PRINT "CONTINUE?(ENTER 1 FOR CONTINUING AND 0 FOR STOP)";
1630 INPUT G7
1635 IF G7=0 THEN 6260
1640 GO TO 1660
1645 READ #1,A1,A2,Z1,Z2
1650 PRINT "X COORD.=";A1,"Y COORD.=";A2
1655 PRINT "X INCREMENT=";Z1,"Y INCREMENT=";Z2
1660 A0(Z5)=A1
1665 B9(Z5)=A2
1670 F8(Z5)=Z1
1675 F9(Z5)=Z2
1680 GO TO 1725
1685 Z5=D3
1690 A1=A0(Z5)
1695 A2=B9(Z5)
1700 Z1=F8(Z5)
1705 Z2=F9(Z5)
1710 GO TO 1725
1715 A1=A7
1720 A2=A8
1725 R=0
1730 PRINT
1735 PRINT
1740 A5=A1
1745 IF D3=0 THEN 1755
1750 PRINT "SEARCH STARTED AT CENTER NO.";D3-Z5+1
1755 A6=A2
1760 Z4=R
1765 Z5=Z5-1
1770 Z3=Z=0
1775 IF Z0=1 THEN 1790
1780 Z8=Z9=0
1785 GO TO 1810
1790 Z8=Z1
1795 Z9=Z2
1800 A9=B2=B3=G6=B5=0
```

```
1805 B4=2
1810 Z3=Z3+1
1815 Z=Z+1
1820 B6=M0=0
1825 FOR E6=1 TO Z3
1830 FOR K0=1 TO Z
1835 K2=0
1840 IF Z0=1 THEN 1850
1845 GO TO 1885
1850 IF B6<>2 THEN 1885
1855 M0=A9=B2=B3=G6=B5=0
1860 B4=2
1865 A1=A5
1870 A2=A6
1875 Z8=Z1
1880 Z9=Z2
1885 X0=A1
1890 Y0=A2
1895 R=Z4
1900 IF Z0<>1 THEN 1940
1905 IF ABS(X0-A5)<20*Z1 THEN 1940
1910 IF ABS(Y0-A6)<20*Z2 THEN 1940
1915 PRINT
1920 PRINT "THE INCREMENTS USED FOR SEARCH ARE TOO SMALL,   OR EQUAL"
1925 PRINT "TO ZERO, SO THE MINIMUM FACTOR OF SAFETY CANNOT BE FOUND"
1930 IF Z5<>0 THEN 2815
1935 STOP
1940 J2=SQR((X0-X1(1,N5))**2+(Y0-Y8(1,N5))**2)
1945 J3=SQR((X0-X1(N6(N5),N5))**2+(Y0-Y8(N6(N5),N5))**2)
1950 IF J3>=J2 THEN 1960
1955 J2=J3
1960 FOR I=1 TO (F6+1)
1965 J4=99999
1970 FOR J=1 TO T8(I)
1975 FOR K=L3(J,I) TO (L4(J,I)-1)
1980 IF X1(K,L2(J,I))=X1(K+1,L2(J,I)) THEN 2035
1985 J8=(X0+S4(K,L2(J,I))*(Y0-Y9(K,L2(J,I))))/(S4(K,L2(J,I))**2+1)
1990 IF J8<X1(K,L2(J,I)) THEN 2000
1995 GO TO 2010
2000 J9=SQR((X0-X1(K,L2(J,I)))**2+(Y0-Y8(K,L2(J,I)))**2)
2005 GO TO 2050
2010 IF J8>X1(K+1,L2(J,I)) THEN 2025
2015 J9=SQR((X0-J8)**2+(Y0-S4(K,L2(J,I))*J8-Y9(K,L2(J,I)))**2)
2020 GO TO 2050
2025 J9=SQR((X0-X1(K+1,L2(J,I)))**2+(Y0-Y8(K+1,L2(J,I)))**2)
2030 GO TO 2050
2035 IF Y0<Y8(K,L2(J,I)) THEN 2000
2040 IF Y0>Y8(K+1,L2(J,I)) THEN 2025
2045 J9=ABS(X0-X1(K,L2(I,J)))
2050 IF J9>=J4 THEN 2060
2055 J4=J9
2060 NEXT K
2065 NEXT J
2070 IF J4<=J2 THEN   2095
2075 PRINT
2080 PRINT"****  WARNING AT NEXT CENTER   ****"
2085 PRINT "MAXIMUM RADIUS IS LIMITED BY END POINT OF GROUND LINE"
2090 J4=J2
2095 R4(I)=J4
2100 NEXT I
2105 R3=R4(F6+1)
2110 FOR L0=1 TO F6
2115 IF N0(L0)=0 THEN 2230
2120 R=R4(L0)
```

```
2125 IF T6(L0)=0 THEN 2135
2130 GO TO 2145
2135 H0=(R4(L0)-R4(L0+1))/N0(L0)
2140 GO TO 2150
2145 H0=T6(L0)
2150 R=R-(T7(L0)-1)*H0
2155 FOR J0=1 TO N0(L0)
2160 IF R>=R4(L0+1) THEN 2200
2165 K2=K2+1
2170 F2(K2)=1E6
2175 R5(K2)=R
2180 IF K2<>1 THEN 2195
2185 F1=1E6
2190 R2=R
2195 GO TO 2235
2200 GOSUB 2925
2205 GOSUB 4480
2210 IF F2(K2)>100 THEN 2235
2215 IF H0<=0 THEN 2230
2220 R=R-H0
2225 NEXT J0
2230 NEXT L0
2235 M0=M0+1
2240 IF Z0<>1 THEN 2255
2245 IF B6-1>0 THEN 2255
2250 GO TO 2260
2255 B6=1
2260 PRINT
2265 PRINT
2270 PRINT
2275 PRINT "AT POINT (";X0;Y0;")";"THE RADIUS AND FACTOR OF SAFETY ARE:"
2280 FOR H4=1 TO K2
2285 PRINT R5(H4),F2(H4)
2290 NEXT H4
2295 PRINT "LOWEST FACTOR OF SAFETY =";F1;"AND OCCURS AT RADIUS =";R2
2300 PRINT
2305 IF Z0-1<0 THEN 2315
2310 GO TO 2360
2315 IF M0=1 THEN 2325
2320 IF F1>=H1 THEN 2345
2325 H1=F1
2330 H2=R2
2335 A7=X0
2340 A8=Y0
2345 A1=A1+Z8
2350 A2=A2+Z9
2355 GO TO 2820
2360 IF M0=1 THEN 2410
2365 A9=A9+1
2370 IF F1-H1<0 THEN 2410
2375 IF G6=1 THEN 2445
2380 IF B3<>0 THEN 2610
2385 IF A9<>1 THEN 2580
2390 A9=0
2395 A1=A1-B4*Z8
2400 B3=1
2405 GO TO 2625
2410 H1=F1
2415 H2=R2
2420 A7=X0
2425 A8=Y0
2430 G1=G3
2435 G2=G4
2440 GO TO 2525
```

```
2445 IF B3<>0 THEN 2475
2450 IF A9<>1 THEN 2495
2455 B3=1
2460 A2=A2-B4*Z9
2465 A9=0
2470 GO TO 2625
2475 IF A9<>1 THEN 2495
2480 IF B2<>1 THEN 2495
2485 B2=2
2490 GO TO 2625
2495 A9=B3=G6=0
2500 B2=M0=1
2505 B4=2
2510 A1=A7+Z8
2515 A2=A8
2520 GO TO 2625
2525 IF G6=1 THEN 2555
2530 IF B3=0 THEN 2545
2535 A1=A1-Z8
2540 GO TO 2625
2545 A1=A1+Z8
2550 GO TO 2625
2555 IF B3=0 THEN 2570
2560 A2=A2-Z9
2565 GO TO 2625
2570 A2=A2+Z9
2575 GO TO 2625
2580 M0=B2=G6=1
2585 A9=B3=0
2590 B4=2
2595 A1=A7
2600 A2=A8+Z9
2605 GO TO 2625
2610 IF A9<>1 THEN 2580
2615 IF B2<>1 THEN 2580
2620 B2=2
2625 R=0
2630 IF B2<>2 THEN 1835
2635 B5=B5+1
2640 IF B5<>1 THEN 2705
2645 IF H1<100 THEN 2660
2650 PRINT "IMPROPER CENTER IS USED FOR SEARCH"
2655 GO TO 2815
2660 Z8=Z1/4
2665 Z9=Z2/4
2670 B2=A9=B3=G6=0
2675 B4=2
2680 M0=1
2685 B5=B5+1
2690 A1=A7+Z8
2695 A2=A8
2700 GO TO 1835
2705 PRINT
2710 PRINT "AT POINT (";A7;A8;")";"RADIUS";H2
2715 PRINT
2720 PRINT
2725 PRINT "THE MINIMUM FACTOR OF SAFETY IS";H1
2730 PRINT
2735 PRINT
2740 PRINT "ANY PLOT?(ENTER 0 FOR NO PLOT AND 1 FOR PLOT)";
2745 INPUT P8
2750 IF P8=0 THEN 2795
2755 PRINT
2760 PRINT "YOU MAY LIKE TO ADVANCE PAPER TO THE TOP OF NEXT PAGE"
```

```
2765 PRINT "SO THE ENTIRE PLOT WILL FIT IN ONE SINGLE PAGE."
2770 PRINT "FOR THE PROGRAM TO PROCEED, HIT THE RETURN KEY."
2775 PRINT "AFTER PLOT, YOU MAY LIKE TO ADVANCE PAPER TO NEXT PAGE"
2780 PRINT "AND HIT THE RETURN KEY AGAIN"
2785 INPUT C$
2790 GOSUB 4725
2795 IF Z0=1 THEN 2875
2800 A1=A7
2805 A2=A8
2810 B6=2
2815 IF Z5<>0 THEN 1560
2820 NEXT K0
2825 A3=A3+Z6
2830 A4=A4+Z7
2835 A1=A3
2840 A2=A4
2845 NEXT E6
2850 PRINT
2855 PRINT "AT POINT (";A7;A8;")";"RADIUS";H2
2860 PRINT
2865 PRINT
2870 PRINT "THE MINIMUM FACTOR OF SAFETY IS";H1
2875 IF Z5<>0 THEN 1560
2880 IF Z0=0 THEN 2890
2885 GO TO 2915
2890 IF Z1<>0 THEN 2905
2895 IF Z2<>0 THEN 2905
2900 GO TO 2915
2905 Z0=Z5=1
2910 GO TO 1715
2915 NEXT P5
2920 GO TO 6260
2925 REM SUBROUTINE FSAFTY
2930 DIM B1(80),C8(80),D6(80),L1(80),S(80),S8(80),T4(80)
2935 DIM W(80),W1(80),X2(80),X4(20),X5(20),Y4(80,20),Y5(80)
2940 IF R>R3 THEN 2955
2945 F7=1E6
2950 GO TO 4475
2955 S7=0
2960 FOR J=1 TO N5
2965 X4(J)=-99999.0
2970 X5(J)=99999.0
2975 N1=N6(J)-1
2980 FOR I=1 TO N1
2985 IF S4(I,J)=99999 THEN 3105
2990 A=1+S4(I,J)**2
2995 B=S4(I,J)*(Y9(I,J)-Y0)-X0
3000 D=(Y9(I,J)-Y0)**2+X0**2-R**2
3005 T2=B**2-A*D
3010 IF ABS(T2)<1 THEN 3110
3015 IF T2>0 THEN 3025
3020 GO TO 3105
3025 X6=(-B-SQR(T2))/A
3030 X7=(-B+SQR(T2))/A
3035 IF X6>=(X1(I,J)-0.01) THEN 3045
3040 GO TO 3070
3045 IF X6<=(X1(I+1,J)+0.01) THEN 3055
3050 GO TO 3070
3055 X4(J)=X6
3060 IF J<>N5 THEN 3070
3065 Y6=S4(I,J)*X4(J)+Y9(I,J)
3070 IF X7>=(X1(I,J)-0.01) THEN 3080
3075 GO TO 3105
3080 IF X7<=(X1(I+1,J)+0.01) THEN 3090
```

```
3085 GO TO 3105
3090 X5(J)=X7
3095 IF J<>N5 THEN 3105
3100 Y1=S4(I,J)*X5(J)+Y9(I,J)
3105 NEXT I
3110 NEXT J
3115 IF X4(N5)=-99999.0 THEN 2945
3120 IF X5(N5)=99999.0 THEN 2945
3125 IF Y0<Y6 THEN 3135
3130 GO TO 3140
3135 IF Y0<Y1 THEN 2945
3140 R1=0
3145 IF Y0>Y6 THEN 3270
3150 X4(N5)=X0-R
3155 X8=X4(N5)
3160 FOR K=1 TO N5
3165 K1=N5+1-K
3170 N1=N6(K1)-1
3175 Y4(1,K1)=-9999.0
3180 FOR I=1 TO N1
3185 IF X1(I,K1)=X1(I+1,K1) THEN 3210
3190 M=I
3195 IF X8>X1(I,K1) THEN 3205
3200 GO TO 3210
3205 IF X8<=X1(I+1,K1) THEN 3220
3210 NEXT I
3215 GO TO 3230
3220 Y4(1,K1)=S4(M,K1)*X8+Y9(M,K1)
3225 IF K=1 THEN 3260
3230 IF (Y4(1,K1)+9999.0)=0 THEN 3265
3235 IF (Y0-Y4(1,K1))>0 THEN 3250
3240 R1=R1+(Y3-Y4(1,K1))*C(K1)
3245 GO TO 3260
3250 R1=R1+(Y3-Y0)*C(K1)
3255 GO TO 3290
3260 Y3=Y4(1,K1)
3265 NEXT K
3270 IF Y0>Y1 THEN 3290
3275 X5(N5)=X0+R
3280 X8=X5(N5)
3285 GO TO 3160
3290 J6=N5-1
3295 N1=N6(N5)-1
3300 R7=(X5(N5)-X4(N5))/P4
3305 P9=0
3310 D7=X4(N5)
3315 FOR I=1 TO P4
3320 D8=0
3325 FOR J=2 TO N1
3330 IF X1(J,N5)<=(D7+0.01) THEN 3350
3335 IF X1(J,N5)>=(D7+R7-0.01) THEN 3350
3340 D8=D8+1
3345 D6(D8)=X1(J,N5)
3350 NEXT J
3355 FOR J=2 TO J6
3360 IF X4(J)<=(D7+0.01) THEN 3385
3365 IF X4(J)>=(D7+R7-0.01) THEN 3385
3370 D8=D8+1
3375 D6(D8)=X4(J)
3380 GO TO 3405
3385 IF X5(J)<=(D7+0.01) THEN 3405
3390 IF X5(J)>=(D7+R7-0.01) THEN 3405
3395 D8=D8+1
3400 D6(D8)=X5(J)
```

```
3405 NEXT J
3410 IF D8<1 THEN 3540
3415 IF D8=1 THEN 3525
3420 E5=D8-1
3425 D9=0
3430 FOR L=1 TO E5
3435 D9=D9+1
3440 E0=D6(D9)
3445 E2=D9+1
3450 E3=D9
3455 FOR K=E2 TO D8
3460 IF D6(K)>=E0 THEN 3475
3465 E0=D6(K)
3470 E3=K
3475 NEXT K
3480 D6(E3)=D6(D9)
3485 D6(D9)=E0
3490 NEXT L
3495 B1(P9+1)=D6(1)-D7
3500 FOR L=2 TO D8
3505 B1(P9+L)=D6(L)-D6(L-1)
3510 NEXT L
3515 B1(P9+D8+1)=D7+R7-D6(D8)
3520 GO TO 3545
3525 B1(P9+1)=D6(1)-D7
3530 B1(P9+2)=D7+R7-D6(1)
3535 GO TO 3545
3540 B1(P9+1)=R7
3545 P9=P9+D8+1
3550 D7=D7+R7
3555 NEXT I
3560 E4=0
3565 FOR J=1 TO P9
3570 IF J>(P9-E4) THEN 3595
3575 IF B1(J)<>0 THEN 3585
3580 E4=E4+1
3585 B1(J)=B1(J+E4)
3590 NEXT J
3595 P9=P9-E4
3600 IF P9<=40 THEN 3605
3605 X2(1)=X4(N5)+B1(1)/2
3610 FOR J=2 TO P9
3615 X2(J)=X2(J-1)+(B1(J)+B1(J-1))/2
3620 NEXT J
3625 FOR J=1 TO P9
3630 Y5(J)=Y0-SQR(R**2-(X2(J)-X0)**2)
3635 S(J)=(X0-X2(J))/R
3640 IF Y1<=Y6 THEN 3650
3645 S(J)=-S(J)
3650 C8(J)=(Y0-Y5(J))/R
3655 NEXT J
3660 FOR J=1 TO P9
3665 I7=0
3670 FOR K=1 TO N5
3675 Y4(J,K)=-9999.0
3680 N1=N6(K)-1
3685 FOR I=1 TO N1
3690 IF X1(I,K)=X1(I+1,K) THEN 3715
3695 M=I
3700 IF X2(J)>X1(I,K) THEN 3710
3705 GO TO 3715
3710 IF X2(J)<=X1(I+1,K) THEN 3725
3715 NEXT I
3720 GO TO 3770
```

```
3725 Y4(J,K)=S4(M,K)*X2(J)+Y9(M,K)
3730 IF I7=0 THEN 3760
3735 IF K=1 THEN 3760
3740 IF Y4(J,K)>(Y3-0.1) THEN 3760
3745 PRINT "BOUNDARY LINE NO.";K;"IS OUT OF PLACE, PLEASE"
3750 PRINT "CHANGE THE INPUT DATA AND RUN THE PROGRAM AGAIN"
3755 STOP
3760 Y3=Y4(J,K)
3765 I7=1
3770 NEXT K
3775 S8(J)=Y4(J,N5)-Y5(J)
3780 IF S8(J)>S7 THEN 3790
3785 GO TO 3795
3790 S7=S8(J)
3795 NEXT J
3800 IF S7<O4 THEN 2945
3805 I6=0
3810 FOR J=1 TO P9
3815 E1=0
3820 W(J)=0
3825 W1(J)=0
3830  FOR K=1 TO N5
3835 IF (Y4(J,K)+9999.0)=0 THEN 3845
3840 GO TO 3855
3845 I6=I6+1
3850 GO TO 4165
3855 S1=Y4(J,N5)-Y4(J,K)
3860 IF S1>=(S8(J)-0.01) THEN 4090
3865 IF E1=1 THEN 3890
3870 PRINT
3875 PRINT "AT POINT (";X0;Y0;")";"WITH RADIUS OF";R
3880 PRINT "THE CIRCLE CUTS INTO ROCK SURFACE"
3885 STOP
3890 IF N9<0 THEN 4080
3895 IF N9=0 THEN 3905
3900 GO TO 4080
3905 Y2=Y5(J)
3910 L1(J)=K-I6
3915 N9=1
3920 IF C(L1(J))=0 THEN 3950
3925 IF J=1 THEN 4105
3930 IF C(L1(J-1))=0 THEN 3940
3935 GO TO 4105
3940 IF T1(L1(J-1))=0 THEN 3960
3945 GO TO 4105
3950 IF T1(L1(J))=0 THEN 3995
3955 GO TO 3925
3960 O2=L1(J-1)
3965 O3=Y4(J-1,N5)
3970 IF Y1<Y6 THEN 3985
3975 O4=-1
3980 GO TO 4045
3985 O4=1
3990 GO TO 4045
3995 IF J=1 THEN 4160
4000 IF C(L1(J-1))<>0 THEN 4015
4005 IF T1(L1(J-1))<>0 THEN 4015
4010 GO TO 4160
4015 O2=L1(J)
4020 O3=Y4(J,N5)
4025 IF Y1<Y6 THEN 4040
4030 O4=1
4035 GO TO 4045
4040 O4=-1
```

```
4045 O5=Y0-SQR(R**2-(X2(J)-B1(J)/2-X0)**2)
4050 O6=O3-O5
4055 W1(J)=O4*0.5*O6**2*G(O2)*(Y0-O5-O6/3)/R
4060 IF C(L1(J-1))=0 THEN 4070
4065 GO TO 4160
4070 IF T1(L1(J-1))=0 THEN 4105
4075 GO TO 4160
4080 Y2=Y4(J,K-I6)
4085 GO TO 4105
4090 N9=0
4095 E1=1
4100 GO TO 4160
4105 W4=B1(J)*(Y4(J,K)-Y2)*G(K-I6)
4110 W(J)=W(J)+W4
4115 W1(J)=W1(J)+W4*S(J)
4120 IF S5(P5)=0 THEN 4160
4125 IF C(K-I6)<>0 THEN 4155
4130 IF T1(K-I6)<>0 THEN 4155
4135 IF Y6>=Y1 THEN 4150
4140 IF X2(J)<X1(2,N5) THEN 4160
4145 GO TO 4155
4150 IF X2(J)>X1(2,N5) THEN 4160
4155 W1(J)=W1(J)+W4*S5(P5)*(Y0-0.5*(Y4(J,K)+Y2))/R
4160 I6=1
4165 NEXT K
4170 NEXT J
4175 IF N3=1 THEN 4205
4180 IF N3=0 THEN 4300
4185 FOR J=1 TO P9
4190 E8(J)=W(J)*(1-R8)
4195 NEXT J
4200 GO TO 4315
4205 N2=N4-1
4210 FOR J=1 TO P9
4215 FOR I=1 TO N2
4220 M=I
4225 IF X2(J)>=X3(I) THEN 4235
4230 GO TO 4240
4235 IF X2(J)<=X3(I+1) THEN 4250
4240 NEXT I
4245 GO TO 4255
4250 T4(J)=((Y7(M+1)-Y7(M))/(X3(M+1)-X3(M)))*(X2(J)-X3(M))+Y7(M)
4255 NEXT J
4260 FOR J=1 TO P9
4265 IF T4(J)-Y5(J)>0 THEN 4275
4270 GO TO 4285
4275 E8(J)=W(J)-(T4(J)-Y5(J))*G5*B1(J)
4280 GO TO 4290
4285 E8(J)=W(J)
4290 NEXT J
4295 GO TO 4315
4300 FOR J=1 TO P9
4305 E8(J)=W(J)
4310 NEXT J
4315 R0=R1
4320 O1=0
4325 FOR J=1 TO P9
4330 R0=R0+C(L1(J))*B1(J)/C8(J)+E8(J)*T1(L1(J))*C8(J)
4335 O1=O1+W1(J)
4340 NEXT J
4345 IF R0=0 THEN 2945
4350 IF O1<1 THEN 2945
4355 F7=R0/O1
4360 IF F7<0 THEN 4455
```

```
4365 IF F7>100 THEN 4475
4370 IF F7>1 THEN 4380
4375 F7=2
4380 I9=0
4385 I9=I9+1
4390 P1=F7
4395 M2=M3=0
4400 FOR J=1 TO P9
4405 M4=C(L1(J))*B1(J)+E8(J)*T1(L1(J))
4410 M5=P1*C8(J)+T1(L1(J))*S(J)
4415 M2=M2+M4/M5
4420 M3=M3+M4*T1(L1(J))*S(J)/M5**2
4425 NEXT J
4430 F7=P1*(1+(R1/P1+M2-O1)/(O1-M3))
4435 IF ABS(F7-P1)/P1>0.0001 THEN 4445
4440 GO TO 4475
4445 IF I9<10 THEN 4385
4450 IF F7>0 THEN 4475
4455 PRINT
4460 PRINT "AT POINT (";X0;Y0;")";"WITH RADIUS OF";R
4465 PRINT "THE FACTOR OF SAFETY IS NEGATIVE"
4470 STOP
4475 RETURN
4480 REM SUBROUTINE SAVE
4485 K2=K2+1
4490 R5(K2)=R
4495 F2(K2)=F7
4500 IF K2<>1 THEN 4530
4505 F1=F2(K2)
4510 R2=R5(K2)
4515 G3=X4(N5)
4520 G4=X5(N5)
4525 GO TO 4720
4530 IF F2(K2)>=F1 THEN 4555
4535 F1=F2(K2)
4540 R2=R5(K2)
4545 G3=X4(N5)
4550 G4=X5(N5)
4555 IF F2(K2)<>1E6 THEN 4570
4560 I3=I8=2
4565 GO TO 4640
4570 IF F6<>1 THEN 4600
4575 IF K2<>N0(1) THEN 4600
4580 IF F2(K2)>F2(K2-1) THEN 4600
4585 I3=I8=3
4590 R5(K2+1)=R3
4595 GO TO 4640
4600 IF K2<3 THEN 4720
4605 IF N7=0 THEN 4720
4610 IF F2(K2)>F2(K2-1) THEN 4620
4615 GO TO 4720
4620 IF F2(K2-1)<F2(K2-2) THEN 4630
4625 GO TO 4720
4630 I3=1
4635 I8=2
4640 D0=R
4645 FOR I0=I3 TO I8
4650 R6=R5(K2-3+I0)
4655 D1=(R5(K2-2+I0)-R5(K2-3+I0))/(N7+1)
4660 FOR N=1 TO N7
4665 R6=R6+D1
4670 R=R6
4675 GOSUB 2925
4680 IF F7>=F1 THEN 4705
```

```
4685 F1=F7
4690 R2=R
4695 G3=X4(N5)
4700 G4=X5(N5)
4705 NEXT N
4710 NEXT I0
4715 R=D0
4720 RETURN
4725 REM SUBROUTINE XYPLOT
4730 DIM B7(50,22),B8(11),I1(51,22),I2(51,22),I5(22),J1(22)
4735 DIM L5(22),L$(51),P$(22),S0(50,22),X9(6)
4740 IF A$="+" THEN 4870
4745 B$=" "
4750 A$="+"
4755 E$="X"
4760 P$(1)="1"
4765 P$(2)="2"
4770 P$(3)="3"
4775 P$(4)="4"
4780 P$(5)="5"
4785 P$(6)="6"
4790 P$(7)="7"
4795 P$(8)="8"
4800 P$(9)="9"
4805 P$(10)="0"
4810 P$(11)="A"
4815 P$(12)="B"
4820 P$(13)="C"
4825 P$(14)="D"
4830 P$(15)="E"
4835 P$(16)="F"
4840 P$(17)="G"
4845 P$(18)="J"
4850 P$(19)="K"
4855 P$(20)="L"
4860 P$(21)="P"
4865 P$(22)="*"
4870 M1=N5
4875 FOR J=1 TO M1
4880 L5(J)=N6(J)
4885 L8=L5(J)
4890 FOR I=1 TO L8
4895 S0(I,J)=X1(I,J)
4900 B7(I,J)=Y8(I,J)
4905 NEXT I
4910 NEXT J
4915 IF N3<>1 THEN 4955
4920 M1=M1+1
4925 L5(M1)=N4
4930 N8=N4
4935 FOR I=1 TO N8
4940 S0(I,M1)=X3(I)
4945 B7(I,M1)=Y7(I)
4950 NEXT I
4955 C1=C2=S0(1,1)
4960 C3=C4=B7(1,1)
4965 FOR J=1 TO M1
4970 L8=L5(J)
4975 FOR I=1 TO L8
4980 IF C2<=S0(I,J) THEN 4990
4985 C2=S0(I,J)
4990 IF C1>=S0(I,J) THEN 5000
4995 C1=S0(I,J)
5000 IF C4<=B7(I,J) THEN 5010
```

```
5005 C4=B7(I,J)
5010 IF C3>=B7(I,J) THEN 5020
5015 C3=B7(I,J)
5020 NEXT I
5025 J1(J)=1
5030 NEXT J
5035 D2=C1-C2
5040 N8=0
5045 IF D2<>0 THEN 5075
5050 D2=0.2*C2
5055 IF C2 <> 0 THEN 5065
5060 D2=0.2
5065 C2=C2-0.5*D2
5070 C1=C1+0.45*D2
5075 FOR I=1 TO 75
5080 IF D2>10 THEN 5105
5085 IF D2>1 THEN 5120
5090 D2=D2*10
5095 N8=-I
5100 GO TO 5115
5105 D2=D2/10
5110 N8=I
5115 NEXT I
5120 IF D2<8 THEN 5135
5125 K7=10
5130 GO TO 5185
5135 IF D2<5 THEN 5150
5140 K7=8
5145 GO TO 5185
5150 IF D2<4 THEN 5165
5155 K7=5
5160 GO TO 5185
5165 IF D2<2 THEN 5180
5170 K7=4
5175 GO TO 5185
5180 K7=2
5185 IF N8>0 THEN 5205
5190 C5=K7
5195 C5=C5/10**(1-N8)
5200 GO TO 5210
5205 C5=K7*10**(N8-1)
5210 C6=0
5215 IF C2>=0 THEN 5235
5220 C6=C6-C5
5225 IF C6>C2 THEN 5220
5230 GO TO 5250
5235 C6=C6+C5
5240 IF C6>C2 THEN 5220
5245 IF C6<C2 THEN 5235
5250 C7=C5*10+C6
5255 IF C7>=C1 THEN 5270
5260 C2=C6-0.001*C5
5265 GO TO 5035
5270 I=INT((C7-C1)/(2*C5))
5275 C6=C6-I*C5
5280 C7=C6+10*C5
5285 D2=C3-C4
5290 N8=0
5295 IF D2<>0 THEN 5325
5300 D2=0.2*C4
5305 IF C4<>0 THEN 5315
5310 C2=0.2
5315 C4=C4-0.5*D2
5320 C3=C3+0.45*D2
```

```
5325 FOR I=1 TO 75
5330 IF D2>10 THEN 5355
5335 IF D2>1 THEN 5370
5340 D2=D2*10
5345 N8=-I
5350 GO TO 5365
5355 D2=D2/10
5360 N8=I
5365 NEXT I
5370 IF D2<8 THEN 5385
5375 K7=10
5380 GO TO 5435
5385 IF D2<5 THEN 5400
5390 K7=8
5395 GO TO 5435
5400 IF D2<4 THEN 5415
5405 K7=5
5410 GO TO 5435
5415 IF D2<2 THEN 5430
5420 K7=4
5425 GO TO 5435
5430 K7=2
5435 IF N8>0 THEN 5455
5440 C0=K7
5445 C0=C0/10**(1-N8)
5450 GO TO 5460
5455 C0=K7*10**(N8-1)
5460 C9=0
5465 IF C4>=0 THEN 5485
5470 C9=C9-C0
5475 IF C9>C4 THEN 5470
5480 GO TO 5500
5485 C9=C9+C0
5490 IF C9>C4 THEN 5470
5495 GO TO 5485
5500 D5=C0*10+C9
5505 IF D5>=C3 THEN 5520
5510 C4=C9-0.001*C0
5515 GO TO 5285
5520 I=INT((D5-C3)/(2*C0))
5525 C9=C9-I*C0
5530 D5=C9+10*C0
5535 M1=M1+1
5540 J1(M1)=1
5545 S0(1,M1)=G1
5550 S0(2,M1)=G2
5555 L5(M1)=2
5560 FOR J=1 TO M1
5565 L8=L5(J)
5570 FOR I=1 TO L8
5575 I1(I,J)=INT(((((S0(I,J)-C6)/(C7-C6))*50)+1.5)
5580 IF J=M1 THEN 5590
5585 I2(I,J)=INT(51.5-(((B7(I,J)-C9)/(D5-C9))*50))
5590 NEXT I
5595 NEXT J
5600 K8=1
5605 J0=51
5610 FOR I=1 TO 10
5615 IF I>5 THEN 5645
5620 IF (C6+C5*(I-1)*2)=0 THEN 5640
5625 IF (C6+C5*(I-1)*2)>0 THEN 5645
5630 K8=10*(I-1)+6
5635 GO TO 5645
5640 K8=10*(I-1)+1
```

```
5645 IF (C9+C0*(I-1)*1.05)>=0 THEN 5655
5650 J0=51-5*I
5655 NEXT I
5660 FOR I=1 TO 11
5665 IF I>6 THEN 5675
5670 X9(I)=C6+C5*(I-1)*2
5675 B8(I)=INT(C9+C0*(11-I))
5680 NEXT I
5685 FOR N8=1 TO M1
5690 IF N8=M1 THEN 5765
5695 N1=L5(N8)-1
5700 H5=L5(N8)
5705 FOR H7=1 TO N1
5710 H3=I1(H7+1,N8)-I1(H7,N8)
5715 H4=I2(H7+1,N8)-I2(H7,N8)
5720 IF H3<2 THEN 5755
5725 H6=H3-1
5730 FOR H8=1 TO H6
5735 I1(H5+H8,N8)=I1(H7,N8)+H8
5740 I2(H5+H8,N8)=I2(H7,N8)+INT(H8*H4/H3)
5745 NEXT H8
5750 H5=H5+H6
5755 NEXT H7
5760 GO TO 5805
5765 H5=I1(2,N8)-I1(1,N8)-1
5770 H9=I1(1,N8)
5775 FOR I=1 TO H5
5780 I1(I,N8)=H9+I
5785 J8=(I1(I,N8)-1.5)*(C7-C6)/50+C6
5790 W2=A8-SQR(H2**2-(J8-A7)**2)
5795 I2(I,N8)=INT(51.5-(W2-C9)/(D5-C9)*50)
5800 NEXT I
5805 FOR I=1 TO H5
5810 FOR J=I TO H5
5815 IF I2(I,N8)<=I2(J,N8) THEN 5850
5820 K=I2(I,N8)
5825 I2(I,N8)=I2(J,N8)
5830 I2(J,N8)=K
5835 K=I1(I,N8)
5840 I1(I,N8)=I1(J,N8)
5845 I1(J,N8)=K
5850 NEXT J
5855 NEXT I
5860 I5(N8)=H5
5865 NEXT N8
5870 PRINT TAB(8);
5875 FOR I=1 TO 72
5880 PRINT T$(I);
5885 NEXT I
5890 PRINT
5895 PRINT "          FOR SEISMIC COEFFICIENT OF";S5(P5)
5900 PRINT "          AT POINT (";A7;A8;")";"RADIUS";H2
5905 PRINT "          THE MINIMUM FACTOR OF SAFETY IS";H1
5910 PRINT
5915 FOR I=1 TO 51
5920 FOR N8=1 TO 51
5925 L$(N8)=B$
5930 NEXT N8
5935 L$(1)=L$(K8)=L$(51)=A$
5940 IF (I-INT(I/5)*5)<>1 THEN 5970
5945 FOR N8=1 TO 51
5950 IF (N8-INT(N8/10)*10)<>1 THEN 5960
5955 L$(N8)=A$
5960 NEXT N8
```

```
5965 L$(1)=L$(K8)=L$(51)=E$
5970 IF I=1 THEN 5990
5975 IF I=J0 THEN 5990
5980 IF I=51 THEN 5990
5985 GO TO 6015
5990 FOR N8=2 TO 50
5995 L$(N8)=A$
6000 IF(N8-INT(N8/10)*10)<>1 THEN 6010
6005 L$(N8)=E$
6010 NEXT N8
6015 FOR J=1 TO M1
6020 IF J1(J)>I5(J) THEN 6145
6025 F3=J1(J)
6030 IF I2(F3,J)<>I THEN 6145
6035 K9=I1(F3,J)
6040 IF L$(K9)=B$ THEN 6055
6045 IF L$(K9)=A$ THEN 6055
6050 GO TO 6095
6055 L7=J
6060 IF J<>M1 THEN 6070
6065 L7=22
6070 IF N3<>1 THEN 6085
6075 IF J<>M1-1 THEN 6085
6080 L7=21
6085 L$(K9)=P$(L7)
6090 GO TO 6135
6095 IF N3=1 THEN 6105
6100 GO TO 6110
6105 IF J>=M1-1 THEN 6135
6110 IF N3<>1 THEN 6120
6115 GO TO 6125
6120 IF J=M1 THEN 6135
6125 IF L$(K9)=P$(J) THEN 6135
6130 L$(K9)=E$
6135 J1(J)=J1(J)+1
6140 GO TO 6020
6145 NEXT J
6150 IF(I-INT(I/5)*5)<>1 THEN 6195
6155 K=INT(0.2*I+1)
6160 PRINT TAB(8);B8(K);TAB(14);
6165 FOR N8=1 TO 51
6170 PRINT L$(N8);
6175 IF N8<>51 THEN 6185
6180 PRINT
6185 NEXT N8
6190 GO TO 6225
6195 PRINT TAB(14);
6200 FOR N8=1 TO 51
6205 PRINT L$(N8);
6210 IF N8<>51 THEN 6220
6215 PRINT
6220 NEXT N8
6225 NEXT I
6230 PRINT USING 6235,X9(1),X9(2),X9(3),X9(4),X9(5),X9(6)
6235 :      ######  ######  ######  ######  ######  ######
6240 PRINT
6245 PRINT
6250 INPUT C$
6255 RETURN
6260 END
```

Index